Membrane Technology

*Edited by
Suzana Pereira Nunes and
Klaus-Viktor Peinemann*

Related Titles

Sammells, A. F., Mundschau, M. V. (eds.)

Nonporous Inorganic Membranes
for Chemical Processing

approx. 380 pages with approx. 150 figures
2006
Hardcover
ISBN 3-527-31342-7

Ohlrogge, K., Ebert, K. (eds.)

Membranen
Grundlagen, Verfahren und industrielle Anwendungen

approx. 592 pages with approx. 270 figures
2006
Hardcover
ISBN 3-527-30979-9

Membrane Technology

in the Chemical Industry

Edited by
Suzana Pereira Nunes and Klaus-Viktor Peinemann

Second, Revised and Extended Edition

WILEY-VCH Verlag GmbH & Co. KGaA

The Editors

Dr. Suzana Pereira Nunes
GKSS Forschungszentrum
Max-Planck-Str. 1
21502 Geesthacht

Dr. Klaus-Viktor Peinemann
GKSS Forschungszentrum
Institut für Chemie
Max-Planck-Str. 1
21502 Geesthacht

■ All books published by Wiley-VCH are carefully produced. Nevertheless, authors, editors, and publisher do not warrant the information contained in these books, including this book, to be free of errors. Readers are advised to keep in mind that statements, data, illustrations, procedural details or other items may inadvertently be inaccurate.

Library of Congress Card No.:
applied for

British Library Cataloguing-in-Publication Data
A catalogue record for this book is available from the British Library.

Bibliographic information published by Die Deutsche Bibliothek
Die Deutsche Bibliothek lists this publication in the Deutsche Nationalbibliografie; detailed bibliographic data is available in the Internet at <http://dnb.ddb.de>

© 2006 WILEY-VCH Verlag GmbH & Co. KGaA, Weinheim, Germany

All rights reserved (including those of translation into other languages). No part of this book may be reproduced in any form – by photoprinting, microfilm, or any other means – nor transmitted or translated into a machine language without written permission from the publishers. Registered names, trademarks, etc. used in this book, even when not specifically marked as such, are not to be considered unprotected by law.

Typesetting K+V Fotosatz GmbH, Beerfelden
Printing betz-druck GmbH, Darmstadt
Binding Litges & Dopf GmbH, Heppenheim
Cover Design Grafik-Design Schulz, Fußgönheim

Printed in the Federal Republic of Germany
Printed on acid-free paper

ISBN-13: 978-3-527-31316-7
ISBN-10: 3-527-31316-8

Contents

Part I	**Membrane Materials and Membrane Preparation**	
	S. P. Nunes and K.-V. Peinemann	
1	**Introduction** *3*	
2	**Membrane Market** *5*	
3	**Membrane Preparation** *9*	
3.1	Phase Inversion *10*	
4	**Presently Available Membranes for Liquid Separation** *15*	
4.1	Membranes for Reverse Osmosis *15*	
4.2	Membranes for Nanofiltration *18*	
4.2.1	Solvent-resistant Membranes for Nanofiltration *20*	
4.2.2	NF Membranes Stable in Extreme pH Conditions *22*	
4.3	Membranes for Ultrafiltration *23*	
4.3.1	Polysulfone and Polyethersulfone *23*	
4.3.2	Poly(vinylidene fluoride) *26*	
4.3.3	Polyetherimide *28*	
4.3.4	Polyacrylonitrile *30*	
4.3.5	Cellulose *32*	
4.3.6	Solvent-resistant Membranes for Ultrafiltration *32*	
4.4	Membranes for Microfiltration *34*	
4.4.1	Polypropylene and Polyethylene *34*	
4.4.2	Poly(tetrafluorethylene) *36*	
4.4.3	Polycarbonate and Poly(ethylene terephthalate) *37*	
5	**Surface Modification of Membranes** *39*	
5.1	Chemical Oxidation *39*	
5.2	Plasma Treatment *40*	
5.3	Classical Organic Reactions *41*	
5.4	Polymer Grafting *41*	

6	**Membranes for Fuel Cells** 45
6.1	Perfluorinated Membranes 46
6.2	Nonfluorinated Membranes 48
6.3	Polymer Membranes for High Temperatures 51
6.4	Organic-Inorganic Membranes for Fuel Cells 52

7	**Gas Separation with Membranes** 53
7.1	Introduction 53
7.2	Materials and Transport Mechanisms 53
7.2.1	Organic Polymers 55
7.2.2	Background 55
7.2.3	Polymers for Commercial Gas-separation Membranes 57
7.2.4	Ultrahigh Free Volume Polymers 58
7.2.5	Inorganic Materials for Gas-separation Membranes 62
7.2.6	Carbon Membranes 62
7.2.7	Perovskite-type Oxide Membranes for Air Separation 64
7.2.8	Mixed-matrix Membranes 67
7.3	Basic Process Design 69
	Acknowledgments 75
	References 75

Part II   Current Application and Perspectives

1	**The Separation of Organic Vapors from Gas Streams by Means of Membranes** 93
	K. Ohlrogge and K. Stürken
	Summary 93
1.1	Introduction 94
1.2	Historical Background 94
1.3	Membranes for Organic Vapor Separation 96
1.3.1	Principles 96
1.3.2	Selectivity 96
1.3.3	Temperature and Pressure 97
1.3.4	Membrane Modules 98
1.4	Applications 100
1.4.1	Design Criteria 100
1.4.2	Off-gas and Process Gas Treatment 102
1.4.2.1	Gasoline Vapor Recovery 103
1.4.2.2	Polyolefin Production Processes 109
1.5	Applications at the Threshold of Commercialization 111
1.5.1	Emission Control at Petrol Stations 111
1.5.2	Natural Gas Treatment 113
1.5.3	Hydrogen/Hydrocarbon Separation 114
1.6	Conclusions and Outlook 116
	References 116

2	**Gas-separation Membrane Applications** *119*
	D. J. Stookey
2.1	Introduction *119*
2.2	Membrane Application Development *120*
2.2.1	Membrane Selection *120*
2.2.2	Membrane Form *123*
2.2.3	Membrane Module Geometry *125*
2.2.4	Compatible Sealing Materials *129*
2.2.5	Module Manufacture *130*
2.2.6	Pilot or Field Demonstration *130*
2.2.7	Process Design *132*
2.2.8	Membrane System *133*
2.2.9	Beta Site *135*
2.2.10	Cost/Performance *136*
2.3	Commercial Gas-separation Membrane Applications *136*
2.3.1	Hydrogen Separations *137*
2.3.2	Helium Separations *140*
2.3.3	Nitrogen Generation *140*
2.3.4	Acid Gas-Separations *143*
2.3.5	Gas Dehydration *144*
2.4	Developing Membrane Applications *146*
2.4.1	Oxygen and Oxygen-enriched Air *146*
2.4.2	Nitrogen Rejection from Natural Gas *147*
2.4.3	Nitrogen-enriched Air (NEA) *147*
	References *147*

3	**State-of-the-Art of Pervaporation Processes in the Chemical Industry** *151*
	H. E. A. Brüschke
3.1	Introduction *151*
3.2	Principles and Calculations *153*
3.2.1	Definitions *153*
3.2.2	Calculation *155*
3.2.3	Permeate-side Conditions *163*
3.2.4	Transport Resistances *166*
3.2.5	Principles of Pervaporation *168*
3.2.6	Principles of Vapor Permeation *171*
3.3	Membranes *175*
3.3.1	Characterization of Membranes *180*
3.4	Modules *182*
3.4.1	Plate Modules *183*
3.4.2	Spiral-wound Modules *185*
3.4.3	"Cushion" Module *185*
3.4.4	Tubular Modules *186*
3.4.5	Other Modules *187*

3.5	Applications 188
3.5.1	Organophilic Membranes 188
3.5.2	Hydrophilic Membranes 189
3.5.2.1	Pervaporation 189
3.5.2.2	Vapor Permeation 191
3.5.3	Removal of Water from Reaction Mixtures 194
3.5.4	Organic–Organic Separation 197
3.6	Conclusion 200
	References 200
4	**Organic Solvent Nanofiltration** 203
	A. G. Livingston, L. G. Peeva and P. Silva
	Summary 203
4.1	Current Applications and Potential 203
4.2	Theoretical Background to Transport Processes 205
4.2.1	Pore-flow Model 205
4.2.2	Solution-Diffusion Model 206
4.2.3	Models Combining Membrane Transport with the Film Theory of Mass Transfer 207
4.3	Transport of Solvent Mixtures 210
4.3.1	Experimental 210
4.3.1.1	Filtration Equipment and Experimental Measurements 210
4.3.2	Results for Binary Solvent Fluxes 210
4.4	Concentration Polarization and Osmotic Pressure 213
4.4.1	Experimental 213
4.4.2	Results for Concentration Polarization and Osmotic Pressure 214
4.4.2.1	Parameter Estimation 214
4.4.2.2	Nanofiltration of Docosane-Toluene Solutions 216
4.4.2.3	Nanofiltration of TOABr-Toluene Solutions 219
4.5	Conclusions 224
	Nomenclature 225
	Greek letters 225
	Subscripts 226
	References 226
5	**Industrial Membrane Reactors** 229
	M. F. Kemmere and J. T. F. Keurentjes
5.1	Introduction 229
5.2	Membrane Functions in Reactors 232
5.2.1	Controlled Introduction of Reactants 232
5.2.2	Separation of Products 238
5.2.3	Catalyst Retention 241
5.3	Applications 242
5.3.1	Pervaporation-assisted Esterification 242
5.3.2	Large-scale Dehydrogenations with Inorganic Membranes 248

5.3.3	OTM Syngas Process 250
5.3.4	Membrane Recycle Reactor for the Acylase Process 251
5.3.5	Membrane Extraction Integrated Systems 253
5.4	Concluding Remarks and Outlook to the Future 254
	References 255

6   Electromembrane Processes   259
T. A. Davis, V. D. Grebenyuk and O. Grebenyuk

6.1	Ion-exchange Membranes 259
6.2	Ion-exchange Membrane Properties 262
6.2.1	Swelling 262
6.2.2	Electrical Conductivity 263
6.2.3	Electrochemical Performance 267
6.2.4	Diffusion Permeability 268
6.2.5	Hydraulic Permeability 269
6.2.6	Osmotic Permeability 269
6.2.7	Electroosmotic Permeability 270
6.2.8	Polarization 271
6.2.9	Chemical and Radiation Stability 273
6.3	Electromembrane Process Application 274
6.3.1	Electrodialysis 274
6.3.2	Electrodeionization 280
6.3.3	Electrochemical Regeneration of Ion-exchange Resin 282
6.3.4	Synthesis of New Substances without Electrode Reaction Participation: Bipolar-membrane Applications 283
6.3.5	Isolation of Chemical Substances from Dilute Solutions 285
6.3.6	Electrodialysis Applications for Chemical-solution Desalination 285
6.4	Electrochemical Processing with Membranes 286
6.4.1	Electrochemistry 286
6.4.2	Chlor-alkali Industry 291
6.4.3	Perfluorinated Membranes 291
6.4.4	Process Conditions 293
6.4.5	Zero-gap Electrode Configurations 294
6.4.6	Other Electrolytic Processes 295
6.4.7	Fuel Cells 297
6.4.8	Electroorganic Synthesis 299
6.4.9	Electrochemical Oxidation of Organic Wastes 300
	Acknowledgments 300
	List of Symbols 300
	References 301

7	**Membrane Technology in the Chemical Industry: Future Directions** *305*
	R. W. Baker
7.1	The Past: Basis for Current Membrane Technology *305*
7.1.1	Ultrathin Membranes *305*
7.1.2	Membrane Modules *306*
7.1.3	Membrane Selectivity *308*
7.2	The Present: Current Status and Potential of the Membrane Industry *309*
7.2.1	Reverse Osmosis *309*
7.2.2	Ultrafiltration *313*
7.2.3	Microfiltration *314*
7.2.4	Gas Separation *315*
7.2.4.1	Refinery Hydrogen Applications *317*
7.2.4.2	Nitrogen (and Oxygen) Separation from Air *319*
7.2.4.3	Natural Gas Separations *323*
7.2.4.4	Vapor/Gas, Vapor/Vapor Separations *326*
7.2.5	Pervaporation *329*
7.2.6	Ion-conducting Membranes *330*
7.3	The Future: Predictions for 2020 *332*
	References *333*

Subject Index *337*

Preface

The idea of the first edition of Membrane Technology in the Chemical Industry was to review the available membranes for the broad variety of separation processes in the chemical sector. A further important decision was to invite well-known membrane scientists with recognized experience in the chemical industry to supply a deep analysis of the main membrane applications in this field. After 5 years the use of membranes is even more widespread, new membranes have entered the market, some are no longer commercialized and some applications have become more relevant, justifying now the second edition of the book. In the first part of this new edition, market statistics have been reviewed, as well as the currently available membranes and membrane materials. A new chapter on Fuel Cells has been added, a field that has grown considerably in recent years. Also, the balance of applications connected to gas separation has been critically reanalyzed. Part II has been reviewed by the various authors. The "future directions" have been reanalyzed. The major change in Part II is the inclusion of a chapter on Organic Solvent Nanofiltration, written by Andrew Livingstone, Ludmila G. Peeva and Pedro Silva, which reflects the rapid development of this field in recent years.

Suzana Pereira Nunes, Klaus-Viktor Peinemann
March 2006

List of Contributors

R. Baker
Membrane Technology & Research
1360 Willow Road
Menlo Park, CA 94025
USA

H. E. A. Brüschke
Kurpfalzstraße 64
D-69226 Nußloch
Germany

T. A. Davis
TAD Consulting
5 Davis Farm Road
Annandale, NJ 08801
USA

O. Grebenyuk
Ionics
65 Grove Street
Watertown, MA 02172
USA

V. D. Grebenyuk
Ionics
65 Grove Street
Watertown, MA 02172
USA

M. F. Kemmere
NIZO Food Research B.V.
P.O. Box 20
6710 BA Ede
The Netherlands

J. T. F. Keurentjes
Eindhoven University of Technology
P.O. Box 513
5600 MB Eindhoven
The Netherlands

A. Livingstone
Department of Chemical Engineering
Imperial College London
Prince Consort Road
South Kensington
London, SW7 2AZ
United Kingdom

S. P. Nunes
GKSS Forschungszentrum
Max-Planck-Str. 1
D-21502 Geesthacht
Germany

K. Ohlrogge
GKSS-Research Center
Max-Planck-Str. 1
D-21502 Geesthacht
Germany

L. G. Peeva
Department of Chemical Engineering
Imperial College London
Prince Consort Road
South Kensington
London, SW7 2AZ
United Kingdom

K.-V. Peinemann
GKSS Forschungszentrum
Max-Planck-Str. 1
D-21502 Geesthacht
Germany

P. Silva
Department of Chemical Engineering
Imperial College London
Prince Consort Road
South Kensington
London, SW7 2AZ
United Kingdom

D. J. Stookey
Elah Strategies
1571 Treherne Court
Chesterfield, MO 63017
USA

K. Stürken
SIHI Anlagentechnik GmbH
Lindenstr. 170
D-25524 Itzehoe
Germany

Part I
Membrane Materials and Membrane Preparation

S. P. Nunes and K.-V. Peinemann

1
Introduction

Membrane technology is presently an established part of several industrial processes. Well known is its relevance in the food industry, in the manufacture of dairy products as well as in the automotive industry for the recovery of electropainting baths. Membranes make possible the water supply for millions of people in the world and care for the survival of the large number of people suffering from kidney disease. The chemical industry is a growing field in the application of membranes, which, however, often requires membrane materials with exceptional stability. The first part of the book will discuss the currently available membranes for different processes, which are suitable for the chemical industry. Information on different methods of membrane preparation will be given. Different materials will be compared, taking into account physical characteristics and chemical stability.

2
Membrane Market

The membranes and module sales in 1998 were estimated at more than US$ 4.4 billion worldwide [1], shared by different applications (Fig. 1.1). If equipment and total membrane systems are also considered, the estimate would be double. At least 40% of the market is in the United States [2, 3], 29% of the market is shared by Europe and the Middle East. The markets in Asia and South America are growing fast. A more recently published study [3] estimates the combined market for membranes used in separation and nonseparation applications to be worth $ 5 billion only in the US, with an annual growth rate of 6.6%. According to another recent study [4], the demand for pure water will drive the market for crossflow membrane equipment and membranes worldwide from $ 6.8 billion in 2005 to $ 9 billion in 2008.

Hemodialysis/hemofiltration alone had sales of over US$ 2200 million in 1998. Reverse osmosis (RO), ultrafiltration (UF) and microfiltration (MF) together accounted for 1.8 billion dollars in sales in 1998. At that time about US$ 400 million worth of membranes and modules were sold each year worldwide for use in reverse osmosis. About 50% of the RO market was controlled by Dow/FilmTec and Hydranautics/Nitto. They were followed by DuPont and Osmonics. Membranes are applied during sea-water desalination, municipal/ brackish water treatment and in the industrial sectors. The market for RO and nanofiltration is growing at a rate higher than 10%/year. The market for desali-

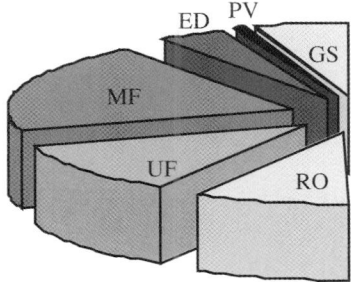
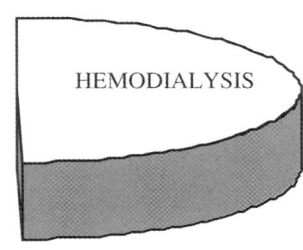

Fig. 1.1 Membrane and module sale for different process applications [1].

Membrane Technology in the Chemical Industry.
Edited by Suzana Pereira Nunes and Klaus-Viktor Peinemann
Copyright © 2006 WILEY-VCH Verlag GmbH & Co. KGaA, Weinheim
ISBN: 3-527-31316-8

nation increased markedly, as did that for process water treatment. Desalination alone is expected to grow to $2 billion in 2008. The USA is the largest purchaser of RO membranes and equipments, while Japan is the second and Saudi Arabia the third. With the increasing demand of reliable municipal water supplies in major cities in China, the market of home RO systems is increasing fast and the RO equipment and membrane market in Asia is expected to exceed $1.8 billion per year in 2008 [4]. General Electric integrated recently Osmonics (and Desal), Ionics and Pall in a large alliance, the GE Infrastructure Water Technology, offering systems for reverse osmosis, nano, micro- and ultrafiltration. Desal commercializes GE and partners recently launched the largest membrane-based water-filtration project, the Sulaibiya Project in the Middle East, to purify more than one hundred million gallons of wastewater each day for agricultural and industrial uses.

Ultrafiltration membranes and modules brought about US$500 million in sales in 1998 with an expected growing rate of 10% a year. Over 58% of the sales are in the US. In contrast to RO, the UF market is shared by a large number of companies, but the leaders are Pall, Amicon/Millipore and Koch. One of the largest industrial sectors for ultrafiltration is still the recovery of electrocoat paints. UF membranes are also to a large degree responsible for supplying pure water for the semiconductor industry. Growing demands of ultrahigh purity chemicals in this sector could also be supplied by UF with the availability of chemical-resistant membranes. Oil/water separation is now a large application for UF in industrial sectors such as metal cleaning and wool scouring and is still growing with the implementation of new environmental legislation. The use of UF in the biotechnology industry is growing even faster than the sector itself. The combined biotechnology and pharmaceutical industries are expected to buy membranes and modules in value of $300 million a year in 2008 and crossflow equipment at a rate of $700 million a year [4].

Sales of microfiltration equipment and membranes are expected to rise from $1.9 billion worldwide in 2005 to $2.5 billion in 2008. The main applications are the production of sterilized water for the pharmaceutical and biotechnology industry.

In the semiconductor industry, MF is used to remove particles from air and to produce pure water. One of the most rapid growths in MF in recent years has been observed in wastewater treatment with membrane bioreactors. Additionally, the MF application has considerably expanded due to the development of new biopharmaceuticals and new research sectors like genomic.

Gas separation (GS) is a relatively young technology and accounted for about US$230 million/year in 1998, but is growing fast with a rate higher than 15% a year. The development of membrane reactors is opening a number of new gas applications. For the electrically driven membrane processes the sales in 1998 were around US$180 million. For pervaporation (PV) in 1996 the market was about US$26 million, with a growth rate of 20%.

The market for nonseparating membranes used in drug delivery, tissue regeneration, batteries, food packaging and high-performance textiles is worth about $2.8 billion, growing at 3.8% a year [3]. High growth rates are expected for fuel-cell membranes, although the market is still quite small.

3
Membrane Preparation

Different methods of polymer membrane preparation have been covered in several reviews [5–9]. Membranes can be classified, according to their morphology as shown in Fig. 3.1.

Dense homogeneous polymer membranes are usually prepared (i) from solution by solvent evaporation only or (ii) by extrusion of the melted polymer. However, dense homogeneous membranes only have a practical meaning when made of highly permeable polymers such as silicone. Usually the permeant flow across the membrane is quite low, since a minimal thickness is required to give the membrane mechanical stability. Most of the presently available membranes are porous or consist of a dense top layer on a porous structure. The preparation of membrane structures with controlled pore size involves several techniques with relatively simple principles, but which are quite tricky.

Commercial membranes were already produced in Germany by Sartorius in the early 1920s. However, they had only a limited application on a laboratory

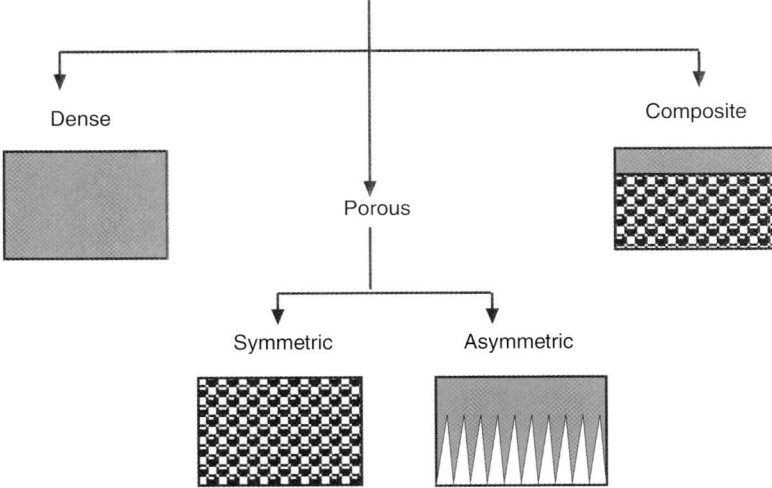

Fig. 3.1 Membrane classification according to the morphology.

Membrane Technology in the Chemical Industry.
Edited by Suzana Pereira Nunes and Klaus-Viktor Peinemann
Copyright © 2006 WILEY-VCH Verlag GmbH & Co. KGaA, Weinheim
ISBN: 3-527-31316-8

scale. The breakthrough of the membrane technology came first in the 1960s with the development of the asymmetric porous membranes by Loeb and Sourirajan [10]. The asymmetric membranes combine high permeant flow, provided by a very thin selective top layer and a reasonable mechanical stability, resulting from the underlying porous structure. An asymmetric structure characterizes most of the presently commercially available membranes, which are now produced from a wide variety of polymers. By far the most common method used in generation of asymmetric structures in membranes is the "phase-inversion" process. Other methods used to form pores in membranes will be discussed in the following sections. Particularly in the case of microfiltration, several techniques other than phase inversion are currently applied in the industry.

3.1
Phase Inversion

The phase-inversion process consists of the induction of phase separation in a previously homogeneous polymer solution either by temperature change, by immersing the solution in a nonsolvent bath (wet process) or exposing it to a nonsolvent atmosphere (dry process).

In the thermal process [11, 12], a low molecular weight component usually acts as a solvent at high temperature and as a nonsolvent at low temperature. It is then removed after formation of the porous structure. Although the thermal process can be applied to a wide range of polymers, it is especially interesting for those with poor solubility, such as polypropylene, which can hardly be manufactured into a porous membrane by other phase-inversion processes. An isotropic microporous structure is usually formed. The isothermal phase inversion is commercially more widespread. Usually the polymer solution is immersed in a nonsolvent bath (wet process) and a solvent–nonsolvent exchange leads to phase separation. The polymer-rich phase forms the porous matrix, while the polymer-poor phase gives rise to the pores. The morphology is usually asymmetric, with a selective skin on the surface, as shown in Fig. 3.2.

The pore structure is generated by phase separation. Phase separation in this case is mainly a liquid–liquid demixing process, although solid–liquid demixing may also play an additional role in systems containing a crystallizable polymer, such as cellulose acetate and poly(vinylidene fluoride). After immersion in a nonsolvent bath, the solvent–nonsolvent exchange brings the initially thermodynamically stable system into a condition for which the minimum Gibb's free energy is attained by separating into two coexisting phases. The predominant mechanism of phase inversion leading to pore formation and the thermodynamics involved are the subject of a fruitful and sometimes controversial discussion in the literature [13–27], as well as the most probable paths in the phase diagram. A simplified diagram is shown in Fig. 3.3. Basically, the mechanism of phase separation depends on the crossing point into the unstable region. If the solvent–nonsolvent exchange brings the system first to a metastable condi-

Fig. 3.2 Asymmetric porous membrane.

tion (Path A), the nucleation and growth mechanism (NG) is favored. A dispersed phase consisting of droplets of a polymer-poor solution is formed in a concentrated matrix. If no additional nonsolvent influx or temperature change in the system were induced, the composition inside the nuclei would, in the early stages, be practically the same as expected at the equilibrium and would, practically, not change with time.

Only the size of the droplets increases with time. If the demixing path crosses the critical point, going directly into the unstable region (Path B), spinodal decomposition (SD) predominates.

A concentration fluctuation appears in the initially homogeneous system and progresses with increasing amplitude, leading to a separation into two cocontinuous phases. Here again, the polymer-poor phase will form the pores. The initial steps of phase separation, either by NG and SD can be relatively well described according to theories of phase separation. However, at later stages, both NG and SD usually progress to a phase coalescence and the final structure can only be predicted with difficulty. At least as important as the starting mechanism of phase separation is the point where the developing structure is fixed. Parallel to demixing, as the concentration of the polymer solution changes, by solvent–nonsolvent exchange, the mobility of the system decreases. Reasons for this may vary from physically unfavorable polymer–solvent (or nonsolvent) interaction, leading to stronger polymer–polymer contacts, to vitrification of the polymer concentrated phase, as the solvent concentration decreases, and also in some cases partial crystallization. If the system gels and solidifies directly after

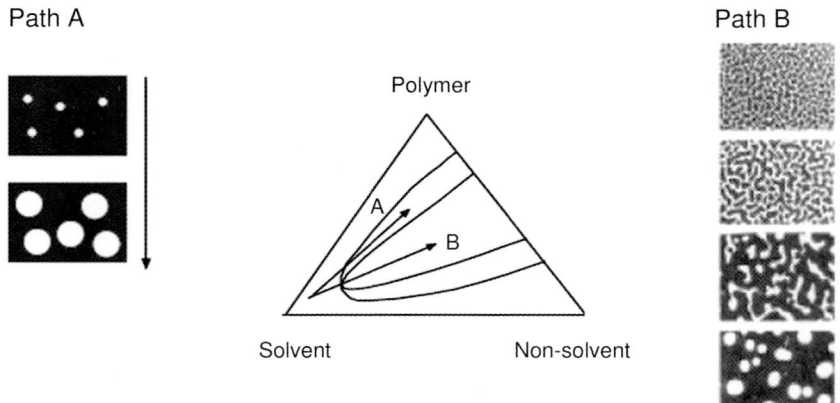

Fig. 3.3 Mechanism of phase separation during membrane formation.

the first steps of phase separation (for instance, at t_2), the membrane will have a fine pore structure, which keeps the original characteristics given by the initial demixing mechanism. If NG demixing stops during the initial stages, a morphology of closed cells would be favored. At later NG stages, the nuclei would grow and touch each other forming interconnected pores. The SD demixing would favor the formation of an interconnected pore structure from the beginning. An asymmetric structure is usually formed across the membrane since the solvent–nonsolvent exchange may lead to different starting conditions for phase separation at layers far from the surface. Besides the NG and SD demixing, other factors influence the morphology. The whole membrane structure usually can be classified as sponge-like or finger-like. Finger-like cavities are formed in many cases, as the nonsolvent enters the polymer solution. This macrovoid structure may contribute to a lack of mechanical stability in membranes to be used at high pressures. A combination of factors is responsible for the formation of macrovoids and this topic has been well reviewed in the literature [28, 29]. For practical purposes, the predominance of a sponge-like or a macrovoid structure can be induced in different ways. Basically, the sponge-like structure is favored by:

1. increasing the polymer concentration of the casting solution
2. increasing the viscosity of the casting solution by adding a crosslinking agent
3. changing the solvent
4. adding solvent to the nonsolvent bath.

The growth of a polymer-poor phase by SD or NG is an isotropical process, which takes place as soon as the solvent–nonsolvent contents supply the thermodynamic condition for demixing. To understand the macrovoid formation, a quite interesting explanation was provided by McKelvey and Koros [28] as de-

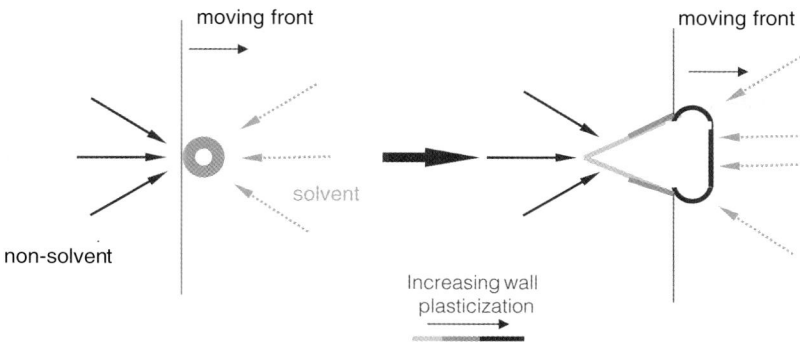

Fig. 3.4 Nonisotropic nucleus growth during macrovoid formation in membranes.

picted in Fig. 3.4. For that, the coupling of the (NG or SD) demixing processes with the rapidly moving front of nonsolvent must be considered. If the nonsolvent diffusion rate into the polymer-poor phase being formed exceeds the rate of outward solvent diffusion, the macrovoid formation is favored. The diffusivity of water is usually expected to be one-to-two orders of magnitude higher than the diffusivity of bulkier organic solvents. The main driving force for the nonsolvent (usually water) influx is the locally generated osmotic pressure. This could be, hypothetically, approximately 100 bar with a difference of only 5 mol% nonsolvent concentration between the initial nucleus and the approaching front. As water moves into a polymer-poor nucleus, its wall is deformed, expanding in the form of a tear. If the walls are fragile, the nucleus may rupture giving rise to macrovoids with unskinned walls. If the walls are stronger, as in the case of nuclei growing in a matrix with higher polymer concentration, the deformation can be restrained or even totally inhibited, giving rise to a macrovoid-free structure.

Increasing the polymer concentration of the casting solution to suppress macrovoids has been well registered in the literature for a wide spectrum of polymers such as cellulose acetate [29], aromatic polyamide [30], and polyetherimide [31]. Other factors such as the addition of crosslinking agents that can improve the strength of the growing nucleus wall also contribute to a macrovoid-free structure. An example is the addition of amines to polyetherimide casting solutions [32]. Another way to suppress the macrovoid formation is to reduce the osmotic pressure between the nonsolvent moving front and the polymer-poor phase inside the nuclei. This can be achieved by adding nonsolvent to the casting solution or adding solvent to the nonsolvent bath. An example is the addition of dioxane to an aqueous coagulation bath for a CA dioxane solution [29].

Changing the solvent may act in different ways. Solvents with higher diffusivity across the nucleus walls would be able to leave the nucleus faster, while the nonsolvent is added in, which does not favor macrovoid formation. But even

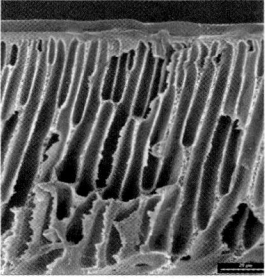  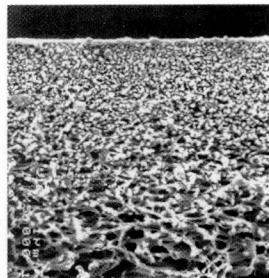

Fig. 3.5 Polyetherimide (PEI) membranes prepared from different casting solutions: (a) 17.5 wt% PEI in dimethylacetamide; (b) 17.5 wt% PEI in 5.5 wt% tetraethoxysilane and 77 wt% dimethylacetamide; (c) 15.5 wt% PEI in 28 wt% THF and 56 wt% g-butyro lactone.

more effective in suppressing macrovoid formation are solvents that increase the solution viscosity or even promote a fast gelation, making the nucleus wall stronger and resistant to deformation. Some examples of solvent influence on membrane morphology are to be found in the literature for polyetherimide [31] (Fig. 3.5) and cellulose acetate [33].

4
Presently Available Membranes for Liquid Separation

4.1
Membranes for Reverse Osmosis

The most common membrane materials for reverse osmosis membranes are cellulose acetate, polyamide and the "thin-film composites", prepared by interfacial polymerization on the surface of a porous support. A review of composite membranes was published by Petersen [34]. Cellulose acetate (CA) was one of the first membrane materials, and it is still being successfully used, especially in water treatment (in spiral-wound modules). They usually allow quite high water flows with low salt solubility. One advantage, when compared to the polyamide-based membranes, is its chlorine tolerance. Also, because of the neutral surface, cellulose acetate membranes usually exhibit a more stable performance than polyamide membranes in applications where the feed water has a high fouling potential, such as in municipal effluents and surface water supplies. However, CA membranes are drastically less stable in organic solvents than polyamide. The recommended pH range is between 3 and 7, they are less resistant to biological attack and the recommended temperature is lower than 50 °C. The susceptibility to hydrolysis increases with temperature and is an inverse function of the degree of acetylation. Aromatic polyamides have a much higher solvent resistance and may be used in a wider pH range (pH 4–11). The main application is the treatment of brackish water and seawater. They can be produced in very thin hollow fibers with large surface area/volume. The membrane top layer is, however, quite thick (>0.2 µm), which leads to relatively low water flows. The main disadvantage is the very low chlorine tolerance.

Integral asymmetric membranes have a relatively low manufacturing cost. CA, in particular, dominates a significant part of the membrane market for water treatment due to its low cost. However, the possibility of expanding the application of reverse osmosis in separations that demand membranes with higher performance came only with the advent of the thin-film composite (TFC) membranes. They consist of an ultrathin layer, usually of polyamide or polyetherurea, which is polymerized *in situ* and crosslinked on an asymmetric porous support, usually polysulfone. Since the dense selective layer is very thin, the membranes can operate at higher flux and lower pressure. The chemical stability is very good, although the chlorine tolerance is low. They are not biode-

Membrane Technology in the Chemical Industry.
Edited by Suzana Pereira Nunes and Klaus-Viktor Peinemann
Copyright © 2006 WILEY-VCH Verlag GmbH & Co. KGaA, Weinheim
ISBN: 3-527-31316-8

gradable and can operate in a pH range of between 2 and 11. The membrane preparation consists of immersing the porous support in an aqueous solution containing a water-soluble monomer. After that the support is immersed in a solution of the second monomer in a nonpolar solvent. Both monomers are only allowed to react at the interface between organic and aqueous solution, forming a thin polymer layer at the surface of the porous support. As soon as the polymer layer is formed it acts as a barrier for the monomer transport and avoids the continuity of the polycondensation. On the other hand, any defect on the polymer layer is immediately repaired, since monomer transport and polycondensation is allowed at that point. One of the most successful TFC membranes is the FT-30, developed by Cadotte et al. [35] in the North Star Laboratories and presently commercialized by Dow. The reaction involved in the preparation of the FT-30 is as follows:

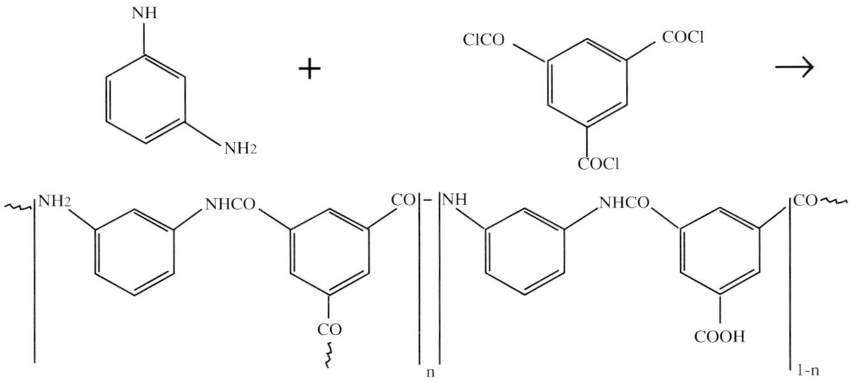

The polyamide layer is formed on an asymmetric microporous polysulfone support cast on a polyester support web. The polyester web gives the major structural support and the polysulfone support with small surface pores with a diameter of ca. 15 nm is the proper substrate for the formation of a 0.2-μm polyamide top layer (Fig. 4.1). The FT-30 has been optimized for different applications, being commercialized [36] as FILMTEC TW-30 (for municipal tap water), BW-30 (for brackish water) and SW-30 (for sea-water conversion to potable water). Salt rejections higher than 99.5% can be obtained with fluxes of 0.6 m^3/m^2 day or rejection. With 0.2% salt feed solution, membranes work at 1.6 MPa with rejections above 98% and fluxes of 1 m^3/m^2 day. This represents a reduction up to 50% in operating pressure for water treatment in comparison to commercial cellulose acetate membranes. The rejection of other solutes by FT 30 is shown in Tab. 4.1. The maximum operating pressure of the FT 30 is about 7 MPa with free chlorine tolerance <0.1 ppm. Recently, Film Tec introduced a low-energy sea-water element, Film Tec SW30XLE-400, claiming a reduction of as much as 20% desalination costs. With an active area of 400 ft^2, the element enables a flow rate of about 34 000 litres per day with 99.7% salt rejection and a maximum pressure of 83 bar.

Table 4.1 Rejection of different solutes by FT30.

Solute	Molecular weight (g/mol)	Rejection (%)
Sodium chloride	58	99
Calcium chloride	111	99
Magnesium sulfate	120	>99
Copper sulfate	160	>99
Formaldehyde	30	35
Methanol	32	25
Ethanol	46	70
Isopropanol	60	90
Urea	60	70
Lactic acid (pH 2)	90	94
Lactic acid (pH 5)	90	99
Sucrose	342	99

2000 ppm solute, 1.6 MPa, 25 °C, pH 7 (unless otherwise noted)

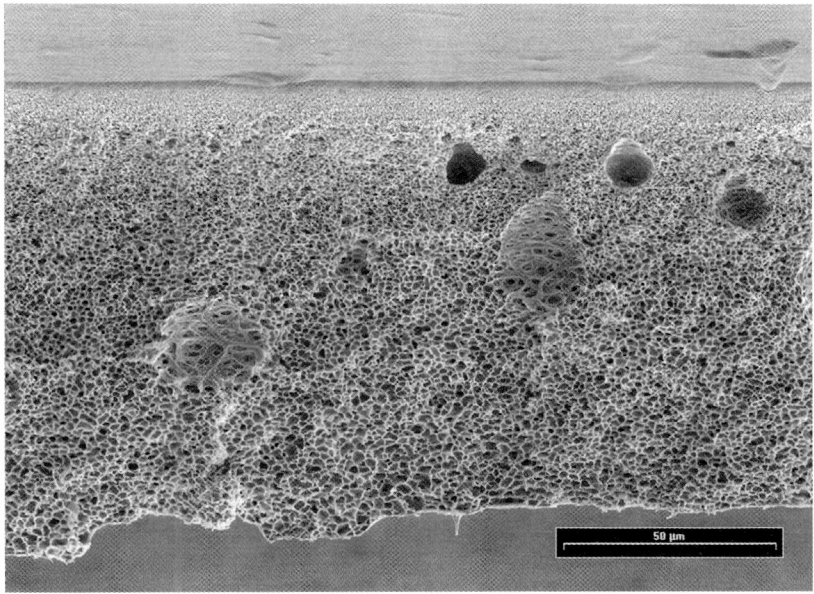

Fig. 4.1 FILMTEC® FT-30.

Another very successful development for reverse osmosis is the energy-saving membrane series produced by Nitto Denko (Tab. 4.2) [37]. The membrane filtrating layer is also an aromatic polyamide. Due to its irregular surface, the actual membrane area available for the permeation is much larger than it would be in the case of a smooth surface on the same porous support. High fluxes are therefore obtained.

Table 4.2 Characteristics of NITTO DENKO RO membranes.

NITTO DENKO Energy Saving ES 10 and ES 15 (Ultralow pressure spiral RO element)
ES10
Applications: manufacture of ultrapure water for electronics industry, manufacture of sterile water for pharmaceuticals, reuse of waste water, desalination of brackish water
Max. Feed Temperature: 40 °C
Max. Operating Pressure: 4.1 MPa
pH: 2–10
Residual chlorine: zero

ES10 operating conditions (2-inch spiral-wound element):
Max. inlet pressure: 4.2 MPa
Standard inlet pressure: 0.75 MPa
NaCl rejection: 99.5% (1500 ppm feed at 7.5 bar, 25 °C, pH 6.5)
Average water flux: 1.7 m^3/day (dimensions: 62.5 mm diameter×1016 mm length)

ES15-D
Applications: primary desalination of ultrapure water manufacturing systems, desalination of brackish water

ES15-D operating conditions (8-inch spiral-wound element):
Max. operating pressure: 2 MPa
NaCl rejection: 99.5% (1500 ppm feed at 7.5 bar, 25° C, pH 7)
Average water flux: 37 m^3/day (surface area 37 m^2)
pH range: 2–10 (permissible range), 1–11 (during cleaning)

ES20-D8
Max. operating pressure: 4.1 MPa
Average water flux: 40 m^3/day (surface area 37 m^2)
pH range: 1–11 (during cleaning)

4.2
Membranes for Nanofiltration

While reverse osmosis and ultrafiltration were being established in several applications, there was a lack of available membranes with cutoffs between 400 and 4000 g/mol. Increasing interest in NF membranes developed in the last decade. An extensive review on principles and applications of nanofiltration has been published recently [38]. Nanofiltration is important for water softening [39] and removal of organic contaminants. In the food industry, nanofiltration can be applied for concentration and demineralization of whey, concentration of sugar and juice. Nanofiltration also finds application in the pulp and paper industry, in the concentration of textile dye effluents and in landfill leachate treatment. The improvement of solvent stability of available NF membranes opens a wide range of potential applications in the chemical and pharmaceutical industry as well as in metal and acid recovery.

FILMTEC NF 90, NF 200 and NF 270 elements are currently commercialized by Dow for the nanofiltration range, for water softening and for organic removal (pesticides and herbicide). Hydranautics commercializes the ESNA membrane series [40], also a thin-film composite with an aromatic polyamide layer, for similar applications. FILMTEC/Dow has commercialized the NF55, NF70 and NF90 (water flow of NF55 > NF70 > NF90) membranes, for the range of nanofiltration, being able to reject at least 95% magnesium sulfate. The chlorine tolerance is lower than 0.1 ppm and the pH range in operation is 3–9. The top layer is a fully aromatic crosslinked polyamide. The exact composition is not completely disclosed. However, one procedure to prepare nanofiltration membranes is the interfacial polymerization between a piperazine or an amine substituted piperidine or cyclohexane and a polyfunctional acyl halide as described in US Patent 4769148 and 4859384 [41, 42]. NF270 is a relatively new membrane composed of a semiaromatic piperazine-based polyamide layer on top of a polysulfone microporous support reinforced with a nonwoven polyester [43]. The membrane is very hydrophilic and when applied to paper-mill waters in typical pH range above 4, its surface is negatively charged leading to the repulsion also of negatively charged solutes.

GE-Osmonics (part of GE Water Technologies) commercializes the Desal™ 5 nanofiltration membranes, used for removal of hardness and other contaminants, alcohol recovery from aqueous solution and removal of salt from salt whey. The membrane has 4 layers, a polyester nonwoven, an asymmetric microporous polysulfone and two proprietary thin films, which might be based on sulfonated polysulfone and polypiperazineamide [34]. A comparison between Desal™ 5 and NF270 for nanofiltration has been reported by [44] (Tab. 4.3).

Nanofiltration membranes can also be obtained by coating ultrafiltration membranes with different polymer solutions. Nitto Denko commercializes NTR-729 HF, a low-pressure spiral element also suitable for nanofiltration of salt and low molecular weight organic compounds. The membrane has a polysulfone porous support coated with a thin layer of polyvinyl alcohol. Analogous procedures have been reported in the literature. Membranes with cutoffs between 800 and 4500 g/mol and water permeabilities of up to 10 l/h m² bar

Table 4.3 Comparison between Desal™ 5 and NF270 membranes for nanofiltration [44].

Membrane	Desal™ 5	NF270
Cutoff (g/mol)	150–300	200–300
Max. operating pressure (bar)	41	41
Retention of 250 ppm glucose (%)	94	90
Max. operating temperature (°C)	50	45
Pure water flux at 40 °C and 6 bar (L/m² h)	65	120
Contact angle (Wilhelmy method)	46	25
pH tolerance	2–11	3–10

could be obtained by coating PVDF membranes with polyether-block-polyamide copolymers [45]. A lower cutoff was obtained by forming a polyamide network dispersed in the block-copolymer layer by reacting a polyether diamine and trimesoyl chloride [46]. Coatings of hydroxyalkyl derivatives of cellulose are used to prepare solvent-resistant membranes [47].

Another way to obtain nanofiltration membranes is the modification of a reverse osmosis membrane, as proposed by Cadotte and Walker [48]. The process involves contacting a crosslinked polyamide selective layer with a strong mineral acid such as phosphoric acid at 100–150 °C, which is then followed by a treatment with a "rejection-enhancing agent" such as tannic acid or water-soluble polymers to selectively plug microscopic leaks and defects. Another procedure [49] to open polyamide RO membranes consists of contacting the membrane with ions to form a membrane ion complex, treating the membrane ion complex with an aqueous solution of alkali metal permanganate to form manganese dioxide crystals in the membrane and finally dissolving the crystals. In another procedure, a reverse osmosis membrane is treated with triethanolamine to open the pores [50]. RO cellulose acetate membranes can be opened by hydrolysis at very high and very low pH. However, it is difficult to control this process.

4.2.1
Solvent-resistant Membranes for Nanofiltration

A reason for the restrained application of membrane technology in the chemical industry, as compared to other fields, is the availability of well-established chemical-resistant membranes that could work in harsh process conditions, eventually at extreme pH conditions or in processes with organic solvents. Although several examples have been described in the patent literature in the last decade and some are commercialized, reasonably low costs associated with a well-documented long-term application are usually required to make them commercially attractive and lower the risks of substituting conventional separation processes. Some solvent stable membranes are discussed here and summarized in Tab. 4.4. Membrane Products Kiryat Weizmann Ltd developed the SelRO® nanofiltration membranes with excellent solvent resistance, which are now commercialized by Koch Membrane Systems. The MPS-44 and -50 [51] claimed to have an excellent stability in alkanes, alcohols, acetates, ketones and aprotic solvents. The MPS 44 is a hydrophilic membrane suitable, for instance, for separation processes in solvent mixtures containing water and organics. Solutes with molecular weights ≥250 g/ml can then be separated or concentrated, while the composition of the solvent mixture does not change through the membrane. The hydrophilic MPS 50 is a nanofiltration membrane for use in a pure organic medium. Some applications include the recovery of antibiotics and peptides from organic solvents, recovery of catalysts from organic medium and recovery of hydrocarbons from cleaning processes.

Although the compositions of the MPS-44 and -50 membranes are not completely open, two patents of Kiryat Weizmann describe interesting procedures

Table 4.4 SelRO® nanofiltration membranes.

Membrane	Cutoff (g/mol)	Stability in water/solvent mixtures (S – stable, LS – limited stability, NS – not stable)	pH range	Maximum temperature (°C)
MPS-44 (hydrophilic)	250	Methanol S Acetone S 2-Propanol S Cyclohexane S Ethanol S MEK S Butanol S MIBK S Pentane S Formaldehyde S Hexane S Ethylene glycol S Dichloroethane S Propylene oxide S Trichloroethane S Nitrobenzene S Methylene chloride S Tetrahydrofuran S Carbon tetrachloride S Acetonitrile S Toluene S Dimethylformamide LS Xylene S N-Methyl Pyrrolidone LS Diethylether S Dimethylacetamide LS Ethylacetate S Dioxane S	3–10	40
MPS-50 (hydrophobic)	700	Methanol S Acetone S 2-Propanol S Cyclohexane S Ethanol S MEK S Butanol S MIBK S Pentane S Formaldehyde S Hexane S Ethylene glycol S Dichloroethane S Propylene oxide S Trichloroethane S Nitrobenzene S Methylene chloride S Tetrahydrofuran S Carbon tetrachloride S Acetonitrile S Toluene S Dimethylformamide LS Xylene S N-Methyl Pyrrolidone LS Diethylether S Dimethylacetamide LS Ethyl acetate S Dioxane S	3–10	40

for the preparation of solvent stable membranes [52, 53]. In the former, porous polyacrylonitrile (PAN) membranes are crosslinked, for instance, by immersing them in a solution containing 1% of metal alkoxides such as sodium ethoxide or 10% of NaOH. The membranes are heated at 110 °C. UF membranes which are not soluble or swellable in DMF, NMP or DMSO are then obtained. If a cutoff in the range of nanofiltration or reverse osmosis is required, the membrane is coated with a hydrophilic polymer, which is later crosslinked, or with polyfunctional reactants, which react forming a crosslinked coating. Some of the described coatings were based on polyethyleneimine and reactive dyes [54]. In a second patent [55], coatings of bromomethylated phenylene oxide were crosslinked with ammonia. An earlier patent [56] discloses the improvement of the solvent resistance of PAN membranes by a reaction with hydroxylamine, followed by treatment with cyanuric chloride and NaOH. The resulting membrane also has an improved resistance to compaction. Carbon membranes are also described in the Kiryat Weizmann patent [52] starting from the same PAN membranes mentioned above. After immersion in an organic or inorganic base solution the membrane is heated at 110–130 °C, below the glass transition temperature, to induce a partial crosslinking and prevent plastic flow during heating at higher temperature. The membrane is then heated further at 250 °C for a few minutes and later in a nonreactive environment at 600–1000 °C for carbonization.

4.2.2
NF Membranes Stable in Extreme pH Conditions

Koch also commercializes SelRO® membranes especially suitable for extreme pH conditions. Some of them are listed in Tab. 4.5.

Such membranes have been applied in the separation of heavy metals from acids and highly alkaline solutions, and in the recovery of alkaline solutions used in cleaning processes. Desal nanofiltration membranes commercialized by GE Osmonics can also work at very low pH levels. They have been used to recover heavy metals and clarify 35% sulfuric acid feed streams or 25% phosphoric acid streams. They have also been applied to permeate boric acid and reject radionuclides at a nuclear power station. Somicon/Nitto Denko announces nanofiltration membranes (NTR-7410, 7430 and 7450 HG) for pH range 1–12, temperatures up to 90 °C and a maximum pressure of 50 bar [37].

Table 4.5 pH Stable SelRO® nanofiltration membranes.

Membrane	Cutoff (g/mol)	pH range	Maximum temperature (°C)	Maximum pressure (bar)
MPT-34	200	0–14	70	35
MPT-36	1000	1–13	70	35

Tanninen et al. [57] reported recently the comparison between the Desal™ 5 membrane with another Desal membrane, Desal KH, the Dow NF270 and two membranes developed by BioPure Technology, Israel, BPT-NF-1 and BPT-NF-2. The BPT membranes were based on a surface layer of melamine polyamine on a polyethersulfone support. For the separation of sulfuric acid from metal salts all the membranes had a good performance, but only Desal KH and BPT-NF-2 maintained their selectivity during 2 months of operation with 8 wt% sulfuric acid at 40 °C.

4.3
Membranes for Ultrafiltration

Table 4.6 lists some commercially available membranes for ultrafiltration. The development of UF membranes form different polymer materials is discussed below.

4.3.1
Polysulfone and Polyethersulfone

UF membranes are usually prepared by phase inversion. The most widely used polymer for the preparation of UF membranes is polysulfone (PSU) or polyethersulfone (PES).

Polysulfone

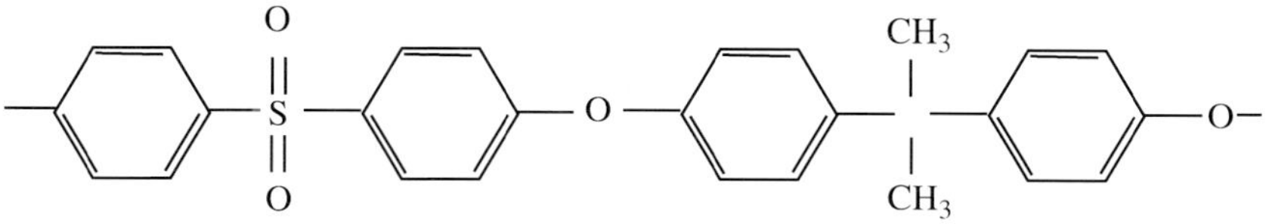

Poly (ether sulfone)

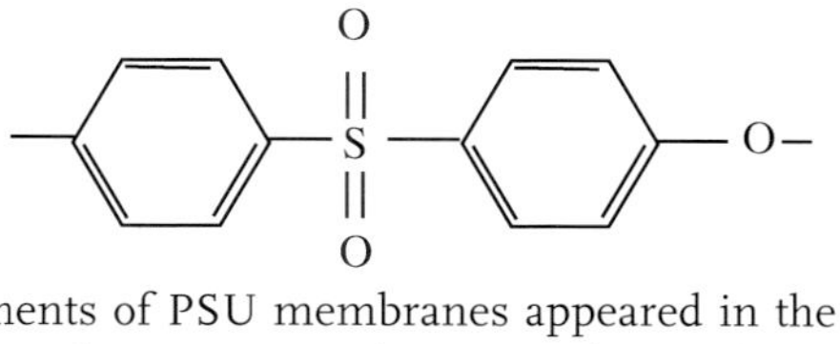

The first developments of PSU membranes appeared in the 1960s as an alternative to cellulosic membranes. Since then several procedures have been described in the literature for PSU membranes [58, 59], in many cases using the high molecular weight polysulfone Udel P-3500 commercialized by Solvay. A great advantage when compared to cellulose acetate is its resistance in extreme pH conditions, as well as its thermal stability. PSU has a T_g of 195 °C and PES even higher, at 230 °C. Both PSU and PES are soluble in chloroform and dimethylformamide, and are easily applied in conventional phase-inversion processes. This high solubility is also the main drawback of PSU as a membrane material, elim-

Table 4.6 Some commercially available membranes for ultrafilration.

Trade name	Membrane material	Cut off (g/mol)	Chlorine tolerance (ppm days)	pH operation range	Max. operating temperature (°C)	Typical operating pressure (psig)	Applications
GE Osmonics							
G-10	Thin film	2500	20–50	2–11	50	70–400	Surface water treatment, color, dye purification
G-20	Thin film	3500	500				
G-50	Thin film	8000	1000			50–200	RO pretreatment, color reduction, colloidal silica and iron removal, oil-water
P-UF membrane elements	PES	10000	5000			80–135	
M-Series	Modified PAN hydrophilic	0.03–1.0 μm pore		2–10	90		Oil–water separation
Millipore							
Biomax®	PES composite	10000		1–14			Concentration of vaccines, clarification, buffer exchange of protein solutions
Amicon® PM	PES on PE						
Ultracel® PLC series	Cellulose on PE	5000–1000000		2–13			Purifying and filtrating therapeutic proteins, DNA, RNA

inating the use of polysulfone-supported membranes in the processing of solvent-based feed solutions. It is also a problem coating the PSU support with polymers, which are only soluble in organic solvents. Another disadvantage of PSU and PES membranes is their hydrophobic character, which prevents spontaneous wetting with aqueous media. Consequently, the membrane must be prevented from drying completely or the membrane must be treated with a hydrophobic agent, glycerin, for example, before drying. Another serious disadvantage of hydrophobic materials consists in the fact that they often possess a powerful nonspecific adsorption capacity. This phenomenon, known as fouling, leads to a rapid deterioration of the membrane permeability.

Suggestions for hydrophilic membranes have already been proposed that do not suffer from these disadvantages. Several procedures have been proposed to make membrane surfaces more hydrophilic and they will be discussed later. One effective way to make polysulfone membranes more hydrophilic is to prepare the membranes from a mixture of sulfonated and nonsulfonated polysulfone [60, 61]. The sulfonation may be controlled to limit the water solubility of the resulting polymer. Insolubilization can also be achieved by crosslinking with additives such as polyols or polyphenols. Sulfonation has been a successful alternative to the incorporation of other hydrophilic polymers, which are soluble in water. Preparation of membranes from polymer blends with hydrophilic polymers has been well described in the literature [59, 62]. Polyvinylpyrrolidone is one of the commonly applied polymers for this purpose [18]. Another common additive is poly(ethylene glycol) [63]. However, hydrophilization of membranes by using large quantities of water-soluble polymers has the disadvantage that the hydrophilic nature of the membrane constantly decreases when they are used in aqueous media, since the water-soluble polymer is washed out. Although not completely described, sulfonated polysulfone probably coats polysulfone supports in the G-series membranes commercialized by Desal.

Sulfonated polysulfone seems to also play an important role in nanofiltration and reverse osmosis membranes commercialized by Desal. According to Petersen [34] the Desal-5 membrane appears to consist of three layers: a microporous polysulfone, a sulfonated overlay and a top ultrathin layer based on polypiperazineamide.

4.3.2
Poly(vinylidene fluoride)

$-(CF_2-CH_2)-$

PVDF

Poly(vinylidene fluoride) (PVDF) is quite interesting in the manufacture of UF membranes due to its chemical resistance. PVDF is resistant to most inorganic and organic acids and can be used in a wide pH range. It is also stable in aromatic hydrocarbons, alcohols, tetrahydrofurane and halogenated solvents. Furthermore it is resistant to oxidizing environments including ozone, which is used in water sterilization. PVDF is semicrystalline with a very low T_g ($-40\,°C$), which makes it quite flexible and suitable for membrane application in temperatures ranging between -50 and $140\,°C$, just prior to its melting temperature. Although stable in most organic solvents, PVDF is soluble in dimethyl formamide, dimethyl acetamide (DMAc), N-methyl pyrrolidone (NMP) and dimethylsulfoxide, making membrane preparation by phase inversion possible. In an early patent [64] on PVDF membranes, solutions containing about 20% PVDF in DMAc were cast and immersed in a methanol bath. Later [65] a preparation was described using a casting solution in DMAc, containing also ca. 17% isopropanol and using an immersion bath with about 40% water, 50% DMAc and 7% isopropanol. Membranes are also prepared from solutions in NMP containing lithium chloride and an immersion bath of methanol. US Patent 4203848 [66] describes the preparation of PVDF membranes by dissolving the polymer in boiling acetone and immersing it in a cold water/acetone bath. Another interesting solvent is triethyl phosphate (TEP) [67], a basic solvent, which complexes with the acidic PVDF [5]. The morphology of a PVDF membrane obtained from a solution in DMAC is shown in Fig. 4.2. Like polysulfone, PVDF is highly hydrophobic and many attempts to make them more hydrophilic have been described in the literature. One procedure is the chemical treatment with a strongly alkaline solution either in the presence of an oxidizing agent [68] or with a polymerization initiator and monomers such as acrylic acid [69].

The membrane surfaces have also been grafted or coated with polyacrylamide, poly(acrylic acid) [70, 71], poly(vinyl alcohol) and cellulose derivatives [72]. Another possibility for improving the membrane properties is the use of polymer blends. Blends of PVDF/PVP [73, 74], PVDF/poly(ethylene glycol) (PEG) [75], PVDF/sulfonated polystyrene [76], PVDF/poly(vinyl acetate) [77] and PVDF/poly(methyl methacrylate) [78] have been used in the preparation of microporous membranes.

4.3 Membranes for Ultrafiltration | 27

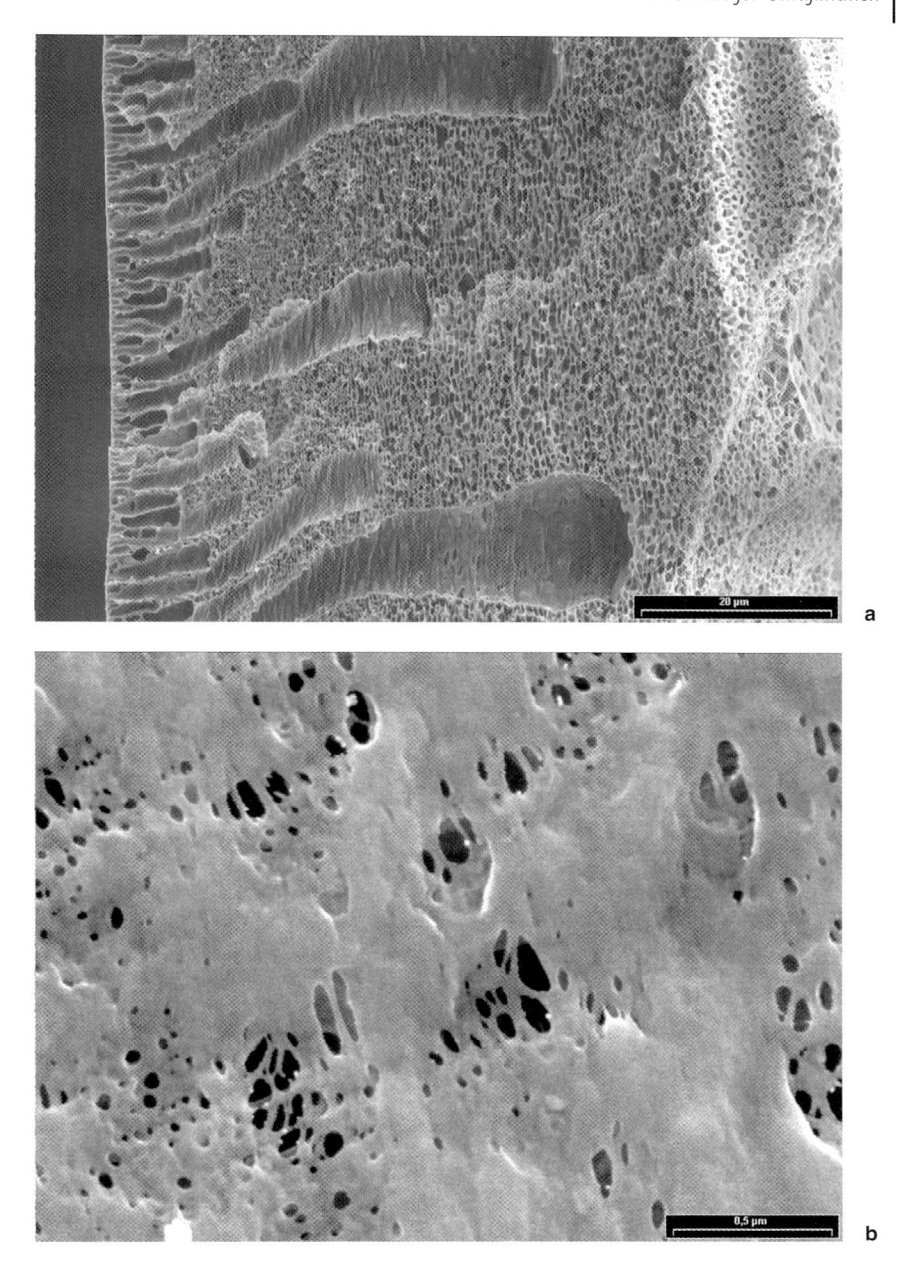

Fig. 4.2 Poly(vinylidene fluoride) membrane: (a) cross section and (b) surface membrane.

4.3.3
Polyetherimide

Polyetherimide (PEI) is an amorphous polymer with T_g near 200 °C. It can be used at temperatures higher than PVDF and it is known for its superior strength. The chemical stability, although much higher than cellulosic polymers, is lower than that of PVDF. PEI can not be used in contact with chloroform and dichloromethane. It is also attacked by tetrahydrofurane. The stability at high pH is poorer than that of PVDF, PSU or PAN. The preparation of membranes from PEI solutions leads to a large variety of porous asymmetric structures, which can be controlled by changing the composition of the solvent mixture. The commercial polymer usually chosen is Ultem® 1000, manufactured by General Electric. Integral asymmetric membranes are quite successful in gas separation [79], particularly in the case of helium recovery. More open asymmetric PEI supports have been used for ultrafiltration or as support for composite membranes. Usually, porosity higher than that of PVDF membranes is obtained with smaller average pore size, as shown in Fig. 4.3. Porous PEI membranes with a dense and thin top layer for gas separation were initially prepared from a solution in a mixture of dichloromethane, 1,1,2-trichloroethane, xylene and acetic acid and coagulated in acetone [79]. Another solvent mixture later allowed the preparation of PEI membranes for gas separation coagulated in water. For that, a mixture of tetrahydrofurane (THF) and gamma-butyrolactone (GBL) was used. Both THF and GBL are nonsolvents for PEI, but due to an effect of cosolvency a stable casting solution is obtained. Membranes with very thin top layers are particularly formed with higher GBL contents. The addition of volatile nonsolvents such as butanol to the casting solution leads to the formation of even thinner top layers. After coagulation in water a sponge-like structure is obtained. The preparation of hollow fibers from PEI solutions in a mixture of NMP and GBL or DMAc and GBL is described by Kneifel and Peinemann [31]. Here, the addition of GBL, a nonsolvent, slightly increases the solution viscosity and favors a sponge-like structure without finger-like cavities. Blends of PEI with other polymers are also reported to improve membrane characteristics [32]. Blends of PEI and poly(ether sulfonamide) were obtained as an attempt to improve the membrane hydrophilicity. In order to make both polymers compatible, 1,3-diamino propanol (DAP) was added to the polymer solution. DAP reacts with PEI and increases the solution viscosity. As a result, besides the compati-

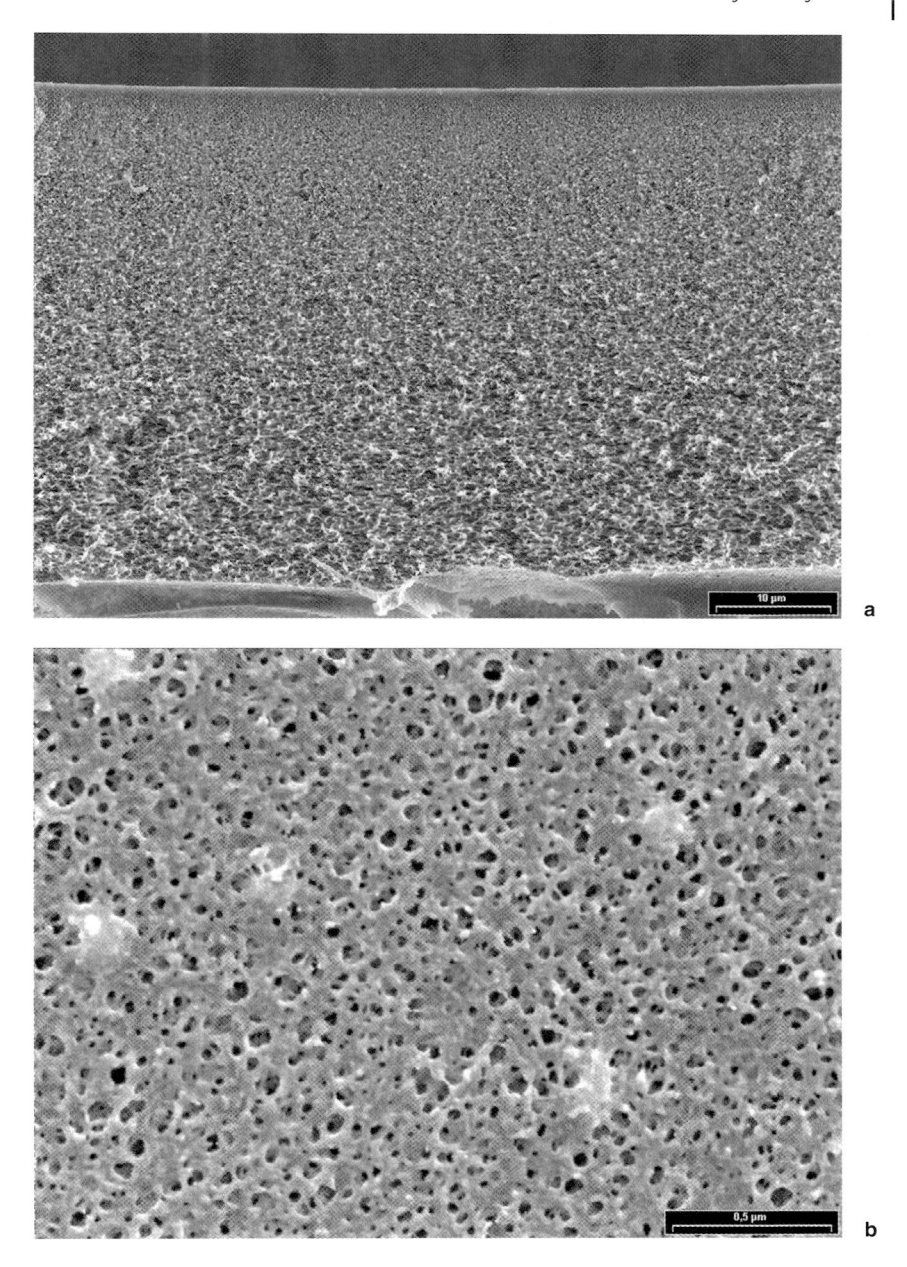

Fig. 4.3 Poly(etherimide) membrane: (a) cross section and (b) surface membrane.

bilization effect, the addition of DAP induces a sponge-like structure, eliminating the finger-like cavities that are usually observed in membranes obtained from PEI or PESA solutions in dimethyl acetamide. PEI is successfully used as a porous support for composite membranes. However, in order to improve their resistance to compaction at extreme conditions of high pressure and hydrocarbon atmosphere, an inorganic polymer was generated in the casting solution by hydrolysis of alkoxy silanes and incorporated in the membrane structure [80]. Here again compatibility was necessary and was achieved with the introduction of amino silanes. Also for the preparation of asymmetric porous membranes a blend of PEI polyimide with phenylindane groups (Matrimid) was reported with the purpose of improving the PEI gas permeation in hollow fibers [81].

4.3.4
Polyacrylonitrile

$$-(CH_2-CH)-$$
$$\quad\quad\quad |$$
$$\quad\quad\quad CN$$

Polyacrylonitrile (PAN) has been used in the preparation of UF membranes for a long time [82, 83] due to its superior resistance to hydrolysis and oxidation. PAN is highly crystalline and relatively hydrophilic and is usually copolymerized with more hydrophilic monomers to improve processability and to make it less brittle. Hollow fibers can be prepared from PAN dissolved in nitric acid [84]. Preparation of PAN membranes by phase inversion from solutions in DMAC, DMF or NMP is also possible. An example is shown in Fig. 4.4. A Sumitomo patent [85] discloses the preparation of membranes from copolymers containing 89% acrylonitrile and 11% ethyl acrylate dissolved in DMF and formamide and coagulated in water. A microporous membrane is obtained. In order to make the membranes suitable for reverse osmosis, they were submitted to a plasma treatment in the presence of 4-vinyl pyridine.

GE Osmonics commercializes a hydrophilic polyacrylonitrile UF membrane, developed and patented by Membrex, Inc., for use in a spinning-disc separation device [86]. The UltraFilic M-series membrane is extremely hydrophilic having water contact angle 4°, much lower than that of the unmodified PAN, which is about 46°. This allows water to pass freely through the membrane, whereas oil, fat and grease are repelled. For ultrafiltration of water produced in oil and gas fields, this is a great advantage. For cleaning the membrane flushing with hot water is sufficient, while for most conventional membranes, complicated cleaning regimens are required.

4.3 Membranes for Ultrafiltration | 31

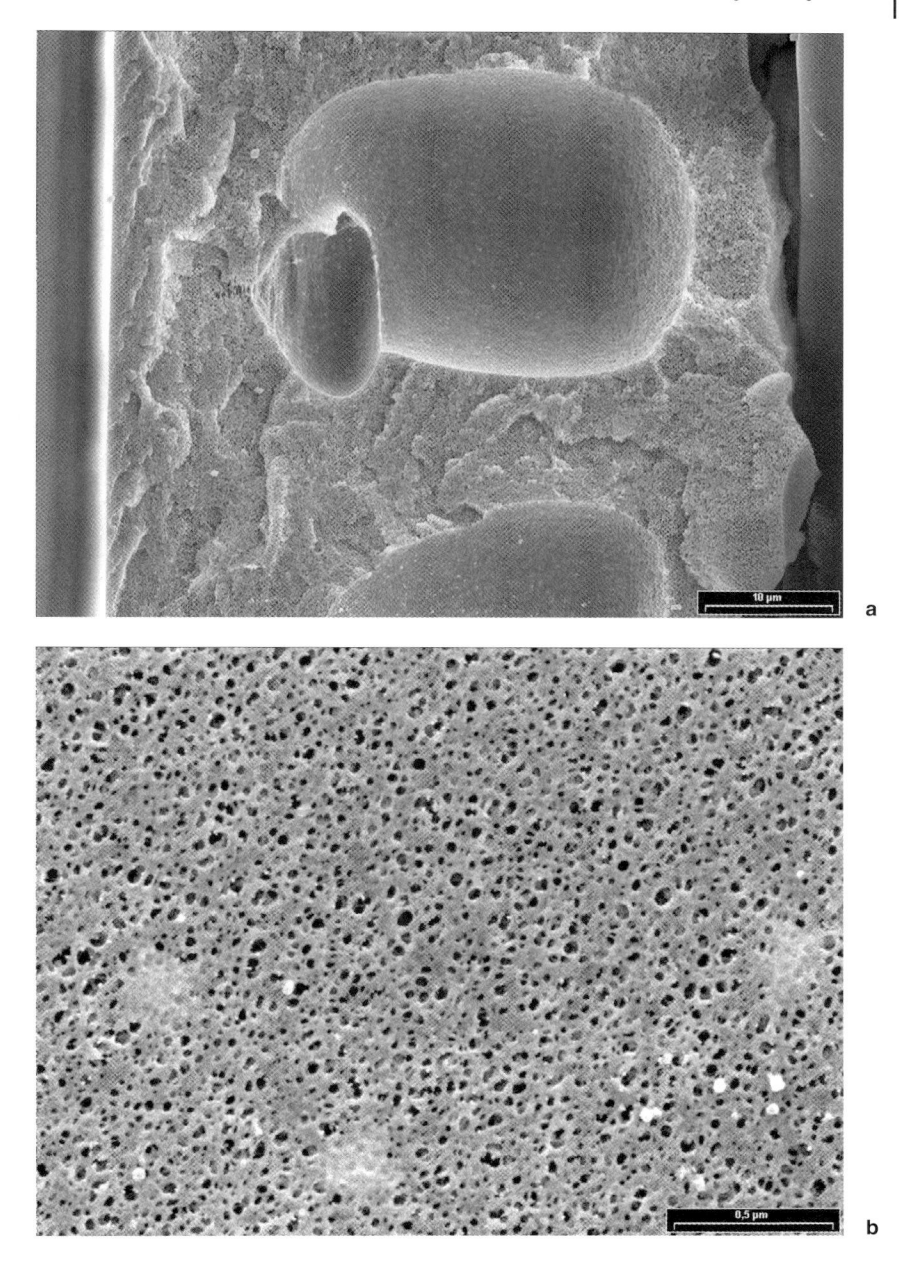

Fig. 4.4 Poly(acrylo nitrile) membrane: (a) cross section and (b) surface membrane.

4.3.5
Cellulose

Cellulose UF membranes are used in applications where low fouling characteristics are required. Cellulose has a very regular structure and is able to form strong intermolecular hydrogen bonds between the several hydroxy groups. As a result, cellulose is practically insoluble in almost all solvents. The only exceptions are dilute solutions in DMAc or NMP with addition of lithium chloride. Cellulose membranes are prepared by methods that basically involve precipitation from a solution of chemically modified native cellulose (from cotton linters, etc.). Until some years ago the three main methods were based on cellophane, cuprophane and cuenophane.

Cellophane membranes are prepared by a viscose process, in which cellulose is regenerated from a cellulose xanthate solution, as described in US Patents 981368 and 991267 [87, 88]. Cuprophane membranes are prepared in a similar way, regenerating cellulose from its soluble copper complex formed by reacting with ammoniacal copper sulfate, as described in US Patent 2067522 [89]. To obtain cuenophane membranes, cellulose is regenerated after dissolving it in cupriethylen diamine. For regeneration of cellulose from solution, a coagulation in strong alkali solutions is usually required. Today, most cellulose membranes are prepared by hydrolysis of asymmetric cellulose acetate membranes [90] also in strong alkali solutions. An alternative method for the preparation of cellulose membranes has been recently proposed; acid hydrolysis of trimethyl silyl cellulose [91].

4.3.6
Solvent-resistant Membranes for Ultrafiltration

Poly(vinylidene fluoride) (PVDF) porous supports coated with cellulose [92] were commercialized by Dow (ETNA membranes). PVDF is soluble only in a few organic solvents such as dimethyl acetamide. Cellulose is very stable in organic and aqueous solvents. However, because of its low solubility, preparing cellulose membranes is not a trivial task. Stengaard [72] proposed the preparation of composite membranes by coating chlorotrifluoroethylene/vinylidene fluoride (CTFE/VF) or PVDF supports with hydroxyethylcellulose and hydroxypropylcellulose, which are water soluble. In order to fixate these materials on the membrane surfaces they have to be crosslinked and/or chemically bonded to the support. PVDF and CTFE/VF copolymer can be made reactive in the following way. Under highly alkaline conditions and elevated temperatures hydrogen fluoride/hydrogen chloride are set free, while reactive groups and double bonds are formed. The support is then treated with a solution of hydroxyalkylcellulose in the presence of NaOH and a crosslinking agent such as 1,3-dichloro-2-propanol, which forms ether bonds with hydroxyalkylcellulose via the OH groups. A stable hydrophilic layer is formed on the top of the PVDF support. The coating may just hydrophilize the UF membrane or close the pores bringing the cutoff into the range of nanofiltration.

A solvent-resistant membrane based on PAN was developed by Hicke et al. [93]. The membrane was prepared from poly(acrylonitrile-coglycidyl methacrylate). The membrane with defined epoxy content could be prepared by a conventional phase-inversion process and stabilized by a post-treatment with a solution containing about 25% ammonia. As a result of the ammoniolysis reaction, a crosslinked membrane with high solvent resistance was obtained.

Polyetherketones are engineering thermoplastics with an exceptional combination of heat and chemical stability. However, their high insolubility in all common solvents, a property that assures the successful application of the membrane in chemical processes, inhibits their production by conventional solution casting methods. In order to bypass this difficulty, alternative preparation methods have been reported in the patent literature. One possibility is the partial sulfonation of PEEK with sulfuric acid and preparation of the membrane by coagulation from the sulfuric acid solution [94]. The final membrane characteristics are, however, affected since the sulfonated PEEK is then soluble in common solvents. Furthermore, sulfonated PEEK swells in aqueous solution, which prejudices the performance of the membrane in water. Membranes have also been prepared from blends of poly(aryl ether ketone) and poly(ether imide) (PEI) submitted to solvent leaching (of PEI) to form the pores [95]. A symmetric membrane with low porosity is formed. Ionics [96] proposed the preparation of membranes by casting from solutions in strongly protic, nonreactive acids such as methanesulfonic acid. Dow described [97] the extrusion of PEEK with a plasticizer, followed by coagulation in a nonsolvent bath and leaching of the plasticizer. Drawing before, during and/or after leaching improves the flux of the membranes.

Poly(phenylene sulfide) (PPS) is another chemically stable polymer. Analogously to the procedure for the preparation of a PEEK membrane, Dow proposed in a recent patent the preparation of PPS membranes by extrusion followed by leaching. Polyimide membranes for ultrafiltration have long been the subject of Nitto patents [98]. Although insoluble in many common solvents such as alcohols, ketones, ethers and esters, solutions in dimethyl formamide may be manufactured into membranes by casting/coagulation procedures. Polyimide membranes with high solvent resistance are claimed by Bend Research, by casting from polyamic acid solutions [99].

Porous polybenzimidazole (PBI) membranes were also developed in an attempt to overcome the temperature and chemical stability limitations of other membranes. Procedures for preparation of PBI membranes by phase inversion are long known [100]. The polymer can be dissolved in dimethyl acetamide and coagulated in a bath containing a mixture of solvent and nonsolvent to hinder the excessive formation of finger-like cavities [101]. Dense PBI membranes, treated with phosphoric acid are now being considered for application in fuel cells [102]. Phosphazenes are also valuable as a membrane material with high chemical stability. They have been explored for ultra- and nanofiltration, pervaporation, gas separation [103] and more recently for application in fuel cells [104]. Although commercial membranes are not available, UF phosphazene membranes have been prepared by phase separation with cutoffs between 70 000 and 500 000 g/mol and water fluxes up to 30 m^3/m^2 day MPa with mean levels of stability in acetone or hexane. Nanofiltration phosphazene membranes on ceramic supports have been successfully used for the separation of organic dyes from isopropanol.

4.4
Membranes for Microfiltration

Several of the polymer solutions mentioned above for ultrafiltration can also be used for the separation of microfiltration membranes by phase inversion. However, for the range of pore size useful for microfiltration, other procedures have also been successfully used for the preparation of commercial membranes. Among these are stretching, track-etching, leaching and sintering. Some commercially available MF membranes are shown in Tab. 4.7. Polypore integrated Celgard, Daramic and Membrana. General Electric integrated Pall and Osmonics (Desal). Microfiltration has been widely applied for water purification with bacteria removal and for obtaining ultrapure water for the electronic industry. A field that is increasing rapidly is the application of microfiltration membranes as battery separators.

4.4.1
Polypropylene and Polyethylene

Stretching is part of the preparation process both of the Celgard® and the Gore-Tex® membranes. Cold drawing was described already in 1969 [105] for membrane preparation starting from crystalline polymers. Another preparation method is solvent stretching [106], where the precursor film is brought into contact with a swelling agent and stretched. The swelling agent is removed while the film is maintained stretched to render the film microporous. Other processes use sequential "cold" and "hot" stretching steps [107]. The Celgard® membrane is made of polypropylene, which is a low-cost and quite inert polymer. It is resistant under extreme pH conditions and is insoluble in most sol-

Table 4.7 Some commercially available membranes for microfiltration.

Membrane	Membrane material	Pore size (µm)	Porosity (%)	Max. operating temperature (°C)
Millipore				
Fluoropore (hydrophobic)	PTFE on HDPE	0.2–3.0	85	130–260
Mitex (hydrophobic)	PTFE	5–10	65	260
LCR	Modified PTFE	0.45	80	130
Omnipore (hydrophilic)	Modified PTFE	0.1–10		
Polypore				
Celgard	Polypropylene	0.1–0.2 × 0.05	37–55	135
Daramic	Ultra high molecular weight polyethylene	0.1–1	60	
Darak	Duroplastic resin in polyester support	0.3–0.7	70	
GE Water Technologies				
Pall Emflon	PTFE	0.02–1.0		
Pall Ultipor N66	Nylon N6,6	0.1–0.2		
Desal J-Series	PVDF	0.3		
Desal E-Series	Polysulfone	0.04		
Desal K-Series	PTFE	0.1–3		

vents at room temperature. It swells, however, in nonpolar solvents such as carbon tetrachloride. No solvent is required for the preparation of the membrane. It involves the extrusion of PP films with high melt stress to align the polymer chains and induce the formation of lamellar microcrystallites when cooling. The film is then 50–300% stretched just below the melting temperature. Under stress, the amorphous phase between the crystallites deforms, giving rise to the slit-like pores of the Celgard® membrane (Fig. 4.5). The film is then cooled under tension. PP is highly hydrophobic and several surface treatments have been proposed to improve the hydrophilicity of PP membranes. Incorporation of surfactants is used to make Celgard® membranes more hydrophilic.

Besides being used in medical applications, Celgard® is used as a separator in batteries. Celgard® now belongs to Polypore, which also integrates Daramic Inc., a leading company in the production of battery separators. The automotive sector of the world market has grown to about 300 million batteries per year. 90% of the batteries are built with microporous pocket separators [108]. Microporous separators based on ultrahigh molecular weight polyethylene have a pore size of about 1 micrometer. The separator acts as an electric insulator between electrodes, while allowing ion migration. Additionally, the use of membranes

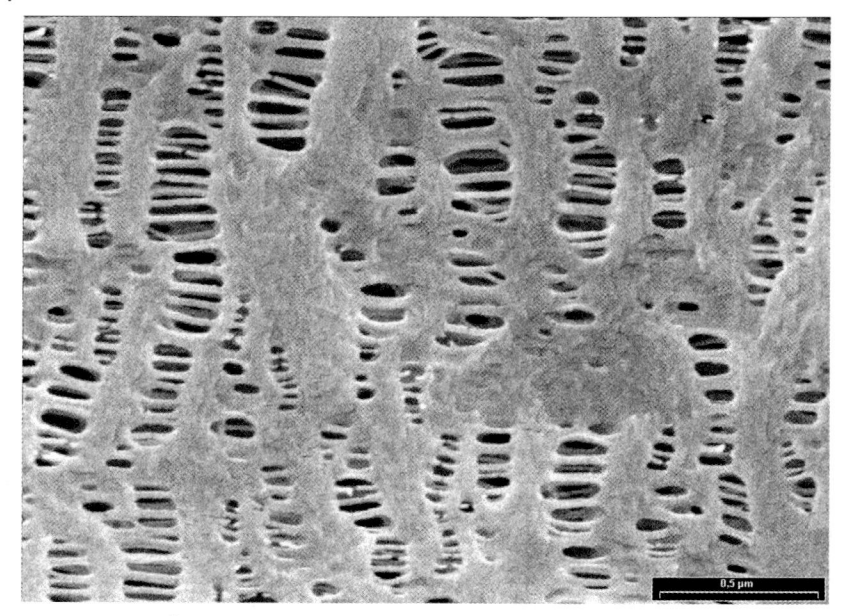

Fig. 4.5 Celgard® membrane.

with pores in the micrometer range allowed the further development of batteries and prevents problems like penetration of lead particles, which precipitate from the electrodes in lead alloy batteries. The increasing demand on chemical stability of separators for operation at more elevated temperatures has also stimulated the optimization of the manufacturing process, leading to membranes like Daramic "High Performance" and "DuraLife" with much higher oxidation stability.

4.4.2
Poly(tetrafluoroethylene)

Another commercial membrane prepared by stretching is GoreTex® (Fig. 4.6). The polymer here is poly(tetrafluorethylene), which makes the membrane extremely inert and thus convenient for processing even harsh streams. Processing PTFE is only possible by paste extrusion. In paste forming the polymer is mixed with a lubricant such as odorless mineral spirits naphtha or kerosene. The lubricant component is removed by heating to 327 °C. Above this temperature, sintering would lead to a dense PTFE film.

After lubricant removal, the PTFE film is submitted to an uniaxial or biaxial stretching, giving rise to an interconnected pore structure. The process was proposed by Gore [109] and the resulting porous film is today a successful product in the membrane and textile industry. For uniaxial stretching, the nonsintered film from the paste extrusion is fed to a machine with heating rollers, where

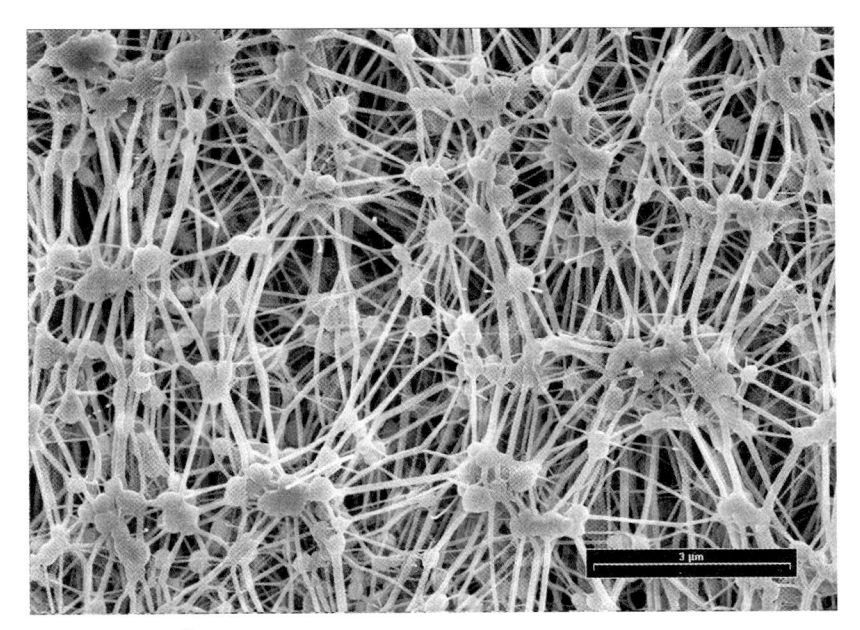

Fig. 4.6 GoreTex® membrane.

one roller is driven faster than the previous one to input stress and induce pore formation. The difference in speed determines the amount of stretch. Additionally in the Gore patent a biaxial stretching is performed using a pantograph. A special characteristic of the Gore membrane is that, since PTFE is very hydrophobic, (liquid) water must not be allowed to wet the membrane and its transport is hindered. On the other hand, water vapor can freely pass through the micropores, making the film suitable for transpirating impermeable cloths. However, due to their inertness, PTFE membranes are also interesting during the processing of aggressive streams. If the goal is the filtration of aqueous wastes the membranes should be modified to become hydrophilic. A solution is disclosed in a later Gore patent [110], mixing PTFE with silica and a surfactant to form the paste for the extrusion. To improve the adhesion of the filler to the PTFE matrix, the membrane, in the last step of preparation, is heated with a solution of dimethyl octadecyl chlorosilane in toluene.

4.4.3
Polycarbonate and Poly(ethylene terephthalate)

Dense films of polycarbonate or poly(ethylene terephtalate) can be transformed into porous microfiltration membranes with very narrow pore-size distribution (Fig. 4.7), by exposing them to fission fragments from radioactive decay with subsequent etching in alkaline solutions. The number of pores can be controlled by the length of exposure to the fission segments. The maximum pore

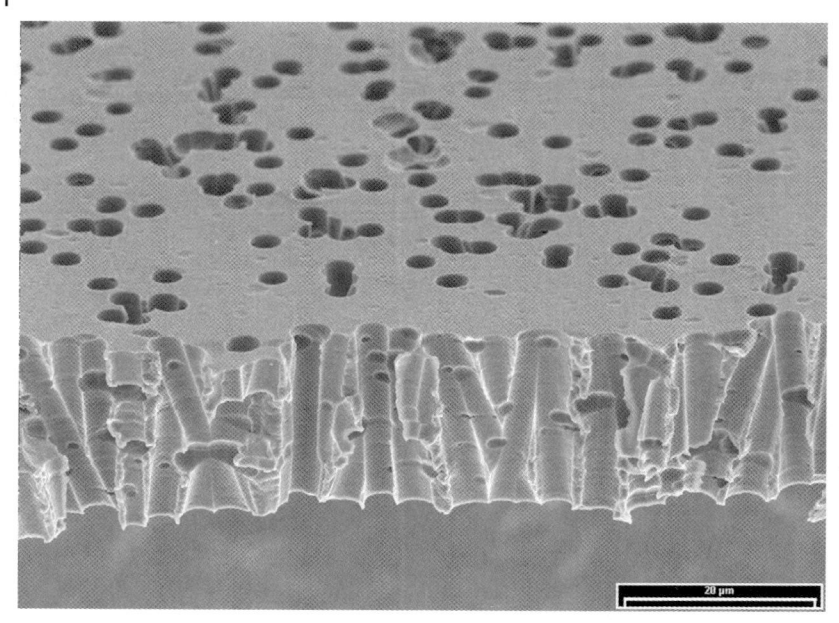

Fig. 4.7 Poly(ethylene terephthalate) membrane.

density is limited by the fact that membranes become excessively brittle and radioactive at very high doses. The size of the pores is controlled by the etching conditions and the length of time in the etching bath. The shape of the nuclear pores (cylindric or conic) is determined by V, the ratio between the rate of etching of the bulk membrane material and the rate of etching of the material along the high-energy particle track. For instance, in poly(ethylene terephthalate) films tracked by argon ions, $V=10–100$ and the pores are less cylindrical than in films tracked by xenon ions, for which $V=100–1000$ [111].

5
Surface Modification of Membranes

Although membrane preparation has been reported concerning a large variety of polymers, it is often not possible to combine the best characteristics needed for the application using just one polymer. It is frequently observed for instance, that polymers with the best solvent resistance or those that provide the most convenient pore structure are too hydrophobic to have an acceptable performance in the filtration of aqueous solutions. On the other hand, chemical modification of the polymer chain prior to membrane formation usually drastically changes the resulting pore structure. Therefore several procedures have been researched to chemically modify the surface of a previously formed porous membrane in order to increase the hydrophilic character or to allow functionalization and incorporation of polymer segments. These may improve selectivity, raise biocompatibility or bond catalytically active groups for membrane reactors. Chemical modification of polymer surfaces has been reviewed [112].

5.1
Chemical Oxidation

In many cases hydrophilicity is the main goal of membrane-surface modification. Surface oxidation is the simplest way to attain this goal. Several methods are available, including glow and corona discharge [113] and surface flaming, which are quite old but still very effective methods particularly in the case of polyolefins. Additionally, one of the earliest surface treatments, still effective, is the exposure to oxidizing chemicals such as chromic acid, nitric acid or potassium permanganate, which lead to the formation of carbonyl, hydroxyl and carboxylic groups on the surface of polyethylene, polypropylene and polyesters. One procedure for the modification of polypropylene membranes [114] consists of reacting the membrane in an aqueous solution of heated potassium peroxy disulfate to produce oxygen-centered radicals that are responsible for introducing hydroxyl groups. If the process is performed in the presence of monomers such as acrylamide, grafting takes place.

5.2
Plasma Treatment

Plasma is a complex gaseous state of matter, consisting of free radicals, electrons, photons, ions, etc. Plasma can be generated by continuous electrical discharge in either an inert or a reactive gas. For membrane application, plasma can be used to improve the characteristics both of porous supports and of polymer films for gas separation. Several examples have been reviewed by Kramer et al. [115]. Porous membranes can be submitted to plasma treatment to achieve the following effects: (i) crosslinking of the top layer and reduction of pore size; (ii) introduction of functional groups to the surface or (iii) grafting and deposition of a thin selective layer on a porous substrate. In the former instance, plasma treatment with inert gases such as argon or helium leads to ablation of the substrate material by the excited plasma molecules and then redeposition of the substrate material as a highly crosslinked layer on the surface. If the time of exposure is limited, a controlled reduction of pore size is obtained and a microfiltration membrane can be transformed into an ultrafiltration or even reverse osmosis membrane. Argon plasma was used to reduce the pores of polyethylene hollow fibers. It has been reported that helium plasma modifies the surface of porous polyacrylonitrile membranes, making them suitable for reverse osmosis. A second application is the incorporation of functional groups. Plasma treatment with air, oxygen and water vapor introduces oxygen-containing functional groups on the surface. Nitrogen, ammonia and alkyl amine plasmas introduce nitrogen-containing functional groups. Treatment of PAN and polysulfone membranes with helium/water plasma, as reported by Belfort and Ulbricht [116, 117], turns them hydrophilic and minimizes fouling. It is interesting to note that amino groups are, for instance, interesting to bind heparin and retard blood coagulation in membranes used in medical applications [115]. Ammonia plasma was also used to improve flux and selectivity of UF-polysulfone membranes [118]. Nitrogen and oxygen plasma have been used to improve the hydrophilicity of poly(vinyl chloride) membranes. Hydrophilization can also be obtained by plasma-induced graft polymerization with the incorporation of hydrophilic monomers. The plasma is then generated from gaseous organic monomers that polymerize and eventually crosslink on the membrane surface. The conditions may also be adjusted to lead to a dense selective polymer coating. One advantage of the plasma treatment is the absence of solvents or any hazardous liquid. A weak point is the complexity of the coating, which is difficult to predict. Several examples are reported in the literature. Ulbricht and Belfort polymerized hydroxy ethyl methacrylate on polysulfone membranes. Acrylic acid was deposited on commercial membranes to improve solute rejection during the ultrafiltration of bleach effluents [119].

5.3
Classical Organic Reactions

The well-defined surface functionalization using classical organic reactions plays an important role in membrane development. In order to be susceptible to reaction, the polymer chain should contain double bonds, hydroxyl groups or benzene rings. An example is the modification of polysulfone by reaction with different chemicals to increase hydrophilicity. The surface modification of polysulfone membranes has been reported by several authors [120, 121].

5.4
Polymer Grafting

The covalent bonding of polymer segments and chains to porous supports can be achieved by polymer graft (Tab. 5.1), which is, to a great degree, based on the free-radical reaction of vinyl or acrylic monomers. Reactive sites on the polymer support, usually in the form of unpaired electrons, can be created by i) UV irradiation in the presence of initiators such as benzophenone and the chosen monomer; ii) reactive sites are also thermally created as a result of the decomposition of organic peroxides; iii) the generation of unpaired electrons by exposure to high-energy radiation, such as gamma or electron irradiation.

Grafting of hydrophilic vinyl monomers, such as hydroxyethyl methacrylate on polysulfone and PAN membranes under UV exposure to make them less susceptible to fouling, has been described in the literature [129, 130].

Photografting has also been reported as a strategy to prepare membranes for affinity separation [145]. Functionalization of PAN has been reported for the manufacture of molecular imprinted membranes for molecular recognition. The membranes are prepared in the presence of selected templates, which are then removed leading to the formation of complementary imprinted sites [146, 147].

Table 5.1 Different monomers and initiation methods for polymer grafting.

Monomer	Substrate	Reference
UV-Photoinitiated		
Acrylic acid	PE	122
	PP	123
	PET	124, 125
Glycidyl acrylate	PE	126, 127
	PS	127
	PP	127
Acrylamide	PE	122
	PP	123, 128
	PET	124, 125
4-Vinyl pyridine	PET	124
2-Hydroxy-ethylmethacrylate	PSU	129, 130
Thermally initiated		
Acrylamide	PE	131
Acrylic acid	PET	132
Methacrylic acid	PET	133
Acrylic acid/maleic anhydride	PP	134
Glycidyl methacrylate	PP	135
Fluorinated acrylic	Polyester	136
High-energy irradiation initiated		
Acrylamide	PE	137 (electron beam)
	LLDPE	138 (electron beam)
	PP	139 (gamma ray)
Acrylic acid	PP	140 (gamma ray, plasma)
		141 (gamma ray)
Fluoroalkyl methacrylate	PDMS	142 (plasma)
Vinyl pyrrolidone	TFB	143 (gamma ray)
Poly(ethylene imine)	PE, PET	144 (plasma)

PE = polyethylene, LLDPE = linear low density polyethylene
PP = polypropylene, PS = polystyrene, PET = poly(ethylene terephthalate),
PSU = polysulfone, TFB = poly(tetrafluoro ethylene-hexafluoropropylene)

The incorporation of positive charges has decreased the fouling susceptibility of membranes even more effectively. This is the principle of the aromatic polyamide membrane series commercialized by Hydranautics as low fouling composite membranes (LFC). Cationic charge-modified nylon membranes are also commercially available from CUNO 3M, under the trademark Zeta Plus®. Pall Corp. sells cationic charge-modified nylon membranes under the trademark N_{66} Posidyne. There are different ways to make the membrane positively charged. A patent from Millipore [148] describes the surface modification of hydrophobic membranes by contacting them with a solution of polyamine epichlorohydrin containing quaternary or ternary ammonium functional groups and acrylate monomers, which can polymerize under UV irradiation onto the surface. A procedure for incorporating positive charges on membranes [149] is disclosed in the US Patent 4473475 [150]. Here a charge modifying agent such as tetraethylene pentamine is bonded to the hydrophilic sites of the polyamide membrane through a polyepoxide crosslinking agent. The membrane surface-modifying polymers, disclosed in the Pall patent [151], are cationic, water-soluble, quaternary ammonium, thermosetting polymers such as the epoxy-functional polyamine epichlorohydrin.

6
Membranes for Fuel Cells

The research on ion-exchange membranes has grown considerably in recent years with the interest in fuel-cell technology for the automotive and portable applications [152]. The most promising fuel-cell technology for low temperature operation makes use of a polymeric membrane separating the anode and cathode of an electrochemical cell.

In the low-temperature fuel cell, the membrane is the solid electrolyte responsible for proton transport and separation of the fuel (hydrogen or methanol) from the oxygen in the cathode, as shown in Fig. 6.1. The requirements for a good membrane are: high chemical stability, high proton conductivity, low fuel and oxygen permeabilities, high electronic resistance and preferentially low cost. The membrane surfaces are in direct contact to a finely dispersed catalyst mainly on Pt base and the electrodes. The catalyst promotes the electrochemical oxidation of hydrogen with proton formation. As soon as the protons are formed, high proton conductivity should assure their fast transport from the an-

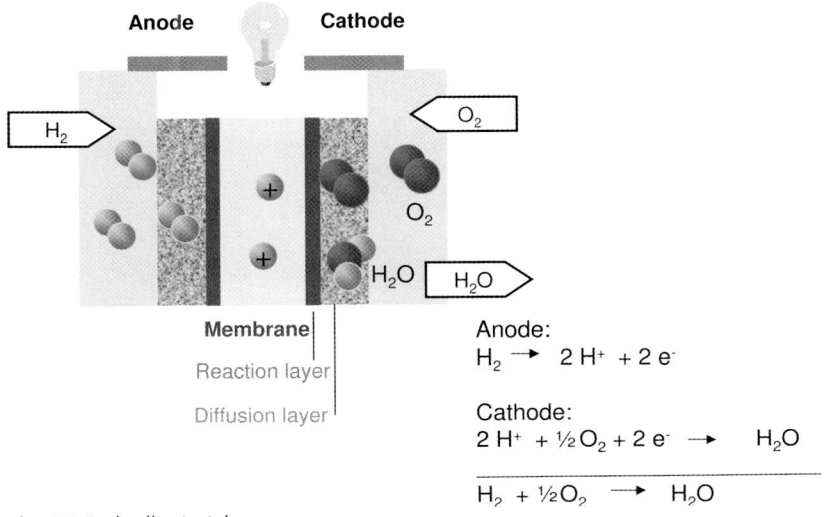

Fig. 6.1 Fuel-cell principle.

Anode:
$H_2 \rightarrow 2 H^+ + 2 e^-$

Cathode:
$2 H^+ + \frac{1}{2} O_2 + 2 e^- \rightarrow H_2O$

$H_2 + \frac{1}{2} O_2 \rightarrow H_2O$

Membrane Technology in the Chemical Industry.
Edited by Suzana Pereira Nunes and Klaus-Viktor Peinemann
Copyright © 2006 WILEY-VCH Verlag GmbH & Co. KGaA, Weinheim
ISBN: 3-527-31316-8

ode to the cathode. In the interface regions, in contrast to the bulk, high electron conductivity is an additional requirement to support the electrochemical reaction. If the cell is fed with a mixture of methanol and water (direct methanol fuel cell, DMFC), an analogous oxidation reaction occurs with formation of protons and CO_2. While water usually favors the proton transport across the membrane, excess water causes cathode flooding, disturbing the catalyst access to oxygen and reducing the whole fuel-cell performance.

The first polymer to be used as fuel cell electrolyte was sulfonated poly(styrenedivinylbenzene) [153], which has a rather poor stability under oxidative operation conditions.

6.1
Perfluorinated Membranes

In 1966 the perfluorinated ionomer membrane from DuPont, Nafion®, was first used for fuel cell, bringing a large improvement as far as stability was concerned. Nafion® or similar perfluorinated membranes, commercialized by Asahi Glass (Flemion®), Asahi Chemical (Aciplex®) or Solvay (Hyflon® Ion) still dominate the market for fuel cells. The ionomer membranes are produced in the sulfonyl fluoride precursor form, a thermoplastic that can be extruded in films. The films are then hydrolyzed and converted to the acid form, which is no longer melt processable, and are insoluble in any solvent at temperatures below 200 °C [154, 155]. The main chain is perfluorinated (x and y only indicate the molar composition and are not related to a sequence length) and long hydrophilic side chains segregate forming clusters immersed in the hydrophobic fluorinated matrix. The clusters act as paths for the proton transport. The difference between the currently available perfluorinated membranes for fuel cells can summarized by listing the functional y monomer unit for each membrane:

Membrane	y monomer unit
Nafion®, Flemion®	$CF_2CFOCF_2CF(CF_3)OCF_2CF_2SO_3H$
Aciplex®	$CF_2CFOCF_2CF(CF_3)OCF_2CF_2CF_2SO_3H$
Dow	$CF_2CFOCF_2CF_2SO_3H$

6.1 Perfluorinated Membranes

The Dow membrane is no longer available. However, Solvay is producing membranes with similar short side chains.

Effectively thin ionomer layers can be obtained on porous polymeric supports giving membranes with good mechanical strength and lower resistance to proton transport. This is the strategy used for the preparation of the Gore-select® [156].

Although having high chemical stability and proton conductivity, Nafion® is expensive and has technical limitations:

1. Decrease of proton conductivity above 100 °C. Not only Nafion® but also most of the sulfonated polymers tend to dehydrate at these temperatures. There are reasons to operate the fuel cell at 130 °C or even higher temperatures: the reaction kinetics would be improved and the contamination of the catalyst by CO and the problems with water drag in the membrane would be minimized.
2. High permeability to methanol and water, which is a disadvantage for direct methanol fuel cells.

α, ββ-Trifluorostyrene

Alternative membrane materials are under investigation by several groups. Examples of alternative fluorinated monomers have been proposed by Ballard. Its 3rd generation membranes (BAM3G) are based on copolymerization of α,β,β-trifluorostyrene with a controlled degree of crosslinking and sulfonation, as described in the US Patents 5773480 and US 3341366 [157, 158].

Radiation grafting has been investigated now for several years as an attempt to balance cost and stability in the development of proton-conductive membranes. Partially fluorinated polymers like perfluoroethylenepropylene (FEP) [159–161] or poly(vinylidene fluoride)(PVDF) [162, 163] are submitted to gamma [159, 160] or electron [162, 163] radiation and immersed in a mixture of styrene and divinyl benzene (for crosslinking):

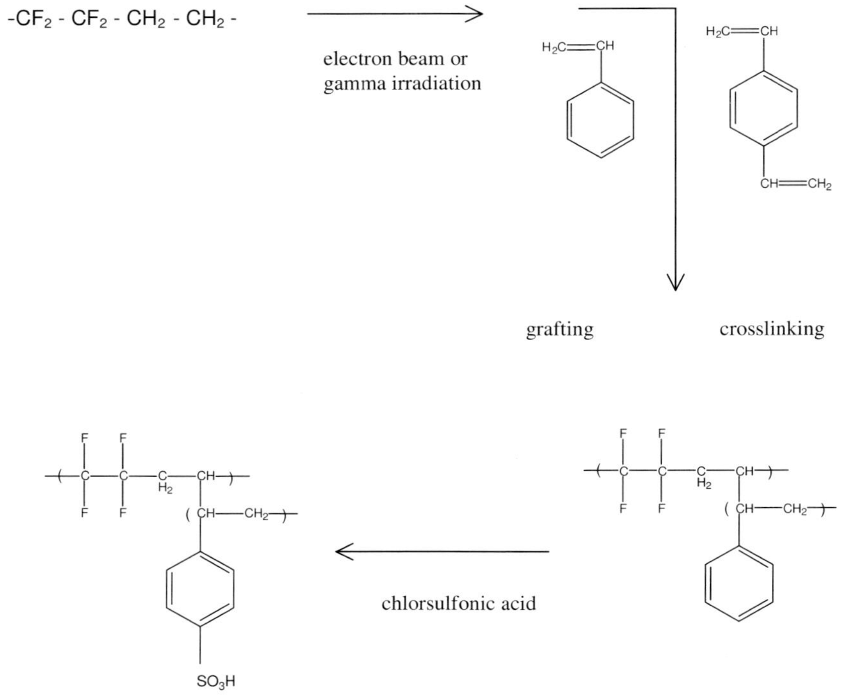

-CF₂ - CF₂ - CH₂ - CH₂ -   →

electron beam or
gamma irradiation

grafting   crosslinking

⟵ chlorsulfonic acid

Ion-exchange membranes based on radiation grafting technology are produced by Solvay. Different fluorinated polymers explored for fuel-cell membranes have been reviewed by Kostova et al. [164].

6.2
Nonfluorinated Membranes

A large number of nonfluorinated polymers are now under investigation. Some of these activities have been reviewed in recent publications [165]. Approaches include the sulfonation of commercial polymers and the polymerization of new functionalized polymers. One of the first polymers chosen for sulfonation was polysulfone [166–174]. More recently the synthesis of stable polysulfones from sulfonated monomers has been explored [175].

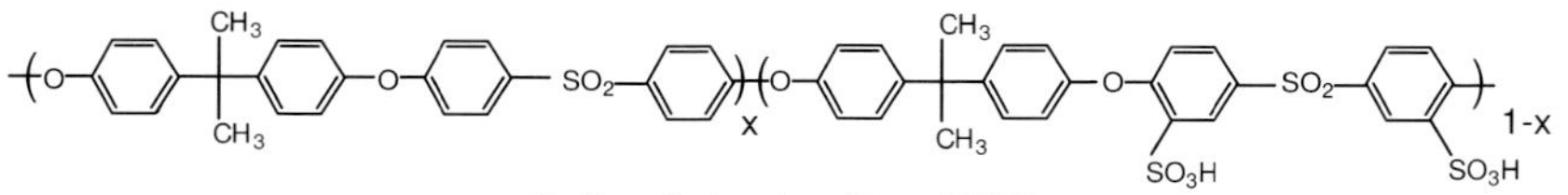

Sulfonated polysulfone [175]

Sulfonated poly(2,6-dimethyl-1,4-phenylene oxide) and copolymers have been considered by some groups [176–178] as proton conductive membranes. The substitution of the methyl groups by phenyl has been tested as one of the initial generations of Ballard membranes in an attempt to increase the stability to oxidation [179]. Even with the modified structure the membranes were unable to provide more than 500 h of continuous running.

Sulfonated polyimide with naphthalenic units

Polyimides have been the topic of investigation for fuel cells for many years [180–184]. By polycondensation of sulfonated amines and phthalic or naphthalenic anhydride it is possible to tailor statistic or block copolymers with good proton conductivity. Polyimides have some susceptibility to hydrolysis, better stability being achieved with naphthalenic structures [185]. Sulfonimide membranes [186, 187] have been investigated due their strong superacidity, water uptake and retention above 80 °C. However, although they are thermally quite stable, the hydrolytic stability is not high.

sPEEK

The investigation of different variants of sulfonated polyetherketones has been widely described in the literature polyetherketone [188], poly(ether ether ketone) [189–191], poly(ether ketone ketone) [192] and poly(ether ether ketone ketone) (sPEEKK) [193, 194], poly(oxa-p-phenylene-3,3-phthalido-p-phenylene-oxa-p-phenyleneoxy-phenylene) (PEEK-WC) [195]. An interesting comparison between different structures in this class of material (sulfonated poly(ether ether ketone) (sPEEK) and poly(phenoxy benzoyl phenylene)) (sPPBP) has been published by Rikukawa and Sanui [196]. Both polymers are isomers. In sPEEK the sulfonic groups are in the main chain, while in sPPBP they hang in a side chain. The water uptake at low relative humidity is higher for sPPBP as well as the conductivity at high temperatures.

sPPBP

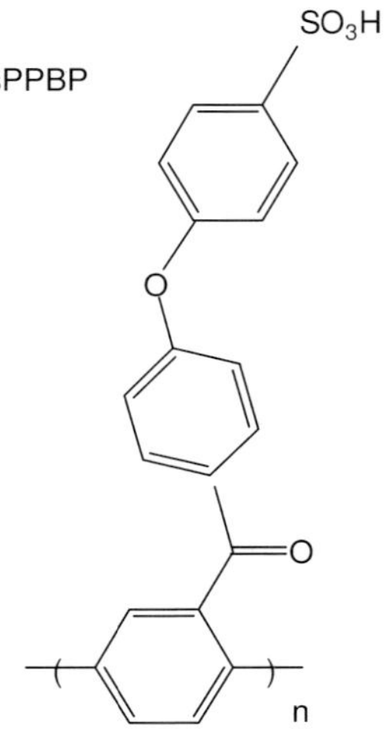

[chemical structure of sPPBP]

A comparison between sulfonated poly(ether ether ketone ketone) and Nafion® taking into account their microstructure and acidity was published by Kreuer [193, 194]. The Nafion® ionic channels are wider and less branched, while sPEEKK has dead-end channels. The hydrophobic matrix of Nafion® is more separated from the ionic phase and the polymer is more acidic. Three recent papers [189–191] report the sulfonation of PEEK and a systematic characterization, describing the influence of sulfonation degree on proton conductivity and swelling. Usually, the range of sulfonation degree useful for fuel-cell application starts about 40%. The proton conductivity and also the swelling increase with the sulfonation degree. Above 90% SD sPEEK is a highly swollen gel in water. Also, the solvent used for the membrane preparation is important. Dimethylformamide is reported to form strong hydrogen bonds between its -CHO groups and the -SO_3H of the sulfonated polymer, decreasing the proton conductivity.

Blends of sulfonated polymers with nonsulfonated polymers or with polymers containing basic groups have been well explored for fuel cells [197, 198].

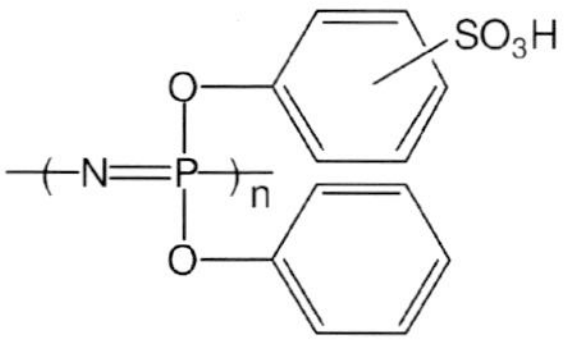

[chemical structure of sulfonated polyphosphazene]

Sulfonated polyphosphazene

Sulfonated poly(phthalazinones) are other polymer classes under investigation for fuel cells [199].

Polyphosphazene sulfonic acids [200, 201] offer a unique combination of inorganic backbone with high stability, high ionic density and structural diversity with crosslinking alternatives. Polyphosphazenes with sulfonamide functionalization have been reported with high proton conductivity [202].

6.3
Polymer Membranes for High Temperatures

A big challenge for fuel-cell membranes is a good performance at temperatures higher than 100 °C. Reviews on materials under investigation to overcome this problem have been published by Alberti and Casciola [203], Li et al. [204] and Hogarth et al. [205].

Polybenzimidazole (PBI) (initially manufactured by Hoechst-Celanese, now PEMEA) is one of the few polymers under consideration for high-temperature operation. The application of PBI [206, 207] and the noncommercial AB-PBI [208] in fuel cells was introduced by Savinell and coworkers. For that, the membrane was immersed in concentrated phosphoric acid to reach the needed proton conductivity. Operation up to 200 °C is reported [209]. A disadvantage of this class of membranes is the acid leaching out during operation, particularly problematic for cells directly fed with liquid fuels. Additionally, the phosphoric acid may adsorb on the platinum surface. A review on membranes for fuel cells operating above 100 °C has been recently published [209].

Phosphonated polymers have been proposed for fuel cells with the expectation of being thermally more stable and better retaining water than sulfonic groups [210, 211]. Phosphonated poly(phenylene oxide) [212], poly(4-phenoxybenzoyl-1,4-phenylene) [213] and polysulfones [214, 215] have been reported. Phosphonated fluoromonomers were polymerized [164]. Characterization of phosphonated films in terms of their proton conductivity has been reported for some of the phosphonated polymers: polyphosphazene [216], trifluoropolystyrene [217], poly(4-phenoxybenzoyl-1,4-phenylene) [218]. Relatively low conductivity values were reported for most of the polymers prepared up to now. The values for polyphosphazene [216] and for perfluorocarbon polymers [219] were quite encouraging. Phosphonated poly(phenylene oxide) [211] was evaluated in fuel cell-tests.

6.4
Organic-Inorganic Membranes for Fuel Cells

The combination of organic and inorganic materials to develop membranes for fuel cells has become a versatile approach. Early reports and patent applications of Stonehart and Watanabe [220] and Antonucci and coworkers [221, 222] claim the advantage of the introduction of small amounts of silica particles to Nafion® to increase the retention of water and improve the membrane performance above 100 °C. The effect is believed to be a result of the water adsorption on the oxide surface. As a consequence, the back-diffusion of the cathode-produced water is enhanced and the water electro-osmotic drag from anode to cathode is reduced [205].

Another important motivation for the development of organic-inorganic membranes for fuel cells is the reduction of methanol and water permeability, which are highly relevant aspects for the establishment of the DMFC technology. The inorganic phase can be included in different ways, starting from the simple dispersion of isotropic particles (SiO_2, ZrO_2, etc.), introduction of fillers with high aspect like layered silicates and the *in situ* generation of an oxide phase in the polymeric matrix by dispersion of inorganic precursors followed by their hydrolysis and polycondensation, as early reported by Mauritz [223]. By introducing silanes to a Nafion® solution and casting films it is possible to change the morphology of the ionic clusters [224]. The reduction of methanol and water permeability of sulfonated poly(ether ether ketone) membranes with the generation of ZrO_2 and SiO_2 in the membrane casting solution was reported in different papers [225–229]. These fillers are isotropic. Inorganic fillers like layered silicates, with high aspect ratios, are expected to lead to a more effective permeability reduction in membranes [230]. Clays and layered silicates have been used by different groups for the development of fuel-cell membranes. Examples are Nafion®/mordenite [231] and Nafion®/montmorillonite [232], Nafion®/sulfonated montmorillonite [233], SPEEK/montmorillonite [234, 235], sSEBS/montmorillonite [236]. At least as important as the aspect ratio is the surface modification of the inorganic filler [230].

Silica and silicates are usually rather passive fillers. Active fillers able to contribute also for the proton conductivity include zirconium [203, 237–250] and boron phosphates [251–253] and heteropolyacids [254–264].

7
Gas Separation with Membranes

7.1
Introduction

The separation of gas mixtures with membranes has emerged from being a laboratory curiosity to becoming a rapidly growing, commercially viable alternative to traditional methods of gas separation within the last two decades. Membrane gas separation has become one of the most significant new unit operations to emerge in the chemical industry in the last 25 years [265]. The gas-separation membrane module business for 2004 is estimated at $ 170 million with an annual growth rate of 8%. Tab. 7.1 shows commercial applications and some of the major suppliers of membrane gas-separation units.

Tab. 7.1 shows the established applications in the field of membrane gas separation. One of the new and currently small applications shown in Tab. 7.1 is natural gas dehydration. Problems related to this separation will be discussed in the last part of this chapter (basic process design considerations).

Besides the well-established applications there are a number of emerging membrane gas separations. These are, for example, natural gas hydrocarbon dewpointing, olefin/paraffin separation and separation of hydrocarbon isomers. These will be addressed in the following material section. The purpose of this chapter is to provide an overview of state-of-the-art and emerging materials for gas-separation membranes, to give some key features of integral asymmetric and composite membranes and finally to explain the influence of basic process parameters.

7.2
Materials and Transport Mechanisms

Organic polymers are the dominating materials for gas-separation membranes. Many polymers exhibit a sufficient gas selectivity and they can be easily processed into membranes. Palladium alloys are the only inorganic materials that are currently used for gas separation (ultrapure hydrogen generation) on a commercial scale. However, during the last decade inorganic materials have been developed with exciting unmatched selectivities for certain gas mixtures and

Membrane Technology in the Chemical Industry.
Edited by Suzana Pereira Nunes and Klaus-Viktor Peinemann
Copyright © 2006 WILEY-VCH Verlag GmbH & Co. KGaA, Weinheim
ISBN: 3-527-31316-8

7 Gas Separation with Membranes

Table 7.1 Gas membrane applications and suppliers.

Gas separation	Application	Supplier
O_2/N_2	Nitrogen generation oxygen enrichment	Permea (Air Products) Generon (IGS), IMS (Praxair) Medal (Air Liquide) Parker Gas Separation, Ube
H_2/hydrocarbons	Refinery hydrogen recovery	Air Products, Air Liquide Praxair, Ube
H_2/CO	Syngas ratio adjustment	as above
H_2/N_2	Ammonia purge gas	as above
CO_2/CH_4	Acid gas treating enhanced oil recovery landfill gas upgrading	Cynara (NATCO), Kvaerner, Air Products Ube
H_2S/hydrocarbon	Sour gas treating	as above
H_2O/hydrocarbon	Natural gas dehydration	Kvaerner, Air Products
H_2O/air	Air dehydration	Air Products, Parker Balston Ultratroc, Praxair
Hydrocarbons/air	Pollution control hydrocarbon recovery	Borsig, MTR, GMT NKK
Hydrocarbons from process streams	Organic solvent recovery monomer recovery	Borsig, MTR, GMT SIHI

Table 7.2 Materials for gas separating membranes.

Organic polymers	Inorganic materials
Polysulfone, polyethersulfone	Carbon molecular sieves
Celluloseacetate	Nanoporous carbon
Polyimide, polyetherimide	Zeolites
Polycarbonate (brominated)	Ultramicroporous amorphous silica
Polyphenyleneoxide	Palladium alloys
Polymethylpentene	Mixed conducting perovskites
Polydimethylsiloxane	
Polyvinyltrimethylsilane	

some of the inorganic membranes described in the scientific literature seem to be on the brink of commercialization. Tab. 7.2 shows relevant membrane materials for gas separation.

7.2.1
Organic Polymers

A number of excellent books and reviews have been published on the subject of polymeric gas separating membranes, which are recommended to the interested reader [266–272]. It is the purpose of this chapter to supply the reader with a basic background that is important for the understanding of the transport mechanism of gases through polymers, to introduce those polymers that are currently of commercial importance and finally to give an outlook on interesting developments in this field.

7.2.2
Background

The simplest model used to explain and predict gas permeation through non-porous polymers is the solution-diffusion model. In this model it is assumed that the gas at the high-pressure side of the membrane dissolves in the polymer and diffuses down a concentration gradient to the low pressure side, where the gas is desorbed. It is further assumed that sorption and desorption at the interfaces is fast compared to the diffusion rate in the polymer. The gas phase on the high- and low-pressure side is in equilibrium with the polymer interface. The combination of Henry's law (solubility) and Fick's law (diffusion) leads to

$$J = \frac{D \times S \times \Delta p}{l} \tag{1}$$

which can be simplified to

$$J = \frac{P \times \Delta p}{l} \tag{2}$$

where D is the diffusion coefficient of the gas in the polymer, S is the gas solubility, Δp is the pressure difference between the high- and low-pressure side, l is the membrane thickness and P is the permeability coefficient[1].

As can be seen from Eqs. (1) and (2) the permeability coefficient P is the product of D (a kinetic term) and S (a thermodynamic term).

$$P = D \times S \tag{3}$$

The selectivity of a polymer to gas A relative to another gas B can be expressed in terms of an ideal selectivity a_{AB} defined by the relation

---

[1] The permeability coefficient is most commonly given in Barrer, defined as 10^{-10} cm³cm/cm²s cm Hg and named after R. Barrer. The corresponding SI unit is kmol · m/m²s kPa (kmol · m/m²s kPa $\times 2.99 \cdot 10^{15}$ = Barrer).

$$\alpha_{AB} = \frac{P_A}{P_B} = \frac{D_A}{D_B} \times \frac{S_A}{S_B} \tag{4}$$

The ratio D_A/D_B can be viewed as mobility selectivity and the ratio S_A/S_B as solubility selectivity. For a given gas pair mobility and solubility selectivity depend on the chemical and physical properties of the polymeric material. Excellent reviews on relationships between polymer structure and transport properties of gases have been given by Stern [270], Pixton and Paul [273] and Freeman [274]. Some general rules are useful for a first understanding. The diffusion coefficient D always decreases with increasing size of the molecule. The extent of this decrease is generally dependent on the flexibility of the polymer backbone. The more rigid the polymer structure the higher the mobility selectivity will be for a given gas pair. The mobility selectivity is dominant for most glassy polymers. Hence, the transport of smaller molecules is favored. On the other hand the solubility of gases generally increases with molecular size, because the intermolecular forces between gas and polymer increase. Most rubbers show a low mobility selectivity due to their flexible polymer chain but their ability to separate gases is dominated by their solubility selectivity. Thus large organic vapor molecules can permeate much faster through some rubbers than smaller gases like oxygen or nitrogen. This is depicted for silicone rubber in Fig. 7.1.

It can be seen that the diffusion coefficient of the large pentane molecule is 3.6 times smaller than the diffusion coefficient of oxygen. However, the solubility of pentane is about 200 times larger than the solubility of oxygen. This solubility selectivity outnumbers the reverse diffusion selectivity. As a result, silicone rubber is much more permeable for pentane than for oxygen.

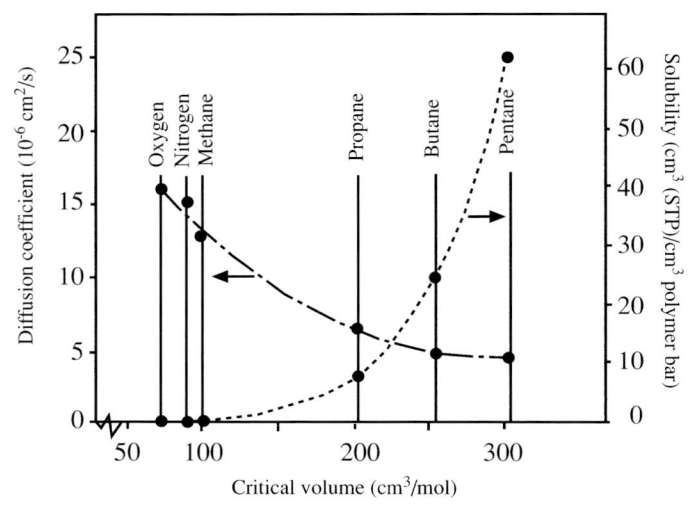

Fig. 7.1 Diffusion and solubility coefficients of different gases in silicone rubber at 30 °C [275–277] vs. the critical volume.

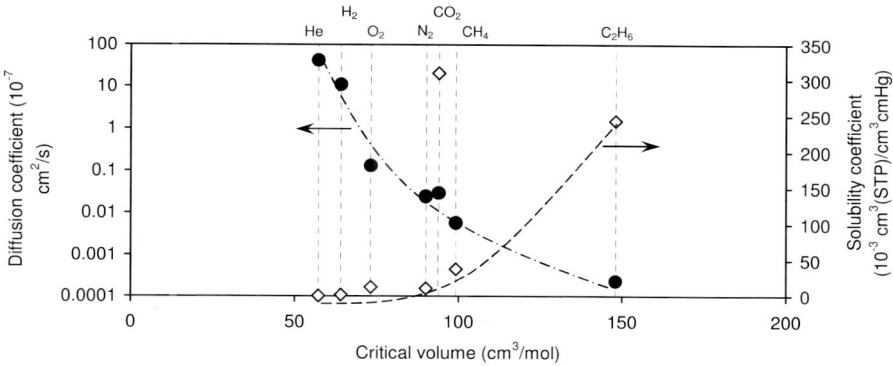

Fig. 7.2 Diffusion and solubility coefficients of different gases in Matrimid at 30 °C vs. the critical volume (data from Shishatskii et al. [278]).

Contrary to rubbers, glassy polymers usually show a preferred permeability to smaller molecules. Their mobility selectivity is much higher than the reverse solubility selectivity. The commercial polyimide Matrimid (Huntsman Advanced Materials), for example, has an impressive hydrogen/methane diffusivity selectivity of 1930. The solubility selectivity, however, favors the larger methane and has a value of 0.055. The resultant permeability selectivity therefore is 109 [278]. It is the dream of membrane developers to uncouple diffusivity and solubility selectivity. The question is, how? Figure 7.2 shows diffusion and solubility coefficients of various gases in Matrimid vs. their critical volume. It is noteworthy that the carbon dioxide solubility does not follow the general trend, but is much higher than predicted from the critical volume.

7.2.3
Polymers for Commercial Gas-separation Membranes

During the last two decades dozens of new polymers that have been developed for gas separation have been described in the literature. The largest group among these are probably polyimides [279]. In spite of these efforts less than 10 polymers are used for industrial gas separations. Nearly all of these are technical polymers developed for totally different applications. "Designer polymers" were either too expensive or their advantage over commercial polymers were not sensational enough or they did not show the expected performance in real applications. The latter is especially true for modified (fluorinated) polyimides. Some of the 6FDA-based polyimides showed tremendous separation abilities in the laboratory but failed in real life due to plasticization or physical aging. Tab. 7.3 gives a list of polymers that are of practical importance for gas separation.

Cellulose acetate, polysulfone and polyimides are by far the most important polymers for gas-separation membranes. When we look at the volume streams

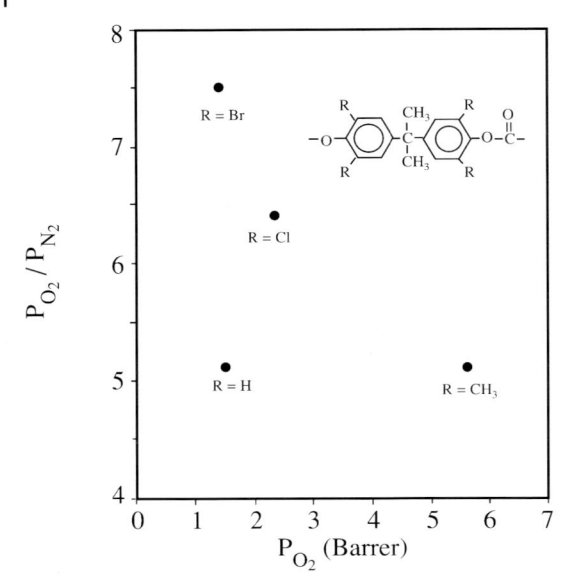

Fig. 7.3 Selectivity vs. permeability for O_2/N_2 for various polycarbonates at 35 °C and 1 bar [280] (reproduced by permission of Elsevier Scientific Publishers).

treated, the old-fashioned cellulose acetate is probably still the dominating polymer. The company Kvaerner Process Systems alone, which has acquired Grace Membrane Systems, has sold or operates membrane plants with CA membranes for carbon dioxide separation for a total stream of more than 5 Mio m³(STP)/day. The only polymer in Tab. 7.3, which is especially designed for gas separation, is the brominated polycarbonate (tetrabromopolycarbonate). It was shown by Paul and coworkers [280] that this relatively simple modification led to an impressive increase in the oxygen/nitrogen separation factor without loss of permeability (see Fig. 7.3).

7.2.4
Ultrahigh Free Volume Polymers

Because the so-called ultrahigh free volume polymers aroused much interest during the last 10 years, they will be briefly described in this introductory chapter. The publication of the physical properties of poly(1-trimethylsilyl-1-propyne) (PTMSP) in 1983 [281] aroused much interest in the field of membrane research. Up to this time it had been believed that the rubbery poly(dimethyl siloxane) has by far the highest gas permeability of all known polymers. Very surprisingly, the glassy PTMSP showed gas permeabilities more than 10 times higher than PDMS. This could be attributed to its very high excess-free volume and the interconnectivity of the free volume elements. Since then a number of

Table 7.3 Gas permeabilities of polymers used for gas-separation membranes.

Polymer	Permeability at 30 °C (Barrer[a])				
	H_2	N_2	O_2	CH_4	CO_2
Cellulose acetate	2.63	0.21	0.59	0.21	6.3
Ethyl cellulose	87	3.2	11	19	26.5
Polycarbonate, brominated		0.18	1.36	0.13	4.23
Polydimethylsiloxane	550	250	500	800	2700
Polyimide (Matrimid)	28.1	0.32	2.13	0.25	10.7
Polymethylpentene	125	6.7	27	14.9	84.6
Polyphenyleneoxide	113	3.81	16.8	11	75.8
Polysulfone	14	0.25	1.4	0.25	5.6
Polyetherimide	7.8	0.047	0.4	0.035	1.32

a) 1 Barrer = 10^{-10} cm^3 cm/cm^2 s cmHg.

high free volume polymers exhibiting strikingly high gas permeabilities have been synthesized.

Two of these are under extensive investigation and are currently being studied for gas separation on a pilot scale. These are DuPont's 2,2-bistrifluoromethyl-4,5-difluoro-1,3-dioxole/tetrafluorethylene copolymer (Teflon AF 2400®) and poly(4-methyl-2-pentyne) (PMP). All three polymers, PTMSP, PMP and Teflon AF2400, are glassy with glass transition above 230 °C and have a very high fractional free volume (FFV). Figure 7.4 shows the chemical structure and fractional free volume of these three polymers.

Table 7.4 shows the oxygen permeability and oxygen/nitrogen selectivity of PTMSP, PMP and Teflon AF2400 in comparison with other polymers.

Although the PMP was already synthesized in 1982 [282], its high gas permeabilities were first published 1996 by Morisato and Pinnau [283] at MTR, Menlo Park. MTR is currently evaluating the performance of PMP membranes for hydrocarbon separation. An attractive application of membranes in this field is natural gas hydrocarbon dewpointing. This means the separation of higher hydrocarbons like butane present in natural gas from methane. Tab. 7.4 shows the performance of PTMSP and PMP membranes for the separation of a methane/butane mixture. Although the permeability and selectivity of PMP is significantly lower compared to PTMSP, its performance for hydrocarbon separation is still superior to all other known polymers. The advantage of PMP lies in its much better chemical stability. In contrast to PTMSP it is not soluble in linear saturated hydrocarbons. With a mixed gas selectivity of butane over methane of 14, the PMP shows a much better performance than poly(dimethyl-siloxane), for which a selectivity of about 5 has been determined under similar conditions [284]. PMP has a poor pure gas selectivity for butane/methane separation, but in the presence of butane the methane permeability drops by a factor of 5. This effect is well known for the gas permeation through nanoporous solids, the con-

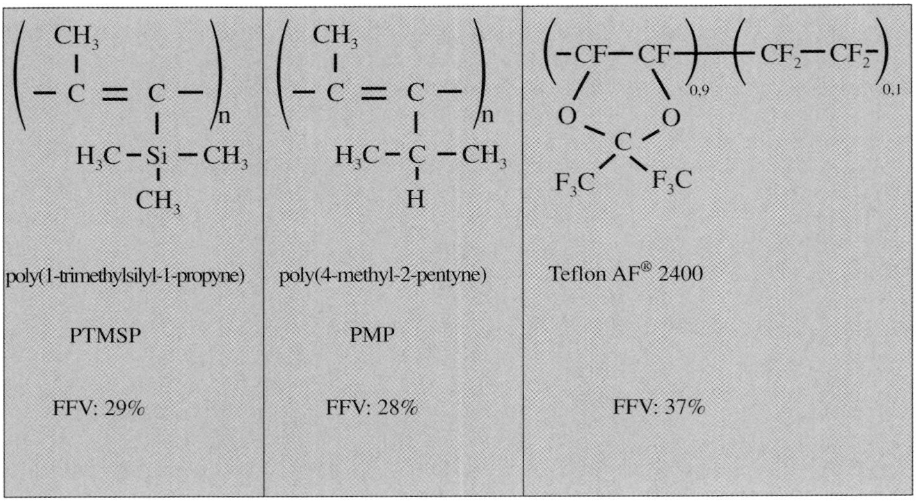

Fig. 7.4 Chemical structure and fractional free volume of PTMSP, PMP and Teflon AF2400.

Table 7.4 Oxygen permeability and oxygen/nitrogen selectivity of high free volume polymers in comparison with conventional polymers.

Polymer	Oxygen permeability (Barrer)	Oxygen/nitrogen selectivity
PTMSP	9700	1.5
PMP	2700	2,0
Teflon AF2400	1300	1.7
PDMS	600	2.2
Polymethylpentene	37	4.2
Polyphenyleneoxide	17	4.4
Ethylcellulose	11	3.4
Polycarbonate	1.4	4.7

densable gas is selectively adsorbed on the pore walls, thus hindering the passage of the smaller molecules. It is believed that in a similar way the extraordinarily high excess free volume of the "superglassy" polymers can be occupied by condensable gases.

It will be quite interesting to observe which class of polymers will finally be applied for the attractive field of hydrocarbon dewpointing of natural gas. The superglasses like PMP show the highest selectivities and fluxes. However, due to their double bonds their chemical stability remains uncertain and they might be prone to physical aging, which is the irreversible absorption of components

Table 7.5 Mixed gas-permeation properties of PTMSP and PMP. Feed: 2% butane in methane, feed pressure: 10 bar, permeate pressure: atmospheric, temperature: 25 °C. From: I. Pinnau et al. In: Polymer Membranes for Gas and Vapor Separation, ACS Symposium Series 733 (1999), 56–67.

Polymer	Permeability (Barrer)		Mixed-gas selectivity $n\text{-}C_4H_{10}/CH_4$	Mixed gas/pure gas CH_4 permeability ratio
	$n\text{-}C_4H_{10}$	CH_4		
PTMSP	53 500	1800	30	0.1
PMP	7500	530	14	0.2

with high boiling points. Rubbery polymers like PDMS on the other hand are stable under natural gas conditions. However, they lose selectivity under high partial pressure of higher hydrocarbons.

It has been reported recently that flux and even selectivity of PMP and PTMSP can be enhanced by the addition of nanoparticles [285, 286]. Merkel et al. [285] added fumed silica to PMP and observed a simultaneous increase of butane flux and butane/methane selectivity. This unusual behavior was explained by fumed-silica-induced disruption of polymer chain packing and an accompanying increase in the size of free volume elements through which molecular transport occurs. Gomes et al. [286] incorporated nanosized silica particles by a sol-gel technique into PTMSP and found also for this polymer a simultaneous increase in flux and selectivity. It has to be studied, if physical aging of the polyacetylenes is reduced by the addition of nanoparticles.

The third polymer listed in Fig. 7.4 has a very different structure in comparison with the polyacetylenes. The Teflon AF2400 is a perfluorinated random copolymer composed of 13 mol% tetrafluoroethylene and 87 mol% 2,2-bis(trifluoromethyl)-4,5-difluoro-1,3-dioxole. Its extraordinarily high gas permeability was first described by Nemser and Roman [287]. Composite membranes fabricated from this polymer are currently being tested on a pilot scale by Compact Membrane Systems, Wilmington. An attractive application seems to be the production of oxygen-enriched or oxygen-depleted air for mobile diesel engines [288] and the separation of supercritical carbondioxide [289].

A new class of polymers with high free volumes has been introduced recently by Budd et al. [290]. The molecular structure of these polymers contains sites of contortion (e.g. spiro-centers) within a rigid backbone (e.g. ladder polymer). A typical chemical structure is shown in Fig. 7.5 A.

The inventors call this polymer class "polymers of intrinsic microporosity (PIMs)", because their porosity arises as a consequence of the molecular structure and is not generated solely through processing. The PIMs can exhibit analogous behavior to that of conventional materials, but, in addition, may be processed into convenient forms for use as membranes [291]. The gas-permeation properties of membranes formed from PIM-1 were investigated at the GKSS

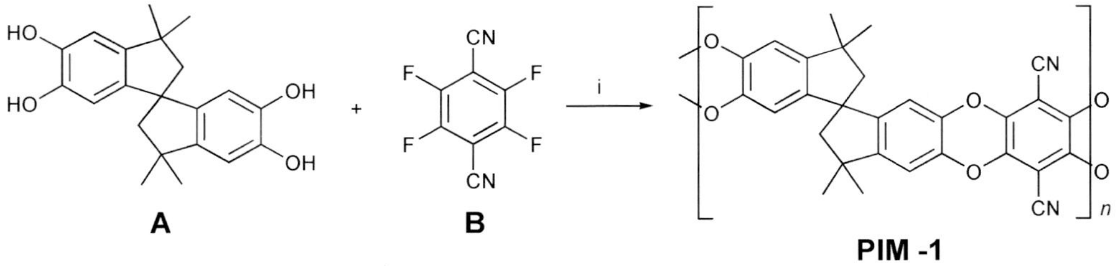

Fig. 7.5 Reaction scheme and structure of PIM I (Budd et al. [290]).

Research Center [292]. With an oxygen permeability of 370 Barrer and an O_2/N_2 selectivity of 4.0 the PIM-1 shows an extraordinary behavior as a gas-separation polymer. However, long-term measurements revealed a physical aging of PIM-1 analogous to PTMSP, which resulted in reduced permeabilities. When the aging problems can be solved, the PIMs will be a highly interesting polymer class for fabrication of gas-separation membranes.

7.2.5
Inorganic Materials for Gas-separation Membranes

The current market for inorganic membranes for gas separation is extremely small. One of the few commercial applications is small-scale palladium membrane systems to produce ultrapure hydrogen for specialized applications. They are marketed by Johnson Matthey and Company. It is not believed that the market share of inorganic membranes will increase significantly in the near future. The main obstacle is their high price and some principle difficulties during reproducible large-scale production. On the other hand, fascinating research results have been published in the past such as a propene/propane mixed gas selectivity of more than 40 with carbon molecular sieve membranes [293] or unmatched selectivities for carbondioxide/methane separation with ceramic membranes [294].

The interested reader is referred to some good reviews on inorganic membrane materials [295–299].

In this chapter, three examples of inorganic membranes or hybrid organic/inorganic membranes for gas separation that might find industrial applications in the future will be discussed.

7.2.6
Carbon Membranes

Since the pioneering paper of Koresh and Soffer on carbon molecular sieve membranes in 1983 [300] much research has been carried out in the field of carbon-based gas-separation membranes. Selectivities and permeabilities far above the performance of the best polymers have been obtained for carbon molecular sieve membranes by many researchers. One example is a recent publication of Koros et al. [301], in which striking selectivities for gas pairs like oxygen/nitrogen and carbon dioxide/methane are described. In spite of these findings carbon molecular sieve membranes have not found their way into industrial separation processes. One reason for this might be the inherent brittleness of carbon materials, high price and aging of the carbon surface by chemical surface reactions are other difficulties.

Very close to commercialization were the nanoporous carbon membranes developed mainly by Air Products in the early 1990s. They were one of the highlights in membrane development for gas separation from 1990 to 2000. They can be produced by different methods. The most advanced membranes of this kind have been produced by Air Products first published in 1993 [302]. Air Products called this membrane a Selective Surface Flow (SSF^{TM}) membrane. It consists of a thin layer (2–3 µm) of a nanoporous carbon matrix (5–7 Å pore diameter) supported on the bore side by a macroporous (<1 µm pore diameter) alumina tube. The membrane is produced by: (a) coating the bore side of the tubular support with a thin uniform layer of a poly(vinylidene chlorideacrylate) terpolymer latex containing small polymer beads in aqueous emulsion, (b) drying the coat under N_2 at 50 °C, (c) heating under a dry N_2 purge to 600 °C for carbonizing the polymer, and (d) finally passivating the nascent carbon film by heating in an oxidizing atmosphere at 200–300 °C. The resulting membranes have quite defined pores in the 5 to 8 Å range. They are especially well suited for the separation of hydrogen/hydrocarbon mixtures. The remarkable point is that they exhibit a preferred permeability for higher hydrocarbons over hydrogen. Tab. 7.6 gives some permeability data from the original paper.

As can be seen from the table the pure gas selectivities of the nanoporous carbon membrane are quite low, e.g. 1.19 for butane/hydrogen. However, for the mixture given in Tab. 7.6 the butane/hydrogen selectivity increases to 94. The reason for this is that the butane is selectively absorbed over hydrogen at the carbon pore wall and because the pores are so small the pathway for hydrogen is blocked. This effect of selective surface flow and pore blocking was first observed by Barrer et al. [303]. Due to its unmatched selectivity the nanoporous

Table 7.6 Gas separation properties of nanoporous carbon membrane.

Gas	Pure gas permeability (Barrer)	Mixed gas permeability	Mixed gas selectivity
H_2	130	1.2	–
CH_4	660	1.3	1.1
C_2H_6	850	7.7	5.1
C_3H_8	290	25	21
C_4H_{10}	155	110	94

41.0% H_2, 20.2% CH_4, 9.5% C_2H_6, 9.4% C_3H_8, 19.9% C_4H_{10}
from: M.B. Rao, S. Sircar, J. Membrane Sci., 85 (1993) 253

carbon membrane looks very attractive for hydrogen enrichment of refinery off-gases with low hydrogen content, e.g. FCC (fluidized catalytic cracker) off-gases. It is much more attractive than hydrogen-selective membranes, because the hydrogen remains on the high-pressure side of the membrane and can be fed into a pressure swing unit for further purification. The drawback of the nanoporous carbon membrane is that water vapor and higher hydrocarbons should be removed before the membranes separation because they adsorb very strongly in the membrane pores. Air Products' SSF membrane was field tested at different refinery sites [304]. Surprisingly, Air Products discontinued the work in this area in 2003. One reason could have been the aging/deactivation of the membranes in the presence of water vapor. In spite of this drawback nanoporous carbon membranes are attractive for future industrial applications.

7.2.7
Perovskite-type Oxide Membranes for Air Separation

It has been known for a long time that certain dense ceramic materials are good conductors of oxygen at elevated temperatures. Oxygen transport through an ionic conductor is a result of oxygen ionic conduction mechanisms that involve oxygen defects such as lattice vacancies. One of the best known ceramic oxygen conductors is yttria-doped zirconia, which is widely used in high-temperature oxygen sensors. Oxygen is transported through these materials as O^{2-} ions. Hence, when oxygen permeates through these materials there must be a flow of electrons in the opposite direction. Oxygen-conducting ceramics like doped zirconia are good oxygen conductors but poor electronic conductors.

The electronic conductivity of yttria-stabilized zirconia is three to four orders of magnitude lower than its ionic conductivity. The oxygen can be pumped through the material by an external electrical field. However, a simple calculation reveals that this is not economic for oxygen separation due to the high electricity consumption. The situation changes when the ceramic material is a good conductor for both oxygen ions and electrons. These materials are referred to as ionic-electronic mixed conductors. With these, a high oxygen flux can be obtained without an external electrical field. Figure 7.6 shows schematically the two types of oxygen ion transport membranes.

One of the first papers that stimulated large interest in this research was published in 1985 by Teraoka et al. [305]. One of the figures of this pioneering paper is given here (Fig. 7.7).

It may be deduced from Fig. 7.7 that, with a specific composition of the ceramic, an oxygen flux of up to 2.4 cm^3/min cm^2 could be obtained at an oxygen partial pressure difference across the membrane of 0.21 bar. The membrane thickness was 1 mm.

With optimized perovskite compositions even much higher fluxes in the range of 6 to 7 cm^3/cm^2 min could be obtained recently [306]. The oxygen flux is not a linear inverse relationship to membrane thickness, because for thin membranes (about 0.3 mm) surface reactions become rate limiting. But even

Fig. 7.6 Types of oxygen ion transport membranes.

Oxygen Ion Transport Membranes

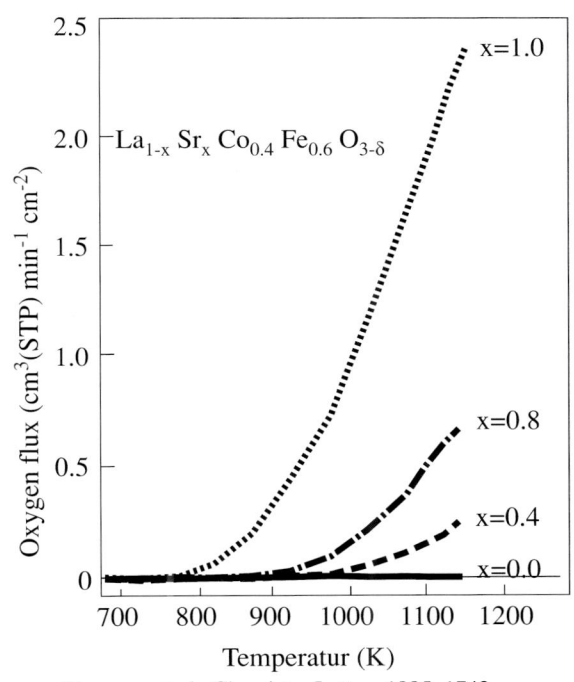

$La_{1-x} Sr_x Co_{0.4} Fe_{0.6} O_{3-\delta}$

Yamazoe et al., Chemistry Letters 1985, 1743

Fig. 7.7 Temperature dependence of the rate of oxygen permeation through perovskite membranes.

0.5-mm thick membranes exhibit an oxygen flux at high temperature, which is orders of magnitude higher than the flux through polymeric membranes.

The promising prospect of these membranes is not in the first place the production of oxygen, but their application in membrane reactors for the partial oxidation of natural gas, which is schematically shown in Fig. 7.8.

The mixed-conducting membrane eliminates the cryogenic air separation plant and it forms a safety barrier between the natural gas and air. The membrane becomes more productive in a configuration like this, because the slow oxygen desorption at the permeate side is enhanced by the chemical reaction.

Another potentially attractive application of the perovskite membranes is the production of hot oxygen for IGCC power plants. An IGCC plant is a power station where a gas turbine is used to generate electricity, and the waste heat from the gas turbine is used to raise steam to generate electricity. When the gas turbine is fired on a gas fuel derived from the gasification of liquid or solid carbonaceous materials, this process is known as an integrated gasification combined cycle.

Perovskite membranes have also been proposed for the production of oxygen enriched air for industrial processes like ammonia synthesis, the Claus process, and the regeneration of the catalyst for the fluid catalytic cracking (FCC) process [307].

The perovskite-type ceramic membranes have attracted much attention from major chemical and petrochemical companies in the USA. Companies currently involved in the development of the mixed-conducting ceramic membranes include Air Products, Praxair, BP and Amoco. The largest currently existing consortium developing this technology is headed by Air Products and sponsored by

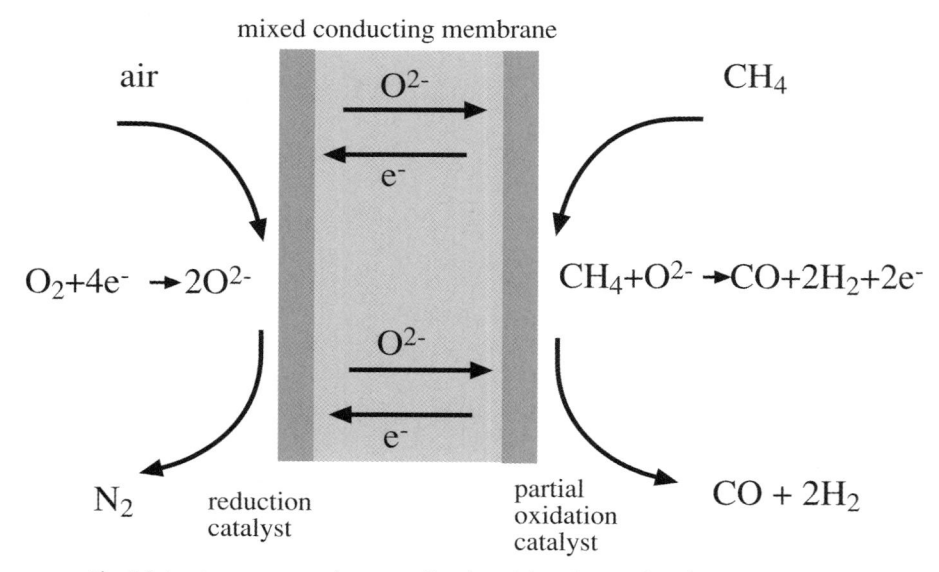

Fig. 7.8 Ion-transport membrane-mediated partial oxidation of methane.

the US Department of Energy. The team includes Ceramatec, Eltron Research, McDermott Technologies, Pennsylvania State University, Siemens Westinghouse and Texaco Gasification. A pilot-scale prototype unit with an oxygen production rate of 5 tons per day is scheduled for the first half of 2006. Many believe that the mixed-conducting membrane technology will represent a major breakthrough in industrial application of inorganic membranes.

7.2.8
Mixed-matrix Membranes

Molecular sieves such as zeolites or carbon molecular sieves show a much higher selectivity for many gas mixtures than polymeric membranes due to their very defined pore sizes. For example it can be calculated from reported sorption and diffusivity data that zeolite 4A has an oxygen permeability of 0.77 Barrer and an O_2/N_2 selectivity of approximately 37 at 35 °C [308].

The preparation of defect-free zeolite layers on a large scale is extremely difficult and it seems doubtful that this will ever be achieved at a competitive price. However, the combination of the superior gas selectivities of molecular sieves with the processibility of polymeric membranes has attracted many researchers. The hybrid membranes consisting of inorganic molecular sieves and polymers are often referred to as mixed-matrix membranes. The term "mixed-matrix membrane" has been introduced by Kulprathipanja et al. [309], who performed pioneering work in the field of polymer/zeolite hybrid membranes. Kulprathipanja showed that the CO_2/H_2 selectivity of cellulose acetate could be reversed by addition of silicalite. The silicalite-CA membrane had a CO_2/H_2 selectivity of 5.1, whereas the pure CA membrane exhibited a selectivity of 0.77.

Hennepe, from the University of Twente, proved for the first time that the incorporation of silicalite in PDMS increased the ethanol/water selectivity significantly under steady-state conditions [310] in pervaporation experiments. Later, it was shown by Jia et al. [311] that using the same approach (silicalite in PDMS) the gas selectivity could also be changed due to a molecular-sieving effect. However, the effects were too small to be of any interest for practical applications. One of the fundamental questions of the mixed-matrix concept is how the permeability of the polymer should match with the molecular-sieve permeability. To answer this question the diffusion and permeation in heterogeneous media has to be modeled. Barrer gave a detailed analysis of this issue [312]. A relatively simple model for transport in heterogeneous media is the Maxwell approach. In the late 1870s Maxwell [313] calculated the electric conductivity of a metal, in which spheres of a second metal are dispersed. The Maxwell equation has been applied by Robeson et al. [314] to calculate the gas permeability of block copolymers. When applied to mixed-matrix membranes we obtain:

$$P_{\text{eff}} = P_c \frac{(P_d + 2P_c - 2\phi(P_c - P_d))}{(P_d + 2P_c + \phi(P_c - P_d))}$$

where P_{eff} is the effective permeability, ϕ is the volume fraction of the dispersed phase, and the subscripts d and c refer to the dispersed and continuous phases, respectively. If this equation is used to calculate the selectivity of a mixed-matrix membrane we obtain

$$a_{eff} = a_c \frac{1 + 2P_{rel} - 2\phi(P_{rel} - 1)}{1 + 2P_{rel} + \phi(P_{rel} - 1)} * \frac{\frac{1}{a_D} + \frac{2P_{rel}}{a_c} + \phi\left(\frac{P_{rel}}{a_c} - \frac{1}{a_D}\right)}{\frac{1}{a_D} + \frac{2P_{rel}}{a_c} - 2\phi\left(\frac{P_{rel}}{a_c} - \frac{1}{a_D}\right)}$$

where a_{eff} is the effective selectivity of the mixed-matrix membrane, P_{rel} is the permeability ratio of continuous phase to dispersed phase for the fast component and a_c and a_d are the selectivities of the continuous and dispersed phase. This equation looks complex at first sight, but it allows the plotting of the selectivity of a mixed-matrix membrane as a function of the permeability ratio of the continuous and dispersed phase. An example is given in Fig. 7.9 with a selectivity of the dispersed phase of 35 and a selectivity of the continuous phase of 7. The volume fraction of the filler is 0.5. The plot reveals that a maximum selectivity of the mixed-matrix material is obtained in this case, when the polymer selectivity is about 3 times lower than the filler permeability.

When the polymer permeability becomes too high the selectivity of the mixed-matrix membrane approaches the polymer selectivity. Hence the above equation gives a theoretical estimation of the selectivity of a mixed-matrix membrane and it gives an idea of how the permeability of molecular sieve and poly-

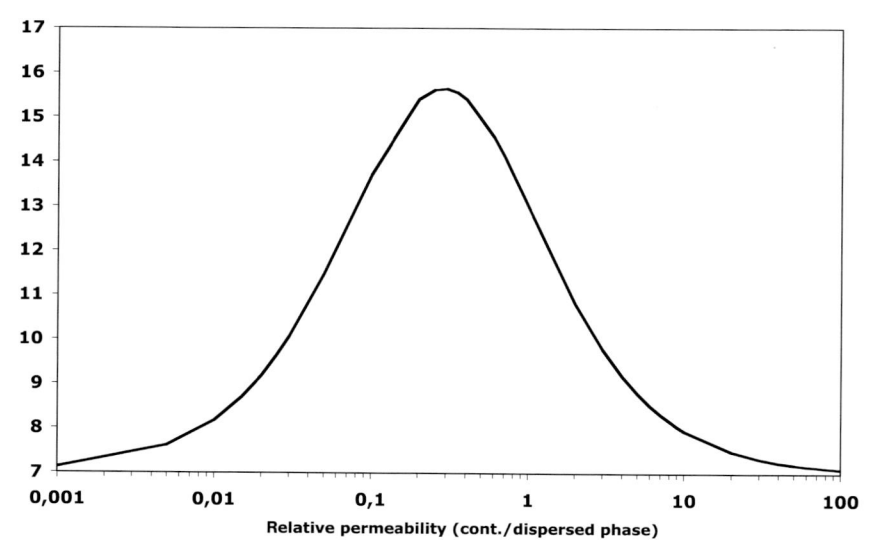

Fig. 7.9 Selectivity of a mixed-matrix membrane vs. permeability ratio of continuous phase to dispersed phase, selectivity of dispersed phase: 35, selectivity of continuous phase: 7.

mer should match. The practical challenge is to improve the compatibility between inorganic molecular sieves and glassy polymers in order to eliminate gas-diffusion pathways at the interface between polymer and zeolite. Additional to voids one might also find regions of reduced polymer permeability close to the filler surface. A number of nonideal effects in mixed-matrix membranes are discussed in a paper of Moore et al. [315]. There are still manufacturing problems to solve, before mixed-matrix membranes will be introduced in commercial gas separation. We see an increasing number of patents filed by big companies active in gas separation [316–319], and it can be concluded that mixed-matrix membranes are on the brink of practical application.

7.3
Basic Process Design

In this chapter some fundamental equations are given that allow a first design of a one-stage membrane separation unit. Questions to be answered are: what is the maximum enrichment that can be achieved with a membrane of a given selectivity? How is the separation performance influenced by feed and permeate pressure? It will be explained why for some applications a high selective membrane will be outperformed by a membrane with a lower selectivity.

For the following we look at a simple gas-separation unit with two components, which is illustrated in Fig. 7.10.

One parameter, which does of course determine the gas enrichment, is the membrane selectivity a, which is a membrane property and defined here as $a = P_2/P_1$ with P_2 and P_1 as permeability coefficients for gas 2 and 1.

Schematic view of binary gas mixture separation

Feed: pressure p', composition c_1', c_2'

Retentate

Permeate: pressure p'', composition c_1'', c_2''

Membrane selectivity $\alpha = \dfrac{P_2}{P_1}$

stage cut $\theta = \dfrac{\text{permeate flow}}{\text{feed flow}}$

pressure ratio $\phi = \dfrac{p'}{p''}$

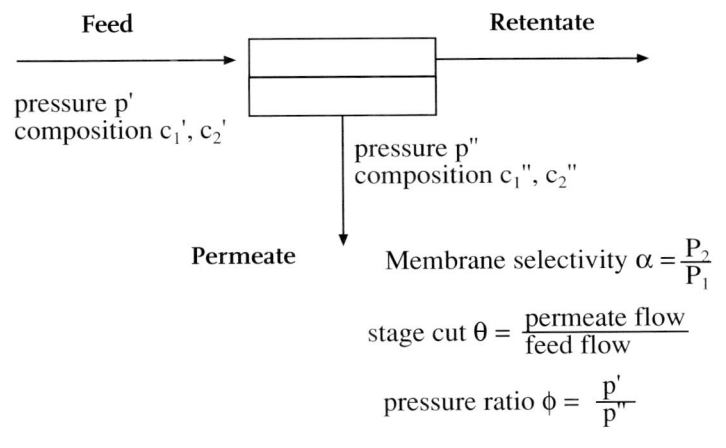

Fig. 7.10 Model gas-separation unit with two components.

Equally important are two process parameters: the stage cut θ and the pressure ratio ϕ. The stage cut is defined as the ratio permeate flow/feed flow and the pressure ratio is the ratio of total feed pressure to total permeate pressure.

For the sake of simplicity we start with a stage cut close to zero, i.e. there is no concentration difference between feed and retentate. The maximum enrichment of the faster component 2 can now be determined easily. For the maximum enrichment the maximum driving force is needed, i.e. the permeate pressure can be neglected when compared to the feed pressure. The flux of component 1 is proportional to its volume fraction on the feed side, for component 2 we have as an additional factor the membrane selectivity.

$$J_1 = \text{const.} \times c_1' = \text{const.} \times (1 - c_2')$$

$$J_2 = \text{const.} \times a \times c_2'$$

The concentration of component 2 on the permeate side must be equal to the flux J_2 of component 2 divided by the total flux J_1+J_2. Combining Eqs. (1) and (2) then yields

$$c_2'' = \frac{c_2' \times a}{1 + c_2'(a-1)} \tag{3}$$

This simple equation gives the maximum possible enrichment of one gas of a two-component mixture when separated by a membrane with a selectivity of a. The equation becomes more complex when the permeate pressure cannot be neglected [320]. Following the simple solution-diffusion model the gas fluxes for gas 1 and 2 through the membrane are

$$J_1 = \frac{P_1(p_1' - p_1'')}{l} \tag{4}$$

and

$$J_2 = \frac{P_2(p_2' - p_2'')}{l} \tag{5}$$

where P_1 and P_2 are the permeabilities of components 1 and 2, l is the membrane thickness, and p_1', p_2' and p_1'', p_2'' are the partial pressures of the two gases in the feed and permeate streams, respectively. The total gas pressure is equal to the sum of the partial pressures, i.e.

$$p' = p_1' + p_2' \tag{6}$$

$$p'' = p_1'' + p_2'' \tag{7}$$

with

$$c_2' = p_2'/p' \quad c_2'' = p_2''/p'' \tag{8}$$

and

$$J_1/J_2 = c_1''/(1-c_1'') = (1-c_2'')/c_2'' \tag{9}$$

Combining Eqs. (4)–(9) then yields the expression

$$c_2'' = 0.5\phi \left[c_2' + \frac{1}{\phi} + \frac{1}{a-1} - \sqrt{\left(c_2' + \frac{1}{\phi} + \frac{1}{a-1}\right) - \frac{4c_2'a}{\phi(a-1)}} \right] \tag{10}$$

Equation (10) gives the concentration of the faster permeating gas in the permeate stream as a function of the membranes selectivity and the pressure ratio across the membrane. It breaks down into two limiting cases. At high driving forces when the pressure ratio is much higher than the selectivity ($\phi \gg a$) Eq. (10) reduces to Eq. (3). We call this a selectivity controlled region, because the enrichment is now independent of the pressure ratio. When on the other hand the pressure ratio becomes much smaller than the selectivity ($\phi \ll a$), Eq. (10) reduces to

$$c_2'' = c_2'\phi \tag{11}$$

The enrichment is now independent of the membranes selectivity. Hence, this is the pressure-ratio-limited region. There is, of course, an intermediate region between these two limiting cases where both the pressure ratio and the membrane selectivity affect the degree of separation. This is illustrated in Fig. 7.11, in which the calculated permeate concentration is plotted versus pressure ratio for a membrane with a selectivity of 200.

For numerous technical applications the pressure ratio does not exceed 10 or 20. An example is the separation of organic vapors, where a typical pressure ratio is about 10. Figure 7.12 shows a plot of permeate concentration versus selectivity at a pressure ratio of 10.

The plot reveals that an increase of the selectivity above 40 does not increase the enrichment significantly. The highest permeate concentration achievable in this example is 5% as predicted by Eq. (11). The process might not only not benefit from a high-selective membrane, but a too-selective membrane might even be a disadvantage. This will be demonstrated with a simple model calculation concerning natural gas dehydration. Natural gas dehydration is one of the emerging applications of membrane-based gas separation. The permeation of water vapor through polymers has one peculiarity. When plotting gas permeability versus selectivity for different polymers a general tendency exists: the more selective a polymer is the lower is its permeability, as shown in the famous Robeson plots [321]. Figure 7.13 displays a plot of water vapor/nitrogen selectivity versus water permeability for a number of polymers.

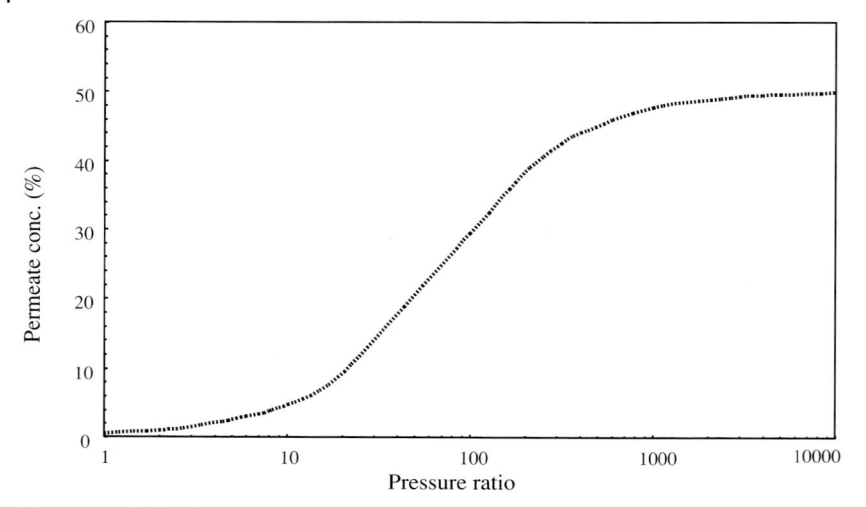

Fig. 7.11 Calculated permeate concentration for a membrane with a selectivity of 200 as a function of pressure ratio. The feed concentration is 0.5%.

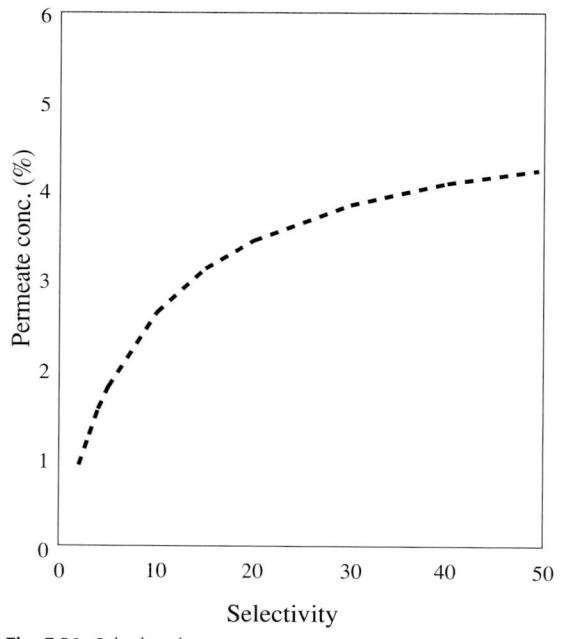

Fig. 7.12 Calculated permeate concentration vs. membrane selectivity for a pressure ratio of 10, feed concentration 0.5%.

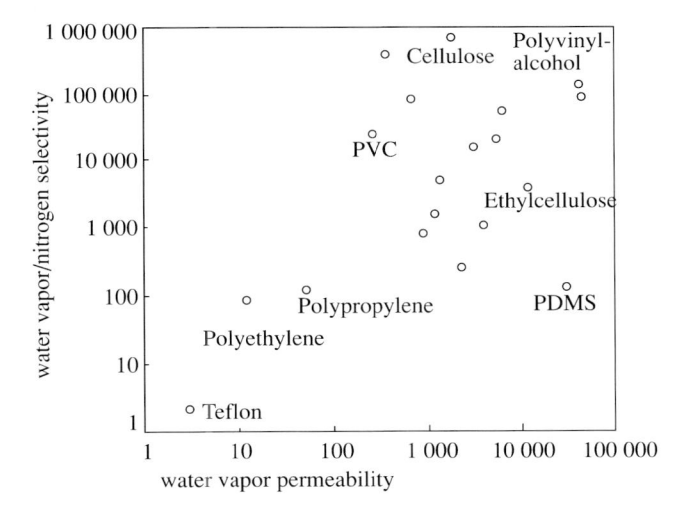

Fig. 7.13 H_2O/N_2 selectivity versus permeability for various polymers.

The aforementioned tendency does not hold here. On the contrary, many very selective polymers also have a very high permeability[2]. Numerous polymers are available with water vapor selectivities of 5000 and more. For the following calculation a two-component water vapor/methane mixture has been assumed with 0.2% water vapor. The membrane unit shall reduce this water content by one order of magnitude; i.e. the retentate water concentration has been set to 0.02%. The simple equation (10) cannot be applied, the equations first derived by Weller and Steiner [322] have been used for the calculation of methane loss and the equation of Saltonstall [323] for calculation of membrane area. The methane loss is simply defined by methane permeate stream divided by methane feed stream times 100. In Fig. 7.14 the methane loss is plotted versus membrane selectivity for two different pressure ratios.

The figure reveals that the methane loss is smaller at the higher pressure ratio. For the pressure ratio of 80 methane loss starts at 5% for a selectivity of 100 and drops to 2.75% for a selectivity of 5000. But even with a very selective membrane it is never smaller than 2.7%. At the high pressure ratio of 1000 the methane loss drops to a favorable 0.17%. However, a pressure ratio of 1000 is quite unrealistic. Additional compression of the feed is out of the question. The permeate pressure could be reduced by vacuum pumps; but this idea is not favored by the gas companies because of cost and safety concerns. If a membrane dehydration system is operated at a pressure ratio of 80, the difference between methane loss from a membrane with a selectivity of 500 (3.2% methane loss) and a membrane with a selectivity of 5000 (2.75% methane loss) is quite small.

---

[2] When looking at the permeabilities in Fig. 7.13 one has to keep in mind, that the water vapor permeability often strongly depends on its vapor pressure. Most permeability data have been generated near saturation pressure.

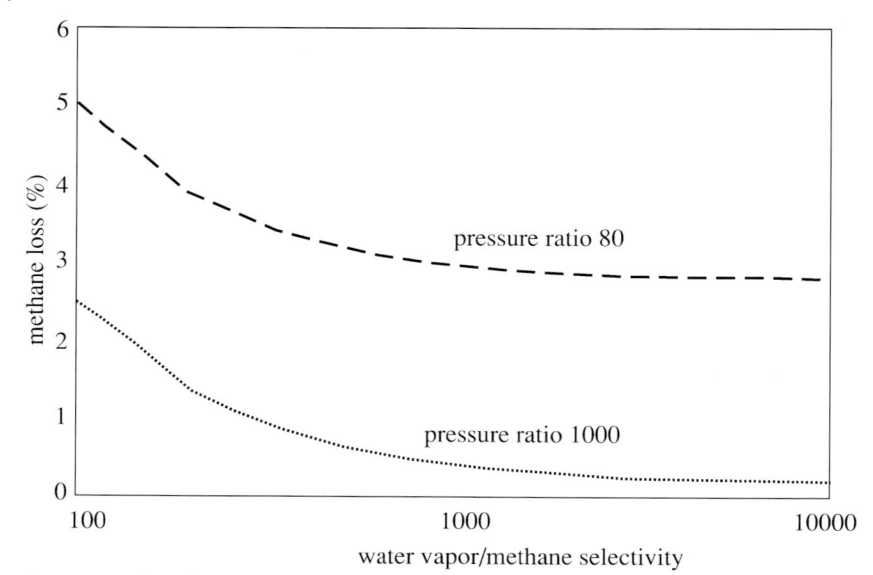

Fig. 7.14 Methane loss vs. selectivity for pressure ratios 80 and 100, volume fraction of methane in feed 0.2%.

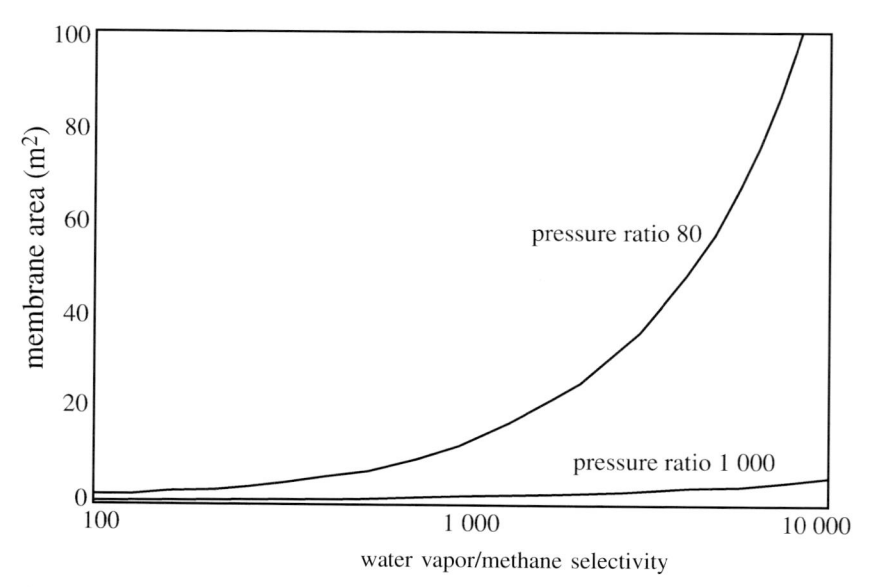

Fig. 7.15 Membrane area versus selectivity at pressure ratios of 80 and 1000, water vapor flux fixed at 1.1×10^{-2} cm^3/cm^2 s cmHg, water vapor feed concentration 0.2%, retentate concentration 0.02%, feed pressure 8 MPa.

Hence, as far as methane loss is concerned there is a small benefit in using the high-selective membrane. The picture changes when the required membrane area is taken into account, which is demonstrated in Fig. 7.15. Membrane area has been calculated for a feed stream of 1000 m^3(STP)/h, the water vapor flux through the membrane has been fixed at 1.1×10^{-2} cm^3/cm^2 s cmHg. Membrane selectivity was adjusted by variation of methane flux.

As the figure shows, the membrane area required increases strongly with selectivity. There is an 8× increase of membrane area, when the membrane selectivity increase from 500 to 5000 for a pressure ratio of 80. This example illustrates that it is not always a good strategy to look for the most selective membrane but that a less-selective membrane may do a better job.

Acknowledgments

The electron micrographs were kindly supplied by Michael Schossig and Marion Aderhold, GKSS-Forschungszentrum Geesthacht.

References

1 H. STRATHMANN, Membrane processes for sustainable industrial growth. *Membrane Technology* **113**, September 1999, p. 9–11.
2 A. W. CRULL, Who's on first. State of the market for RO, MF, UF. Filtration News, Sept/October 1997, p. 38.
3 New Era, Business Communications Co, USA; Membrane Technology, October 2005, p. 4.
4 RO, UF, MF, World Markets, McIlvaine Co, Membrane Technology, September 2005, p. 4.
5 R. E. KESTING, Synthetic polymeric membranes. A structural perspective. John Wiley & Sons, New York, Chichester, Brisbane, Toronto, Singapore, 1985.
6 H. STRATHMANN, Review – Membrane Separation Processes. *Journal of Membrane Science* **9** (1981) 121.
7 H. K. LONSDALE, Review – The Growth of Membrane Technology. *Journal of Membrane Science* **10** (1982) 81.
8 W. S. W. HO, K. K. SIRKAR, Eds. Membrane Handbook, Van Nostrand Reinhold, 1992.
9 R. W. BAKER, E. L. CUSSLER, W. EYKAMP, W. J. KOROS, R. L. RILEY, H. STRATHMANN, Membrane Separation Systems: Recent Developments and Future Directions. Noyes Publishing, Park Ridge, NJ, USA.
10 S. LOEB, S. SOURIRAJAN, Sea water demineralization by means of an osmotic membrane. *Advances in Chemistry Series* **38** (1962), 117.
11 A. CASTRO (Akzona) Methods for making microporous products. US Patent 4247498, January 1981.
12 D. R. LLOYD, S. S. KIM, K. E. KINZER, Microporous membrane formation via thermally-induced phase separation. II. Liquid-liquid phase separation. *Journal of Membrane Science* **64** (1991) 1.
13 S. P. NUNES, Recent advances in the controlled formation of pores in membranes. *Trends in Polymer Science* **5** (1997) 187.
14 W. J. KOROS, I. PINNAU, in Polymeric Gas Separation Membranes (D. R. Paul, Y. Yampol'skii, Eds.) pp. 209–271, CRC Press 1994.
15 A. J. REUVERS, C. A. SMOLDERS, Formation of membranes by means of immersion precipitation. Part II The mechanism of formation of membranes prepared from the system cellulose acetate-acetone-water. *Journal of Membrane Science* **34** (1987) 45, 67.
16 A. M. W. BULTE, B. FOLKERS, M. H. V. MULDER, C. A. SMOLDERS, Membranes of semicrystalline aliphatic polyamide Ny-

lon 4,6: Formation by diffusion-induced phase separation. *Journal of Applied Polymer Science* **50** (1993) 13.
17 G. E. Gaides, A. J. McHugh, Gelation in an amorphous polymer: a discussion of its relation to membrane formation. *Polymer* **30** (1989) 2118.
18 F. J. Tsai, J. M. Torkelson, Roles of phase separation mechanism and coarsening in the formation of poly(methyl methacrylate) asymmetric membranes, *Macromolecules* **23** (1990) 775.
19 S. P. Nunes, T. Inoue, Evidence for spinodal decomposition and nucleation and growth mechanisms during membrane formation. *Journal of Membrane Science* **111** (1996) 93.
20 R. M. Boom, T. van den Boomgaard, C. A. Smolders, Mass transfer and thermodynamics during immersion precipitation for a two-polymer system. Evaluation with the system PES-PVP-NMP-water. *Journal of Membrane Science* **90** (1994) 231.
21 J. G. Wijmans, J. P. B. Baaij, C. A. Smolders, The Mechanism of Formation of Microporous or Skinned Membranes Produced by Immersion Precipitation. *Journal of Membrane Science* **14** (1983) 263.
22 L. Zeman, T. Fraser, Formation of air-cast cellulose acetate membranes. Part I. Study of macrovoid formation. *Journal of Membrane Science* **84** (1993) 93.
23 I. Pinnau, J. Koros, A. qualitative skin layer formation mechanism for membranes made by dry/wet phase inversion. *Journal of Polymer Science B: Polymer Physics* **31** (1993) 419.
24 M. Mulder, Basic Principles of Membrane Technology. Kluwer Academic, 1991.
25 K. V. Peinemann, J. F. Maggioni, S. P. Nunes, Poly (ether imide) membranes obtained from solution in cosolvent mixtures. *Polymer* **39** (1998) 3411.
26 J. Y. Kim, H. K. Lee, K. J. Baik, S. C. Kim, Liquid-liquid phase separation in polysulfone/solvent/water systems. *Journal of Applied Polymer Science* **65** (1997) 2643.
27 J. Y. Kim, Y. D. Kim, T. Kanamori, H. K. Lee, K. J. Baik, S. C. Kim, Vitrification phenomena in polysulfone/NMP/water system. *Journal of Applied Polymer Science* **71** (1999) 431.
28 S. A. McKelvey, W. Koros, Phase separation, vitrification and the manifestation of macrovoids in polymeric asymmetric membranes. *Journal of Membrane Science* **112** (1996) 29.
29 C. A. Smolders, A. J. Reuvers, R. M. Boom, I. M. Wienk, Microstructures in phase-inversion membranes. Part 1. Formation of macrovoids. *Journal of Membrane Science* **73** (1992) 259.
30 H. Strathmann, K. Koch, P. Amar, R. W. Baker, The formation mechanism of asymmetric membranes. *Desalination* **16** (1975) 179.
31 K. Kneifel, K. V. Peinemann, Preparation of hollow fiber membranes from polyetherimide for gas separation. *Journal of Membrane Science* **65** (1992) 295.
32 C. Blicke, K. V. Peinemann, S. P. Nunes, Ultrafiltration membranes of PESA/PEI. *Journal of Membrane Science* **79** (1993) 83.
33 S. P. Nunes, F. Galembeck, N. Barelli, Cellulose Acetate Membranes for Osmo-sedimentation: Performance and Morphological Dependence on Preparation Conditions. *Polymer* **27** (1986) 937–943.
34 R. J. Petersen, Composite reverse osmosis and nanofiltration membranes. *Journal of Membrane Science* **83** (1993) 81.
35 J. E. Cadotte, R. J. Petersen, R. E. Larson, E. E. Erickson, A new thin-film composite seawater reverse osmosis membrane. *Desalination* **32** (1980) 25.
36 FT-30 Catalogue; M. Busch, W. E. Mickols, Reducing energy consumption in seawater desalination. *Desalination* **165** (2004) 299–312.
37 Nitto Denko Catalogue; http://www.somicon.com/products/Products_Nitto_files/nitto.html.
38 A. I. Schäfer, A. G. Fane, T. D. Waite, Eds., Nanofiltration – Principles and Applications, Elsevier, 2005.
39 W. J. Conlon, S. A. McClellan, Membrane softening: a treatment process comes to age. *Journal of American Water Works Association* **81** (1989) 47.
40 Hydranautics Catalogue.
41 R. F. Fibiger, J. Koo, D. J. Forgach, R. J. Petersen, D. L. Schmidt, R. A. Wessling, T. F. Stocker, (Dow Chemical). Novel polyamide reverse osmosis mem-

branes US Patent 4,769,148, September, 1988.
42 R. F. Fibiger, J. Koo; D. J. Forgach, R. J. Petersen, D. L. Schmidt, R. A. Wessling, T. F. Stocker (Filmtec, Dow Chemical) Novel polyamide reverse osmosis membranes. US Patent 4859384, August, 1989.
43 M. Mänttäri, T. Pekuri, M. Nyström, NF270, A new membrane having promising characteristics and being suitable for treatment of dilute effluents from the paper industry. *Journal of Membrane Science* **242** (2004) 107–116.
44 M. Nyström, S. Butylina, S. Platt, NF retention and critical flux of small hydrophilic/hydrophobic molecules. *Membrane Technology*, October 2004, p. 5–8.
45 S. P. Nunes, M. L. Sforça, K. V. Peinemann, Dense hydrophilic composite membranes for ultrafiltration. *Journal of Membrane Science* **106** (1995) 49.
46 M. L. Sforça, S. P. Nunes, K. V. Peinemann, Composite nanofiltration membranes prepared by in situ polycondensation of amines in a poly(ethylene oxide-b-amide) layer. *Journal of Membrane Science* **135** (1997) 179.
47 M. Schmidt, K. V. Peinemann, D. Paul, H. Rödicker, Cellulose ether als Trennschichten hydrophiler Polymermembranen. *Die Angewandte Makromolekulare Chemie* **249** (1997) 11.
48 J. E. Cadotte, D. R. Walker (Filmtec). Novel water softening membranes. US Patent 4812270, March 1989.
49 J. W. Wheeler (DuPont) Process for opening reverse osmosis membranes. US Patent 5262054, November 1993.
50 W. E. Mickols (Dow Chemical) Method of treating polyamide membranes to increase flux. US Patent 5755964, May 1998.
51 Koch Catalogue.
52 C. Linder, M. Perry, M. Nemas, R. Katraro (Aligena) Solvent stable membranes. US Patent 5039421, August 1991.
53 M. Perry, H. Yacubowicz, C. Linder, M. Nemas, R. Katraro (Membrane Products Kiryat Weizmann) Polyphenylene oxide-derived membranes for separation in organic solvents. US Patent 5151182, September 1992.
54 C. Linder, M. Perry, M. Nemas, R. Katraro (Aligena) Solvent stable membranes. US Patent 5039421, August 1991.
55 M. Perry, H. Yacubowicz, C. Linder, M. Nemas, R. Katraro (Membrane Products Kiryat Weizmann) Polyphenylene oxide-derived membranes for separation in organic solvents. US Patent 5151182, September 1992.
56 C. Linder, G. Aviv, M. Perry, R. Kotraro (Aligena) Modified acrylonitrile polymer containing semipermeable membranes. US Patent 4477634, October 1984.
57 J. Tanninen, S. Platt, A. Weis, M. Nyström, Long-term acid resistance and selectivity of NF membranes in very acid conditions. *Journal of Membrane Science* **240** (2004) 11–18.
58 W. J. Wrasidlo (Brunswick) Asymmetric membranes. US Patent 4629563, December 1986.
59 M. Kraus, M. Heisler, I. Katsnelson, D. Velazques (Gelman) Filtration membranes and method of making the same. US Patent 4900449, February 1990.
60 K. Ikeda, S. Yamamoto, H. Ito (Nitto) Sulfonated polysulfone composite semipermeable membranes and process for producing the same. US Patent 4818387, April 1989.
61 W. Loffelmann, J. Passlack, H. Schmitt, H. D. Sluma, M. Schmitt (Akzo Nobel) Polysulfone membrane and method for its manufacture. US Patent 5879554, March 1999.
62 H. D. W. Roesink, D. M. Koenhen, M. H. V. Mulder, C. A. Smolders (X-Flow) Process for preparing a microporous membrane and such a membrane. US Patent 5076925, December, 1991.
63 M. A. Kraus, M. D. Heisler, I. Katsnelson, D. J. Velazquez, Filtration membrane and method of making the same. EP A 0228072. B1, August 1991.
64 J. L. Bailey, R. F. McCune (Polaroid). Microporous vinylidene fluoride polymer and process of making same. US Patent 3642668, February 1972.

65 P. J. Degen, I. P. Sipsas, G. C. Rapisarda, J. Gregg. Polyvinylidenfluorid-Membran. DE 4445973. A1, June 1995.

66 J. D. Grandine, ii (Millipore) Processes of making a porous membrane material from polyvinylidene fluoride, and products. US Patent 4203848, May 1980.

67 W. D. Benzinger, D. N. Robinson (Pennwalt). Porous vinylidene fluoride polymer membrane and process for its preparation. US Patent 4384047, May 1983.

68 M. Onishi, Y. Seita, N. Koyama (Terumo) Hydrophilic, porous poly(vinylidene fluoride) membrane process for its preparation. EP 0344312 A1, December 1989.

69 I. B. Joffee, P. J. Degen, F. A. Baltusis (Pall Corporation) Microporous membrane structure and method of making. EP 0245000A2, November 1987.

70 H. Iwata, T. Matsuda, Preparation and properties of novel environment sensitive membranes prepared by graft polymerization onto a porous membrane. *Journal of Membrane Science* **38** (1988) 185.

71 Y. M. Lee, J. K. Shim, Plasma surface graft of acrylic acid onto a porous PVDF membrane and its riboflavin permeation. *Journal of Applied Polymer Science* **61** (1996) 1245.

72 F. F. Stengaard (Dow) Permeable, porous, polymeric membrane with hydrophilic character, methods for preparing said membranes, and their use. EP 0257635 B1, March 1993.

73 J. Sasaki, K. Naruo, Kyoichi (Fuji Photo Film). Asymmetric microporous membrane containing a layer of minimum size pores below the surface thereof. US Patent 4933081, June 1990.

74 I. F. Wang, J. F. Ditter, R. Zepf (USF Filtration and Separations) Highly porous polyvinylidene difluoride membranes. US Patent 5834107, November 1998.

75 A. Bottino In: Drioli, Nagaki, *Membranes and Membrane Processes*, 1986, pp. 163.

76 T. Uragami, M. Fujimoto, M. Sugikara, studies on syntheses and permeabilities of special polymer membranes. 28. permeation characteristics and structure of interpolymer membranes from poly (vinylidene fluoride) and poly(styrene sulfonic acid). *desalination* **34** (1980) 311.

77 M. Kasi, N. Koyama (Terumo) Method for production of porous membrane US Patent 4772440, September 1988.

78 S. P. Nunes, K. V. Peinemann, Ultrafiltration membranes from PVDF/PMMA blends. *Journal of Membrane Science* **73** (1992) 25.

79 K. V. Peinemann, K. Fink, P. Witt, Asymmetric Polyetherimide Membranes for Helium Separation, *Journal of Membrane Science* **27** (1986) 215.

80 S. P. Nunes, K. V. Peinemann, K. Ohlrogge, A. Alpers, M. Keller, A. T. N. Pires, Membranes of poly(ether imide) and nanodispersed silica. *Journal of Membrane Science*, **157** (1999) 219.

81 J. T. Macheras, B. Bikson, J. K. Nelson (Praxair) Method of preparing membranes from blends of polyetherimide and polyimide polymers. US Patent 5443728, August 1995.

82 A. Goetz, Method of making microporous filter film. US Patent 2926104, February 1960.

83 Y. Hashino, M. Yoshino, H. Sawabu, S. Kawashima (Asahi), Membranes of acrylonitrile polymers for ultrafilter and method for producing the same. US Patent 3933653, January 1976.

84 Y. Hashino, M. Yoshino, H. Sawabu, T. Konno (Asahi) Method for producing hollow fibers of acrylonitrile polymers for ultrafilter. US Patent 4181694, January 1980.

85 T. Sano, T. Shimomura, M. Sasaki, I. Murase, Ichiki (Sumitomo) Process for producing semipermeable membranes,. US Patent 4107049, 1978.

86 GE Water Technologies Catalogue. (http://www. gewater. com).

87 E. Brandenberger, Process for the continuous manufacture of cellulose films. US Patent 981368, January 1911.

88 E. Brandenberger. Apparatus for the continuous manufacture of cellulose films. US Patent 991267, May 1911.

89 R. Etzkorn, E. Knehe (I. P. Bemberg). Method of producing cellulosic films. US Patent 2067522, January 1937.

90 R. Tuccelli, P. V. McGrath (Millipore) Cellulosic ultrafiltration membrane. US Patent 5522991, June 1996.

91 K. V. Peinemann, S. P. Nunes, J. Timmermann (GKSS) Kompositmembran aus einer Mikroporösen Trägermembran und Trennschicht aus Regenerierter Zellulose sowie ein Verfahren zu ihrer Herstellung. German Patent DE59911447D, February 2005.

92 K. B. Hvid, P. S. Nielsen, F. F. Stengaard Preparation and characterization of a new ultrafiltration membrane. *Journal of Membrane Science* **53** (1990) 189.

93 H. G. Hicke, I. Lehmann, G. Malsch, M. Ulbricht, M. Becker, Preparation and characterization of a novel solvent-resistant and autoclavable polymer membrane. *Journal of Membrane Science* **198** (2002) 187–196.

94 P. Zschocke, D. Quellmatz (Berghof). Integralasymmetrische, lösungsmittelbeständige Ultrafiltrationsmembran aus partiell sulfoniertem, aromatischen Polyetheretherketon. German Patent DE 3321860, June 1992.

95 R. S. Dubrow, M. F. Froix (Raychem). Polymeric articles and methods of manufacture thereof. US Patent 4721732, January 1988.

96 L. C. Costa (Ionics) Asymmetric semipermeable poly(aryletherketone) membranes and method of producing same. US Patent 5089192, February 1992.

97 H. N. Beck, R. A. Lundgard, R. D. Mahoney (Dow Chemical) Microporous membranes from poly(etheretherketone)-type polymers. US Patent 5200078, April 1993.

98 A. Iwama, Y. Kihara, M. Abe, Y. Kazuse (Nitto) Process for preparing selective permeable membrane. US Patent 4410568, October 1983.

99 W. K. Miller, S. B. McCray, D. T. Friesen (Bend Research). Solvent resistant microporous polymide membranes. US Patent 5725769, March 1998.

100 W. C. Brinegar. Reverse osmosis process employing polybenzimidazole membranes. US Patent 3720607, March 1973.

101 M. J. Sansone (Celanese). Process for the production of polybenzimidazole ultrafiltration membranes. US Patent 4693824, September 1987.

102 R. F. Savinell, M. H. Litt, Proton conducting polymers prepared by direct acid casting. US Patent 5716727, February 1998.

103 G. Golemme, E. Drioli, Polyphosphazene membrane separations –Review. *Journal of Inorganic and Organometallic Polymers* **6** (1996) 341.

104 Q. Guo, P. N. Pintauro, H. Tang, S. O'Connor, Sulfonated and crosslinked polyphosphazene-based proton-exchange membranes. *Journal of Membrane Science* **154** (1999) 175.

105 H. S. Bierenbaum, R. B. Isaacson, P. R. Lantos (Celanese) Breathable medical dressing. US Patent 3426754, February 1969.

106 J. W. Soehngen, K. Ostrander (Celanese) Solvent stretch process for preparing a microporous film. US Patent 4257997, March 1981.

107 H. M. Fisher, D. E. Leone (Hoechst Celanese) Microporous membranes having increased pore densities and process for making the same. EP Patent 0342026 B1, 1989.

108 W. Böhnstedt, A review of future directions in automotive battery separators. *Journal of Power Sources* **133** (2004) 59–66.

109 R. W. Gore, W. L. Gore (Gore) Process for producing porous products. US Patent 3953566, April 1976.

110 J. L. Dennison, C. B. Jones, W. P. Mortimer, E. K. Probst, W. L. Gore (Gore) Improved filled porous polymers with surface active agents and methods of making same. WO97/28898, August 1997.

111 B. V. McHedlishvili, V. V. Beryozkin, V. A. Oleinikov, A. I. Vilensky, A. B. Vasilyev, Structure, physical and chemical properties and applications of nuclear filters as a new class of membranes. *Journal of Membrane Science* **79** (1993) 285.

112 L. S. Penn, H. Wang, Chemical modification of polymer surfaces: A review. *Polymers for Advanced Technologies* **5** (1994) 809.

113 J. Sutherland, R. Popat, D.M. Brewis, Corona discharge treatment of Polyolefins. *Journal of Adhesion* **46** (1994) 79.

114 C.H. Bamford, K.G. Al-lamee, (University of Liverpool) Functionalisation of polymers. US Patent 5618887, April 1997.

115 P.W. Kramer, Y.-S. Yeh, H. Yasuda, Low temperature plasma for the preparation of separation membranes. *Journal of Membrane Science* **46** (1989) 1.

116 G. Belfort, M. Ulbricht, Surface modification of ultrafiltration membranes by low temperature plasma. I. Treatment of polyacrylonitrile. *Journal of Applied Polymer Science* **56** (1995) 325.

117 M. Ulbricht, G. Belfort, Surface modification of ultrafiltration membranes by low temperature plasma. II. Graft polymerization onto polyacrylonitrile and polysulfone. *Journal of Membrane Science* **111** (1996) 193.

118 G. Clarotti, F. Schue, J. Sledz, K.E. Geckler, W. Göpel, A. Orsetti, Plasma deposition of thin fluorocarbon films for increased membrane hemocompatibility. *Journal of Membrane Science* **61** (1991) 289.

119 D.L. Cho, O. Ekengren, Composite membranes formed by plasmapolymerized acrylic acid for ultrafiltration of bleach effluent. *Journal of Applied Polymer Science*

120 A. Higuchi, N. Iwata, T. Nakagawa, Surface-Modified Polysulfone Hollow Fibers. *Journal of Applied Polymer Science* **40** (1990) 709.

121 M.D. Guiver, P. Black, C.M. Tam, Y. Deslandes, Functionalized PSU membranes by heterogeneous lithiation. *Journal of Applied Polymer Science* **48** (1993) 1597.

122 Z. Feng, B. Ranby, Photoinitiated surface grafting of synthetic fibers. I. Photoinitiated surface grafting of ultrahigh strength polyethylene fibers. *Angewandte Makromolekulare Chemie* **195** (1992) 17.

123 P.Y. Zhang, B. Ranby, Surface modification by continuous graft copolymerization. IV. Photoinitiated graft copolymerization onto polypropylene fiber surface. *Journal of Applied Polymer Science* **41** (1990) 1469.

124 Z. Feng, B. Ranby, Photoinitiated surface grafting of synthetic fibers. III. Photoinitiated surface grafting of poly(ethylene terephthalate) fibers. *Angewandte Makromolekulare Chemie* **196** (1992) 113.

125 P.Y. Zhang, B. Ranby, Surface modification by continuous graft copolymerization. III. Photoinitiated graft copolymerization onto poly(ethylene terephthalate) fiber surface. *Journal of Applied Polymer Science* **41** (1990) 1459.

126 K. Allmer, A. Hult, B. Ranby, Surface Modification of Polymers. II. Grafting with Glycidyl Acrylates and the Reactions of the Grafted Surfaces with Amines. *Journal of Polymer Science, Polymer Chemistry Edition* **27** (1989) 1641.

127 K. Allmer, A. Hult, B. Ranby, Surface Modification of Polymers. III. Grafting of Stabilizers onto Polymer Films. *Journal of Polymer Science, Polymer Chemistry Edition* **27** (1989) 3405.

128 P.Y. Zhang, B. Ranby, Surface modification by continuous graft copolymerization. II. Photoinitiated graft copolymerization onto polypropylene film surface acrylamide. *Journal of Applied Polymer Science* **43** (1991) 621.

129 J.V. Crivello, G. Belfort, H. Yamagishi (Rensselaer Polytechnic Institute) Low fouling ultrafiltration and microfiltration aryl polysulfone. US Patent 5468390, November 1995.

130 M. Ulbricht, H. Matuschewski, A. Oechel, H.-G. Hicke, Photo-induced graft polymerization surface modifications for the preparation of hydrophilic and low-protein-adsorbing ultrafiltration membranes. *Journal of Membrane Science* **115** (1996) 31.

131 K. Kildal, K. Olafsen, A. Stoni, Peroxide-initiated grafting of acrylamide onto polyethylene surfaces. *Journal of Applied Polymer Science* **44** (1992) 1893.

132 I.F. Osipenko, V.I. Martinovicz, Grafting of the Acrylic Acid on Poly(ethylene Terephthalate). *Journal of Applied Polymer Science* **39** (1990) 935.

133 R. Çokun, M. Yigitoclu, M. Saçak, Adsorption behavior of copper (II) ion from aqueous solution on methacrylic acid-grafted PET fibers. *Journal of Applied Polymer Science* **75** (2000) 766.

134 A. R. Oromehie, S. A. Hashemi, D. N. Waters, Functionalisation of Polypropylene with Maleic Anhydride and Acrylic Acid for Compatibilising Blends of Polyproylene with Poly(ethylene terephthalate). *Polymer International* **42** (1997) 117.

135 Y. Pan, J. Ruan, D. Zhou, Solid-Phase Grafting of Glycidyl Methacrylate onto Polypropylene. *Journal of Applied Polymer Science* **65** (1997) 1905.

136 A. Ghenaim, A. Elachari, M. Louati, C. Caze, Surface energy analysis of polyester fibers modified by graft fluorination. *Journal of Applied Polymer Science* **75** (2000) 10.

137 A. Wirsen, K. T. Lindberg, A. C. Albertsson, Graft polymerization of acrylamide onto linear low-density polyethylene film by electron beam pre-irradiation in air or argon: 3. Morphology. *Polymer* **37** (1996) 761.

138 A. Wirsen, A. C. Albertsson, Graft Polymerization of Acrylamide onto LLDPE Film by Electron Beam Pre-Irradiation in Air and Argon. II. Influence of Mahr's Salt. *Journal of Polymer Science. Part A, Polymer Chemistry* **33** (1995) 2049.

139 A. M. Dessouki, N. H. Taher, M. B. El Arnaouty, Gamma Ray Induced Graft Polymerization of N-Vinylpyrrolidone. Acrylamide and Their Mixtures Onto Polypropylene Films. *Polymer International* **45** (1998) 1.

140 I. L. J. Dogue, N. Mermilliod, R. Foerch, Grafting of acrylic acid onto polypropylene – comparison of two pretreatments: gamma-irradiation and argon plasma. *Nuclear Instruments & Methods in Physics Research* **105** (1995) 164.

141 I. L. J. Dogue, N. Mermilliod, A. Gandini, Modification of Industrial Polypropylene Film by Grafting of Poly(acrylic acid). *Journal of Applied Polymer Science* **56** (1995) 33.

142 S. Mishima, T. Nakagawa, Sorption and diffusion of volatile organic compounds in fluoroalkyl methacrylate grafted PDMS membrane. *Journal of Applied Polymer Science* **75** (2000) 773.

143 A. M. Dessouki, N. H. Taher, M. B. El Arnaouty, Synthesis of permselective membrane induced grafting of N-vinylpyrrolidone onto poly(tetrafluoroethylene-hexafluoropropylene-vinylidene fluoride) (TFB) films. *Polymer International* **48** (1999) 92, 67.

144 B. Zhu, H. Iwata, Y. Ikada, Immobilization of poly(ethylene imine) onto polymer films pretreated with plasma. *Journal of Applied Polymer Science* **75** (2000) 576.

145 H. Borcherding, H. G. Hicke, D. Jorcke, M. Ulbrich, Surface functionalized microfiltration membranes for affinity separation. *Desalination* **149** (2002) 297–302.

146 T. Kobayashi, H. Y. Wang, N. Fujii, Molecular imprinting of theophylline in acrylonitrile-acrylic acid co-polymer membrane. *Chemistry Letters* **10** (1995) 927.

147 F. Trotta, C. Baggiani, M. P. Luda, E. Drioli, T. Massari, A molecular imprinted membrane for molecular discrimination of tetracycline hydrochloride. *Journal of Membrane Science* **254** (2005) 13–19.

148 D. Wang (Millipore) Hydrophobic membrane having hydrophilic and charged surface and process. US Patent 5137633, August 1992.

149 C. J. Hou, J. Disbrow, K. C. Hou (Cuno) Low protein binding membrane. US Patent 4921654, May 1990.

150 R. G. Barnes, Jr., C. Chu, G. Emond, A. K. Roy (AMF) Charge modified microporous membrane, process for charge modifying said membrane, and process for filtration of fluids. US Patent 4473475, September 1984.

151 P. J. Degen, I. B. Joffee, T. C. Gsell (Pall), Charge modified polyamide membrane. US Patent 4702840, October 1987.

152 W. Vielstich, A. Lamm, H. A. Gasteiger, Eds. Handbook of Fuel Cells. John Wiley & Sons, New York, 2003.

153 W. T. Grubb, Fuel Cell. US Patent 2913511, Nov 17, 1959.
154 M. Doyle, G. Rajendran, Perfluorinated membranes. In W. Vielstich, A. Lamm, H. A. Gasteiger, Eds. Handbook of Fuel Cells. John Wiley & Sons, New York, 2003, p. 351.
155 M. Doyle, M. E. Lewittes, M. G. Roelofs, S. A. Perusich, R. E. Lowrey, *Journal of Membrane Science* **184** (2001) 257.
156 B. Bahar, A. R. Hobson (Gore) Ultrathin integral composite membrane. US Patent 5.547.551, Aug 20, 1996.
157 C. Stone, A. E. Steck, J. Wei (Ballard), Trifluorostyrene and substituted trifluorostyrene copolymeric compositions and ion-exchange membranes formed therefrom. US Patent 5773480, Jun 30, 1998.
158 R. B. Hodgdon, J. F. Enos, E. J. Aiken, Sulfonated polymers of α,β,β-trifluorostyrene with applications to structures and cells. US Patent 3341366, Sept 12, 1967.
159 B. Gupta, F. N. Buchi, G. G. Scherer, A. Chapiro, Materials research aspects of organic-solid proton conductors. *Solid State Ionics* **61** (1993) 213–218.
160 T. R. Dargavillea, G. A. Georgeb, D. J. T. Hilla, A. K. Whittaker, High energy radiation grafting of fluoropolymers. *Progress in Polymer Science* **28** (2003) 1355–1376.
161 J. Huslage, T. Rager, B. Schnyder, A. Tsukada, Radiation-grafted membrane/electrode assemblies with improved interface. *Electrochimica Acta* **48** (2002) 247–254.
162 S. D. Flint, C. T. Slade, Investigation of radiation-grafted PVDF-γ-polystyrene-sulfonic-acid ion exchange membranes for use in hydrogen oxygen fuel cells. *Solid State Ionics* **97** (1997) 299–307.
163 T. Lehtinen, G. Sundholm, S. Holmberg, P. Björnbom, M. Bursell, *Electrochimica Acta* **43** (1998) 1881–1890.
164 G. Kostova, B. Ameduri, B. Boutevin. New approaches to the synthesis of functionalized fluorine-containing polymers. *Journal of Fluorine Chemistry* **114** (2002) 171–176.

165 M. A. Hickner, H. Ghassemi, Y. S. Kim, B. R. Einsla, J. E. McGrath, Alternative Polymer Systems for Proton Exchange Membranes (PEMs) *Chemical Reviews* **104** (2004) 4587–4612.
166 J. B. Rose (Imperial Chemical Industries), Sulfonated polyarylethersulfone copolymers US Patent 4273903, Jun 16, 1981.
167 M. J. Coplan, G. Götz (Albany International Corporation), Heterogeneous sulfonation process for difficulty sulfonatable poly(ether sulfone). US Patent 4413106, Nov 1, 1983.
168 L. E. Karlson, P. Jannasch, Polysulfone ionomers for proton-conducting fuel-cell membranes: sulfoalkylated polysulfones. *Journal of Membrane Science* **230** (2004) 61–70.
169 C.-M. Bell, R. Deppisch, H. J. Golh, (Gambro Dialysatoren) Membrane and process for the production thereof. US Patent 5401410, Mar 28, 1995.
170 H. S. Chao, D. R. Kelsey, (Union Carbide Corporation) Process for preparing sulfonated poly(aryl ether) resins US Patent 4625000, Nov 25, 1986.
171 P. Genova-Dimitrova, B. Baradie, D. Foscallo, C. Poinsignon, J. Y. Sanchez, Ionomeric membranes for proton exchange membrane fuel cell (PEMFC): sulfonated polysulfone associated with phosphatoantimonic acid. *Journal of Membrane Science* **185** (2001) 59–71.
172 A. Dyck, D. Fritsch, S. P. Nunes, Proton-Conductive Membranes of Sulfonated Polyphenylsulfone. *Journal of Applied Polymer Science* **86** (2002) 2820–2827.
173 W. Schnurberger, J. Kerres, S. Reichle, G. Eigenberger (DLR) Modifying thermoplastic polymers insoluble in organic dispersants. German Patent DE19622338, December 1997.
174 M. D. Guiver, G. P. Robertson, M. Yoshikawa, C. M. Tam, In: I. Pinnau, B. Freeman, Membrane Preparation and Modification. Oxford University Press, 2000.
175 W. L. Harrison, F. Wang, J. B. Mecham, V. A. Bhanu, M. Hill, Y. S. Kim, J. E. McGrath, Influence of the

Bisphenol Structure on the Direct Synthesis of Sulfonated Poly(arylene ether) Copolymers. I. *Journal of Polymer Science: Part A: Polymer Chemistry* **41** (2003) 2264–2276.

176 B. VISHNUPRIYA, K. RAMYA, K. S. DHATHATHREYAN. Synthesis and characterization of sulfonated poly(phenylene oxides) as membranes for polymer electrolyte membrane fuel cells. *Journal of Applied Polymer Science* **83** (2002) 1792–1798.

177 K. RAMYA, K. S. DHATHATHREYAN, Poly(phenylene oxide)-Based Polymer Electrolyte Membranes for Fuel-Cell Applications. *Journal of Applied Polymer Science* **88** (2003) 307–311.

178 K. RICHAU, V. KUDELA, J. SCHAUER, R. MOHR, Electrochemical Characterization of ionically conductive polymer membranes. *Macromolecular Symposium* **188** (2002) 73–89.

179 O. SAVADOGO, Emerging membrane for electrochemical systems: (I) solid polymer electrolyte membranes for fuel cell systems. *Journal of New Materials for Electrochemical Systems* **1** (1998) 47–66.

180 E. VALLEJO, G. POURCELLY, C. GAVACH, R. MERCIER, M. PINERI, Sulfonated polyimides as proton conductor exchange membranes. Physicochemical properties and separation H^+/Mz^+ by electrodialysis comparison with a perfluorosulfonic membrane. *Journal of Membrane Science* **160** (1999).

181 C. GENIES, R. MERCIER, B. SILLION, R. PETIAU, N. CORNET, G. GEBEL, M. PINERI, Stability study of sulfonated phthalic and naphthalenic polyimide structures in aqueous medium *Polymer* **42** (2001) 5097–5105.

182 W. ESSAFI, G. GEBEL, R. MERCIER, Sulfonated Polyimide Ionomers: A Structural Study. *Macromolecules* **37** (2004) 1431–1440.

183 H. J. KIM, M. H. LITT, S. Y. NAM, E. M. SHIN, Synthesis and characterization of sulfonated polyimide polymer electrolyte membranes. *Macromolecular Research* **11** (2003) 458–466.

184 Y. YIN, J. FANG, Y. CUI, K. TANAKA, H. KITA, K. OKAMOTO, Synthesis, proton conductivity and methanol permeability of a novel sulfonated polyimide from 3-(20,40-diaminophenoxy)propane sulfonic acid. *Polymer* **44** (2003) 4509–4518.

185 C. GENIES, R. MERCIER, B. SILLION, R. PETIAUD, N. CORNET, G. GEBEL, M. PINERI, Stability study of sulfonated phthalic and naphthalenic polyimide structures in aqueous medium. *Polymer* **42** (2001) 5097–5105.

186 D. D. DESMARTEAU, L. M. YAGUPOLSKII, Y. L. YAGUPOLSKII, N. V. IGNAT'EV, N. V. KONDRATENKO, A. Y. VOLKONSKII, V. M. SLASOV, R. NOTARIO, P. C. MARIA, *Journal of American Chemical Society* **116** (1994) 3047.

187 K. RAHMAN, G. AIBA, A. B. HASAN, S. Y. SASAYA, K. OTA, M. WATANABE, Synthesis, Characterization, and Copolymerization of a Series of Novel Acid Monomers Based on Sulfonimides for Proton Conducting Membranes. *Macromolecules* **37** (2004) 5572–5577.

188 P. CHAMOCK, D. J. KEMMISH, P. A. STANIL, B. WILSON, (Victrex Manufacturing) Ion-exchange polymers. US Patent 6969755. November 2005.

189 S. KALIAGUINE, S. D. MIKHAILENKO, K. P. WANG, P. XING, G. ROBERTSON, M. GUIVER, Properties of SPEEK based PEMs for fuel-cell application. *Catalysis Today* **82** (2003) 213–222.

190 P. XING, G. P. ROBERTSON, M. D. GUIVER, S. D. MIKHAILENKO, K. WANG, S. KALIAGUINE, Synthesis and characterization of sulfonated poly(ether ether ketone) for proton exchange membranes. *Journal of Membrane science* **229** (2004) 95–106.

191 L. LI, J. ZHANG, Y. WANG, Sulfonated poly(ether ether ketone) membranes for direct methanol fuel cell. *Journal of Membrane Science* **226** (2003) 159–167.

192 B. BAUER, German Patent DE 10116391 A1, Oct 10, 2002.

193 K. D. KREUER, On the development of proton conducting polymer membranes for hydrogen and methanol fuel cells. *Journal of Membrane Science* **185** (2001) 29–39.

194 K. D. KREUER, Hydrocarbon Membranes, In: W. VIELSTICH, A. LAMM, H. A. GASTEIGER, Eds. Handbook of Fuel

Cells. John Wiley & Sons, New York, 2003.

195 E. Drioli, A. Regina, M. Casciola, A. Oliveti, F. Trotta, T. Massari, Sulfonated PEEK-WC membranes for possible fuel-cell applications. *Journal of Membrane Science* **228** (2004) 139–148.

196 M. Rikukawa, K. Sanui, Proton-conducting polymer electrolyte membranes based on hydrocarbon polymers. *Progress in Polymer Science* **25** (2000) 1463–1502.

197 J. Kerres, Development of ionomer membranes for fuel cells. *Journal of Membrane Science* **185** (2001) 3–27.

198 B. Kosmala, J. Schauer, Ion-exchange membranes prepared by blending sulfonated poly(2,6-dimethyl-1,4-phenylene oxide) with polybenzimidazole. *Journal of Applied Polymer Science* **85** (2002) 1118–1127.

199 Y. Gao, G. P. Robertson, M. D. Guiver, X. Jian, S. D. Mikhailenko, K. Wang, S. Kaliaguine. Sulfonation of poly(phthalazinones) with fuming sulfuric acid mixtures for proton exchange membrane materials. *Journal of Membrane Science* **227** (2003) 39–50.

200 X. Zhou, J. Weston, E. Chalkova, M. A. Hofmann, C. M. Ambler, H. R. Allcock, S. N. Lvov, High temperature transport properties of polyphosphazene membranes for direct methanol fuel cells. *Electrochimica Acta* **48** (2003) 2173–2180.

201 A. K. Rianov, A. Marin, J. Chen, J. Sargent, N. Corbett, Novel Route to Sulfonated Polyphosphazenes: Single-Step Synthesis Using "Noncovalent Protection" of Sulfonic Acid Functionality. *Macromolecules* **37** (2004) 4075–4080.

202 M. A. Hofmann, C. M. Ambler, A. E. Maher, E. Chalkova, X. Y. Zhou, S. N. Lvov, H. R. Allcock, Synthesis of polyphosphazenes with sulfonimide side groups. *Macromolecules* **35** (2002) 6490.

203 G. Alberti, M. Casciola, Solid state protonic conductors, present main applications and future prospects. *Solid State Ionics* **33** (2003) 3–16.

204 Q. Li, R. He, J. O. Jens, N. J. Bjerrum, Approaches and Recent Development of Polymer Electrolyte Membranes for Fuel Cells Operating above 100 °C. *Chemistry Materials* **15** (2003) 4896–4915.

205 W. H. J. Hogarth, J. C. Diniz da Costa, G. Q. (Max) Lu, Solid acid membranes for high temperature (>140 °C) proton exchange membrane fuel cells. *J. Power Sources* **142** (2005) 223–237.

206 R. F. Savinell, M. H. Litt (Case Western University) Proton conducting polymers used as membranes. US Patent 5525436, Jun 11, 1996.

207 R. F. Savinell, M. H. Litt, Proton conducting polymers prepared by direct acid casting. US Patent 5716727, Feb 10, 1998.

208 J. S. Wainright, M. H. Litt, R. F. Savinell, High-temperature membranes. In: Handbook of Fuel Cells. John Wiley & Sons, New York, 2003, p. 436–446.

209 O. Savadogo, Emerging membranes for electrochemical systems Part II. High temperature composite membranes for polymer electrolyte fuel cell (PEFC) applications. *Journal of Power Sources* **127** (2004) 135–161.

210 I. Cabasso, J. Jagur-Grodzinski, D. Vofsi, Synthesis and characterization of polymers with pendent phosphonate groups. *Journal of Applied Polymer Science* **18** (1974) 1969–1986.

211 X. Xu, I. Cabasso, Preliminary study of phosphonate ion exchange membranes for PEM fuel cells. *Journal of Polymer Material Science* **120** (1993) 68.

212 A. K. Bhatacharya, G. Thyagarajan, The Michaelis-Arbuzov Rearrangement. *Chemical Reviews* **81** (1981) 415–430, 16.

213 S. Yanagimachi, K. Kaneko, Y. Takeoka, M. Rikukawa, Synthesis and evaluation of phosphonated poly(4-phenoxybenzoyl-1,4-phenylene). *Synthetic Metals* **135/136** (2003) 69–70.

214 K. Jakoby, K. V. Peinemann, S. P. Nunes, Palladium-catalyzed phosphonation of polyphenylsulfone. *Macromolecular Chemical Physics* **204** (2003) 61–67.

215 B. Lafitte, P. Jannasch, Phosphonation of polysulfones via lithiation and

reaction with chlorophosphonic acid esters, *Journal of Polymer Science: Part A: Polymer Chemistry* **43** (2005) 273–286.
216 H. R. Allcock, M. A. Hofmann, C. M. Ambler, S. N. Lvov, X. Y. Y. Zhou, E. Chalkova, J. Weston, Phenyl phosphonic acid functionalized poly(aryloxyphosphazenes) as proton-conducting membranes for direct methanol fuel cells. *Journal of Membrane Science* **201** (2002) 47–54.
217 C. Stone, T. S. Daynard, L. Q. Hu, C. Mah, A. E. Steck, Phosphonic acid functionalized proton exchange membranes for PEM fuel cells. *Journal of New Materials Electrochemical Systems* **3** (2000) 43–50.
218 S. Yanagimachi, K. Kaneko, Y. Takeoka, M. Rikukawa, Synthesis and evaluation of phosphonated poly(4-phenoxybenzoyl-1,4-phenylene). *Synthetic Metals* **135/136** (2003) 69–70.
219 S. V. Kotov, S. Pedersen, W. Qiu, Z. M. Qiu, D. J. Burton, Preparation of perfluorocarbon polymers containing phosphonic acid groups. *Journal of Fluorine Chemistry* **82** (1997) 13–19.
220 P. Stonehart, M. Watanabe, Polymer solid-electrolyte composition and electrochemical cell using the composition. US Patent 5523181, June 4, 1996.
221 V. Antonucci, A. Arico (De Nora) Polymeric membrane electrochemical cell operating at temperatures. EP 0926754A1, June 1999.
222 A. S. Arico, V. Baglio, A. di Blasia, P. Creti, P. L. Antonucci, V. Antonucci, Influence of the acid–base characteristics of inorganic fillers on the high temperature performance of composite membranes in direct methanol fuel cells. *Solid State Ionics* **161** (2003) 251–265.
223 K. A. Mauritz, Organic-inorganic hybrid materials: perfluorinated ionomers as sol-gel polymerization templates for inorganic alkoxides. *Materials Science and Engineering* **C6** (1998) 121–133.
224 R. A. Zoppi, I. V. P. Yoshida, S. P. Nunes, Hybrids of perfluorosulfonic acid ionomer and silicon oxide by sol-gel reaction from solution: Morphology and thermal analysis. *Polymer* **39** (1998) 1309–1315.
225 S. P. Nunes, E. Rikowski, A. Dyck, D. Fritsch, M. Schossig-Tiedemann, K. V. Peinemann, Euromembrane 2000, Jerusalem, Proceedings p. 279–280.
226 D. J. Jones, J. Roziere, in: Handbook of Fuel Cells. W. Vielstich, A. Lamm, H. A. Gasteiger, Eds. Wiley 2003, vol. 3, p. 447.
227 S. P. Nunes, B. Ruffmann, E. Rikowski, S. Vetter, K. Richau, Inorganic modification of proton conductive polymer membranes for direct methanol fuel cells. *Journal of Membrane Science* **203** (2002) 215–225.
228 V. S. Silva, B. Ruffmann, H. Silva, Y. A. Gallego, A. Mendes, L. M. Madeira, S. P. Nunes, Proton electrolyte membrane properties and direct methanol fuel-cell performance I. Characterization of hybrid sulfonated poly-(ether ether ketone)/zirconium oxide membranes. *Journal of Power Sources* **140** (2005) 34–40.
229 V. S. Silva, J. Schirmer, R. Reissner, B. Ruffmann, H. Silva, A. Mendes, L. M. Madeira, S. P. Nunes, Performance and efficiency of a DMFC using non-fluorinated composite membranes operating at low/medium temperatures. *Journal of Power Sources* **140** (2005) 41–49.
230 C. S. Karthikeyan, S. P. Nunes, L. A. S. A. Prado, M. L. Ponce, H. Silva, B. Ruffmann, K. Schulte, Polymer nanocomposite membranes for DMFC application. *Journal of Membrane Science* **254** (2005) 139–146.
231 S.-H. Kwak, T.-H. Yang, C.-S. Kim, K. H. Yoon, Nafion/mordenite hybrid membrane for high-temperature operation of polymer electrolyte membrane fuel cell. *Solid State Ionics* **160** (2003) 309–315.
232 D. H. Jung, S. Y. Cho, D. H. Peck, D. R. Shin, J. S. Kim, Preparation and performance of a Nafion®/montmorillonite nanocomposite membrane for direct methanol fuel cell. *Journal of Power Sources* **118** (2003) 205–211.
233 C. H. Rhee, H. K. Kim, H. Chang, J. S. Lee, Nafion/Sulfonated Montmorillon-

234 C. S. Karthikeyan, S. P. Nunes, K. Schulte, Ionomer-silicates composite membranes: Permeability and conductivity studies. *European Polymer Journal* **41** (2005) 1350–1356.

235 Z. Gaowen, Z. Zhentao, Organic/inorganic composite membranes for application in DMFC. *Journal of Membrane Science* **261** (2005) 107–113.

236 J. Won, Y. S. Kang, Proton-conducting polymer membranes for direct methanol fuel cells. *Macromolecular Symposium* **204** (2003) 79–91.

237 G. Alberti, M. Casciola, L. Massineli, B. Bauer, Polymeric proton conducting membranes for medium temperature fuel cells (110–160 °C). *Journal of Membrane Science* **185** (2001) 73–81.

238 G. Alberti, M. Casciola, U. Constantino, R. Vivani, Layered and pillared metal (IV) phosphates and phosphonates. *Advanced Materials* **8** (1996) 291–303.

239 G. Alberti, M. Casciola, R. Palombari, Inorgano-organic proton conducting membranes for fuel cells and sensors at medium temperatures. *Journal of Membrane Science* **172** (2000) 233–239.

240 L. Tchicaya-Bouckary, D. Jones, J. Roziere, Hybrid polyaryletherketone membranes for fuel-cell applications. *Fuel Cells* **2** (2002) 40.

241 W. G. Grot, G. Rajendran (E. I. du Pont de Nemours and Company) Membranes containing inorganic fillers and membrane and electrode assemblies and electrochemical cells employing same. US Patent 5919583, July 1999.

242 C. Yang, P. Costamagna, S. Srinivasan, J. Benziger, A. B. Bocarsly, Approaches and technical challenges to high temperature operation of proton exchange membrane fuel cells. *Journal Power Sources* **103** (2001) 1–9.

243 F. Damay, L. C. Klein, Transport properties of Nafion™ composite membranes for proton-exchange membranes fuel cells. *Solid State Ionics* **162/163** (2003) 261–267.

244 C. Yang, S. Srinivasan, A. B. Bocarsly, S. Tulyani, J. B. Benziger, A comparison of physical properties and fuel-cell performance of Nafion and zirconium phosphate/Nafion composite membranes. *Journal of Membrane Science* **237** (2004) 145–161.

245 F. Bauer, M. Willert-Porada, Microstructural characterization of Zr-phosphate-Nafion® membranes for direct methanol fuel-cell applications. *Journal of Membrane Science* **233** (2004) 141–149.

246 B. Ruffmann, S. P. Nunes, Organic/inorganic composite membranes for application in DMFC. *Solid State Ionics* **162/163** (2003) 269–275.

247 P. Costamagna, C. Yang, A. B. Bocarsly, S. Srinivasan, Nafion® 115/zirconium phosphate composite membranes for operation of PEMFCs above 100 °C. *Electrochimica Acta* **47** (2002) 1023–1033.

248 L. A. S. A. Prado, H. Wittich, K. Schulte, G. Goerigk, V. M. Garamus, R. Willumeit, S. Vetter, B. Ruffmann, S. P. Nunes, Anomalous Small-Angle X-Ray Scattering Characterization of Composites Based on Sulfonated Poly(ether ether ketone), Zirconium Phosphates, and Zirconium Oxide. *Journal of Polymer Science-Physics* **42** (2003) 567–575.

249 R. He, Q. Li, G. Xiao, N. J. Bjerrum, Proton conductivity of phosphoric acid doped polybenzimidazole and its composites with inorganic proton conductors. *Journal of Membrane Science* **226**, (2003) 169–184.

250 M. Y. Jang, Y. Yamazaki, Preparation, characterization and proton conductivity of membrane based on zirconium tricarboxybutylphosphonate and polybenzimidazole for fuel cells. *Solid State Ionics* **167** (2004) 107–112.

251 S. M. J. Zaidi, Preparation and characterization of composite membranes using blends of SPEEK/PBI with boron phosphate. *Electrochimica Acta* **50** (2005) 4771–4777.

252 S. M. J. Zaidi, S. D. Mikhailenko, S. Kaliaguine, Electrical properties of sulfonated poly ether ketone/polyetherimide blend membranes doped with inorganic acids. *Journal of Polymer Science: Part B: Polymer Physics* **38** (2000) 1386–1395.

253 K. Palanichamy, J. S. Park, T. H. Yang, Y. G. Yoon, W. Y. Lee, C. S. Kim, Proceedings of the ICOM 2005, Seoul, Korea, August 21–26, p. 98.

254 B. Tazi, O. Savadogo, Parameters of PEM fuel cells based on new membranes fabricated from Nafion®, silicotungstic acid and thiophene. *Electrochemical Acta* **45** (2000) 4329–4339.

255 S. Zaidi, S. Mikhailenko, G. Robertson, M. Guiver, S. Kaliaguine, Proton conducting composite membranes from polyether ether ketone and heteropolyacids for fuel-cell applications. *Journal of Membrane Science* **173** (2000) 17–34.

256 P. Staiti, M. Minutoli, S. Hocevar, Membranes based on phosphotungstic acid and polybenzimidazole for fuel-cell application. *Journal of Power Sources* **90** (2000) 231–235.

257 P. Staiti, M. Minutoli, Influence of composition and acid treatment on proton conduction of composite polybenzimidazole membranes. *Journal of Power Sources* **94** (2001) 9–13.

258 P. Staiti, Proton conductive membranes constituted of silicotungstic acid anchored to silica-polybenzimidazole matrices, *Journal of New Materials for Electrochemical Systems* **4** (2001) 181–186.

259 Y. S. Kim, F. Wang, M. Hickner, T. Zawodzinski, J. Mc grath, Fabrication and characterization of heteropolyacid (H3PW12O40)/directly polymerized sulfonated poly(arylene ether sulfone) copolymer composite membranes for higher temperature fuel-cell applications. *Journal of Membrane Science* **212** (2003) 263–282.

260 I. Honma, Y. Takeda, J. M. Bae, Protonic conducting properties of sol-gel derived organic/inorganic nanocomposite membranes doped with acidic functional molecules. *Solid State Ionics* **120** (1999) 255–264.

261 I. Ahmed, S. M. J. Zaidi, S. U. Rahman, Proceedings of the *ICOM 2005*, Seoul, Korea, August 21–26, p. 102.

262 M. L. Ponce, L. Prado, B. Ruffmann, K. Richau, R. Mohr, S. P. Nunes, Reduction of methanol permeability in polyetherketone-heteropolyacid membranes. *Journal of Membrane Science* **217** (2003) 5–15.

263 M L. Ponce, L. A. S. A. Prado, V. Silva, S. P. Nunes, Membranes for direct methanol fuel cell based on modified heteropolyacids. *Desalination* **162** (2004) 383–391.

264 P. Staiti, A. S. Arico, V. Baglio, F. Lufrano, E. Passalacqua, V. Antonucci, Hybrid Nafion–silica membranes doped with heteropolyacids for application in direct methanol fuel cells. *Solid State Ionics* **145** (2001) 101–107.

265 R. Prasad, R. L. Shaner, K. J. Doshi, Comparison of membranes with other gas separation technologies. In: Polymeric Gas-separation Membranes, D. R. Paul, Y. P. Yampolskii (Eds.). CRC Press, Boca Raton, 1994.

266 N. Toshima (Ed.) Polymers for Gas Separation, Wiley-VCH, New York, Weinheim, Cambridge, 1992.

267 R. E. Kesting, A. K. Fritzsche (Eds.) Polymeric Gas-separation Membranes. John Wiley & Sons, New York 1993.

268 W. J. Koros, G. K. Fleming, Membrane-based gas separation. *Journal of Membrane Science* **83** (1993) 1–80.

269 D. R. Paul, Y. P. Yampolskii (Eds.) Polymeric Gas-separation Membranes. CRC Press, Boca Raton 1994.

270 A. Stern, Polymers for gas separations: The next decade. *Journal of Membrane Science* **94** (1994) 1–65.

271 B. D. Freeman, I. Pinnau (Eds.), Polymer membranes for gas and vapor separation, ACS Symposium Series 733, Washington 1999.

272 R. W. Baker, Membrane Technology and Applications, 2nd edn, Chapter 8. John Wiley & Sons, Chichester 2004.

273 M. R. Pixton, D. R. Paul, Relationship between structure and transport prop-

erties for polymers with aromatic backbones. In 146, p. 83–153.
274 B. Freeman, Basis of permeability/selectivity trade-off relations in polymeric gas-separation membranes. *Macromolecules* **32** (1999) 375.
275 W. L. Robb, Thin silicone membranes – their permeation properties and some applications. *Annals New York Academy of Sciences* **146** (1986) 119.
276 J. A. Barrie, K. Munday, Gas transport in heterogeneous polymer blends. *Journal of Membrane Science* **13** (1983) 175.
277 R. M. Barrer, J. A. Barrie, N. K. Raman, Solution and diffusion in silicone rubber. *Polymer* **3** (1962) 595.
278 S. Shishatskiy, C. Nistor, M. Popa, S. P. Nunes, K.-V. Peinemann, Polyimide asymmetric membranes for hydrogen separation: influence of formation conditions on gas transport properties. Submitted to *Advanced Engineering Materials*.
279 M. Langsam, Polyimides for gas separation. In: Polyimides – Fundamentals and Applications, M.K. Gosh, K.L. Mittal (Eds.) Marcel Dekker, New York.
280 N. Muruganamda, D. R. Paul, Evaluation of substituted polycarbonates and a blend with polystyrene as gas-separation membranes. *Journal of Membrane Science* **34** (1987) 185.
281 T. Masuda, E. Isobe, T. Higashimura, K. Takada, *J. Am. Chem. Soc.* 1983, 7473.
282 T. Masuda, M. Kawasaki, Y. Okano, T. Higashimura, *Polym. J.* (Tokyo) **14** (1982) 371.
283 A. Morisato, I. Pinnau, Synthesis and gas-permeation properties of poly(4-methyl-2-pentyne). *Journal of Membrane Science* **121** (1996) 243.
284 J. Schultz, K.-V. Peinemann, Membranes for separation of higher hydrocarbons from methane. *Journal of Membrane Science* **110** (1996) 37.
285 T. C. Merkel, B. D. Freeman, R. J. Spontak, Z. He, I. Pinnau, P. Meakin, A. J. Hill, Ultrapermeable, reverse-selective nanocomposite membranes. *Science* 2002, 296 (5567), 519–552.
286 D. Gomes, S. P. Nunes, K. V. Peinemann, Membranes for gas separation based on poly(1-trimethylsilyl-1-propyne)-silica nanocomposites. *Journal of Membrane Science* 2005, 246 (1), 13–25.
287 S. M. Nemser, I. C. Roman, Perfluorodioxole membranes, US Patent 5051114 (1991).
288 S. M. Nemser, K. P. Callaghan, T. C. Reppert, Combustion engine air supply. US Patent 5960777.
289 K.-V. Peinemann, M. Schossig-Tiedemann, L. Sartorelli, W. Kuhlcke, G. Brunner, Verfahren zur Hochdruck-Gastrennung, German Patent DE 10030643. C1 (2002).
290 P. M. Budd, B. S. Ghanem, S. Makhseed, N. B. McKeown, K. J. Msayib, C. E. Tattershall, Polymers of intrinsic microporosity (PIMs): robust, solution-processable, organic nanoporous materials. *Chemical Communications* 2004, (2), 230–231.
291 N. B. McKeown, P. M. Budd, K. J. Msayib, B. S. Ghanem, H. J. Kingston, C. E. Tattershall, S. Makhseed, K. J. Reynolds, D. Fritsch, Polymers of intrinsic microporosity (PIMs). *Chemistry-A European Journal* 2005, 11, (9), 2610–2620.
292 P. M. Budd, K. J. Msayib, C. E. Tattershall, B. S. Ghanem, K. J. Reynolds, N. B. McKeown, D. Fritsch, Gas separation membranes from polymers of intrinsic microporosity. *Journal of Membrane Science* 2005, 251, (1–2), 263–269.
293 K. Okamoto, S. Kawamura, M. Yoshino, H. Kita, Y. Hirayama, N. Tanihara, Y. Kusuki, Olefin/Paraffin separation through carbonized membranes derived from an asymmetric polyimide hollow fiber membrane, *Ind. Eng. Chem. Res.* **38** (1999) 4424.
294 C. Tsai, S. Tam, Y. Lu, C. J. Brinker, Dual layer asymmetric microporous silica membranes, *Journal of Membrane Science* **169** (2000) 255.
295 R. R. Bhave, Inorganic membranes: characteristics and applications, Van Nostrand Reinhold, New York 1991.
296 H. B. Hsieh, Inorganic membranes for separation and reaction. Elsevier, 1996.

297 A. J. Burggraaf (Ed.), Fundamentals of inorganic membranes. Elsevier, 1996.

298 J. Caro, M. Noack, P. Kölsch, Zeolite membranes: from the laboratory scale to technical applications. *Adsorption* **11** (2005) 215.

299 A. F. Sammells, M. V. Mundschau (Eds.) Nonporous inorganic membranes for chemical processing, Wiley-VCH, New York, Weinheim, Cambridge 2006.

300 J. E. Koresh, A. Sofer, Molecular-Sieve Carbon Permselective Membrane. 1. Presentation of a New Device for Gas-Mixture Separation. *Separation Science and Technology* **1983**, 18 (8), 723–734.

301 K. M. Steel, W. J. Koros, An investigation of the effects of pyrolysis parameters on gas separation properties of carbon materials. *Carbon* **2005**, 43 (9), 1843–1856.

302 M. B. Rao, S. Sircar, Nanoporous carbon membranes for separation of gas mixtures by selective surface flow. *Journal of Membrane Science* **85** (1993) 253.

303 R. M. Barrer, R. Ash, C. G. Pope, Flow of adsorbable gases and vapors in a microporous medium. Part II. Binary mixtures. *Procceedings of the Royal Society* **A271** (1963) 19.

304 S. Sircar, M. B. Rao, M. A. Thaeron, Selective surface flow membrane for gas separation. *Separation Science & Technology* **34** (1999) 2081.

305 Y. Teraoka, H. Zhang, S. Furukawa, N. Yamazoe, Oxygen permeation through perovskite-type oxides, *Chemical Letters* (1985) 1743.

306 H. H. Wang, S. Werth, T. Schiestel, A. Caro, Perovskite hollow–fiber membranes for the production of oxygen-enriched air. *Angewandte Chemie-International Edition* **2005**, 44 (42), 6906–6909.

307 T. Schiestel, M. Kilgus, S. Peter, K. J. Caspary, H. Wang, J. Caro, Hollow fibre perovskite membranes for oxygen separation. *Journal of Membrane Science* **2005**, 258 (1/2), 1–4.

308 C. M. Zimmermann, A. Singh, W. J. Koros, Tailoring mixed matrix composite membranes for gas separations. *Journal of Membrane Science* **137** (1997) 145.

309 S. Kulprathipanja, R. W. Neuzil, N. N. Li, Separation of fluids by means of mixed-matrix membranes, US Patent 4740219 (1988).

310 H. J. C. Te Hennepe, D. Bargeman, M. H. V. Mulder, C. A. Smolders, Zeolite-filled silicone rubber membranes. Part 1. Membrane pervaporation and pervaporation results. *Journal of Membrane Science* **35** (1987) 39.

311 M. Jia, K.-V. Peinemann, R.-D. Behling, Molecular sieving effect of the zeolite-filled silicone rubber membranes in gas permeation, *Journal of Membrane Science* **57** (1991) 289.

312 R. M. Barrer, Diffusion and permeation in heterogeneous media. In: J. Crank, G. S. Park, Diffusion in Polymers. Academic Press, London, New York 1968.

313 J. C. Maxwell, A Treatise on electricity and magnetism, Vol 1. p. 440. Dover Publications, New York 1954 (first published by Clarendon Press in 1891).

314 L. M. Robeson, A. Noshay, M. Matzner, C. N. Merriam, Physical property characteristics of polysulfone/poly(dimethylsiloxane) block copolymers. *Angew. Makromol. Chemie* **29/30** (1973) 47.

315 T. T. Moore, W. J. Koros, Non-ideal effects in organic-inorganic materials for gas-separation membranes. *Journal of Molecular Structure* **2005**, 739 (1–3), 87–98.

316 S. J. Miller, A. Kuperman, Q. de Vu, Mixed-matrix membranes with small pore molecular sieves and methods for making and using membranes, US Patent Application 20050139066 (2005).

317 W. J. Koros, D. Wallace, J. D. Wind, S. J. Miller, C, Staudt-Bickel, Q. de Vu, Crosslinked and crosslinkable hollow fiber mixed-matrix membranes and method of making them. US Patent 6755900 (2004).

318 S. Kulprathipanja, J. Charoenphol, Mixed-matrix membrane for separation of gases. US Patent 6726744 (2004).

319 O. M. Ekiner, S. S. Kulkarni, Process for making hollow fiber mixed-matrix membrane, US Patent 6663805 (2003).

320 K.-V. Peinemann, J. Mohr, R. W. Baker, The separation of organic vapors from air. In: *AIChE Symposium Series* **82** (1987) 250, 14–26.

321 L. Robeson, Correlation of separation factor versus permeability for polymeric membranes. *Journal of Membrane Science* **62** (1991) 165.

322 S. Weller, W. A. Steiner, Separation of gases by fractional permeation through membranes. *Journal of Applied Physics* **21** (1950) 279.

323 C. W. Saltonstall, Calculation of the membrane area required for gas separation, *Journal of Membrane Science* **32** (1987) 185.

Part II
Current Application and Perspectives

1
The Separation of Organic Vapors from Gas Streams by Means of Membranes

K. Ohlrogge and K. Stürken

Summary

The use of membranes to separate and recover organic vapors from off-gas and process gas streams has grown from an outsider application to an accepted and established technology. At the end of the 1980s and the beginning of the 1990s the first applications to recover gasoline vapors or solvents were designed and commissioned in an industrial scale.

The membranes, which are used to separate organic vapors, are thin-film composite membranes with a rubbery polymer as a permselective layer. The commonly used membrane type in industrial applications is a flat-sheet membrane converted into spiral-wound modules [1, 2] or membrane envelopes that are introduced into the GKSS GS membrane module. Gas streams with a high loading of organic vapors are a favorable application. The membrane selectivity of various organic compounds over nitrogen is typically 10 to over 100. A sufficient membrane selectivity, the simple modular design, the ease of operation and advantages in investment and operating costs have led the membrane technology to separate organic vapors to an established position in competition to adsorption, absorption or only condensation processes.

Membrane separation is applied in off-gas treatment with the goal to recover organic compounds and to separate contaminants in accordance with stipulated clean air regulations. The other main application in production plants is the recovery of valuable compounds from process gas streams, e.g. vinylchloride monomer or propylene. The size of the membrane separation units ranges from some m^3/h to more than 4000 m^2/h. The small scale units are often used to treat the off-gases from dryers or pump exhausts, the big units are installed at gasoline loading facilities to treat the off-gases generated by truck, rail road tanker or ship loading.

Meanwhile, approximately 400 membrane units to separate organic vapors have been sold or commissioned by GKSS licensees. The units have been sold to nearly all industrially developed countries in the world. Approx. 40% of the units are installed in Germany. This is because of the stringent air pollution regulations. Major developments to introduce this technology have been performed in the US by MTR, in Europe by GKSS Research Center and its licensees and in Japan by Nitto Denko as a membrane producer and NKK as a plant manufacturer. Based on the growing success new applications have been developed and are at the threshold of becoming commercial. These are hydrocarbon vapor separation from hydrogen in refineries, hydrocarbon dewpointing of natural gas and the control of the methane number of the fuel gas of gas engines and gas turbines.

1.1
Introduction

Many effluents of processes in the chemical, petrochemical and pharmaceutical industry contain volatile organic vapors (VOCs). In addition to this, VOC emissions are generated by handling, storage and distribution of solvents and gasoline products. The emissions, which are allowed to be vented into the atmosphere, are governed by certain governmental regulations. The introduction of new clean air acts at the end of the 1980s in Germany [3] and the beginning of 1990 in the US [4, 5] generated a driving force to develop new technologies to meet the new and more stringent emission standards. In the beginning, membrane technology occupied niches in effluent gas treatment. The growing acceptance based on the confirmed performance, reliability and economic efficiency leads to new and bigger applications to use membranes in production processes to recover valuable compounds and to control concentrations in process gas streams. Membrane technology is now, in a wide range of applications, a serious competitor to established technologies like adsorption, absorption or condensation.

1.2
Historical Background

Rubbery films play an important rule in exploring and understanding gas permeation through dense films. The publications of Mitchell [6] in 1831 and Sir Thomas Graham in 1866 [7] are focused on gas absorption in rubbery material and on the first quantitative measurements of gas permeation rates. A very early patent to use rubbery membranes to separate different hydrocarbons was filed by Frederik E. Frey from Phillips Petroleum Company. The US Patent 2159434 "Process for concentrating hydrocarbons" was granted in 1939. Jean P. Jones filed a second patent on a similar application from Phillips Petroleum titled: "Separation of hydrocarbons from nonhydrocarbons by diffusion". This patent was granted in 1952 as US Patent 2167493. G.J. van Amerongen [8] and R.M. Barrer [9] have provided very important contributions to the knowledge of permeation polymers. In 1981, Roger W. Fenstermaker filed the US Patent 4370150 "Engine performance operating on filed gas engine fuel". This patent describes a membrane separator based on silicone membranes to separate hydrogen sulfide and heavier hydrocarbons from natural gas to up grade the fuel gas of gas engines. One substantial drawback to convert the available knowledge and the inventions into economic technical feasible operations was the lack of commercially available membranes with acceptable fluxes and selectivities. Activities to develop suitable membranes were started in the 1980s in the US, Japan and Germany. The work was pioneered by MTR by evaluating different elastomers as selective layers of thin-film flat-sheet composite membranes [10]. The favored module configuration was the spiral-wound module. Membrane and module development has been performed in a cooperation with Nitto Denko. GKSS started its activities in the separation of volatile organic compounds in the mid-1980s. Filing process patents associates the developments. Richard Baker, MTR, filed his US Patent 4553983 "Process for recovering organic vapors from air" in 1985. This patent is focused on the treatment of effluents from dryers with a maximum solvent concentration of 2 vol%. Kato et al. from Nippon Kokan Kabushiki Kaisha (NKK) invented, in 1986, a process to separate organic vapors from off-gases as a combination of membrane separation and absorption. The key feature of the GKSS patent "Method for Extracting Organic Compounds from Air/Permanent Gas Mixtures" is the compression of organic-vapor-laden gas stream in order to enhance the recovery of organic vapors by condensation or absorption and to optimize the economics of the membrane stage. This patent was filed in February 1988. A process to treat the effluents of a vacuum pump was invented by Gerhard Hauk in 1989 [11]. The vacuum pump was used to supply the feed stream to the membrane module and to provide the vacuum to assist the permeation through the membranes. This patent is owned by Sterling-SIHI.

The introduction of industrial plants in the US, in Japan and Germany occurred at the end of the 1980s. MTR built plants to treat refrigerant vent streams in 1989; NKK has commissioned a gasoline vapor-recovery unit in 1988 and the first gasoline vapor-recovery unit built by the GKSS licensee Aluminum Rheinfelden (now Borsig Membrane Gas Processing) was commissioned in 1989. Meanwhile more than 400 [12–15] membrane-based vapor-separation units are in operation. It concludes all kind of common solvents and organic vapors. The customers include the chemical, petrochemical and pharmaceutical industries. A well-explored market is the treatment of gasoline vapor at tank farms and loading facilities for ships, railroad tankers and trucks. A potential market is the emission reduction at petrol stations.

1.3
Membranes for Organic Vapor Separation

1.3.1
Principles

The membranes, which are being used for organic vapor separations, are thin-film composite membranes. Typically the membrane consists of a three layer structure: a nonwoven material, e.g. polyester, to provide the mechanical strength, a microporous substrate made from polysulfone, polyimide, polyetherimide, polyacrylonitrile or polyvinylidenefluoride and a thin pore-free coating of a rubbery polymer as a selective layer. A common selective coating is polydimethylsiloxane (PDMS) which has a high flux and a sufficient selectivity for many organic vapors. Some specific separation problems require higher selectivities as provided by PDMS. In this case the use of a polyoctylmethylsiloxane (POMS) is favorable. This polymer shows higher selectivities but lower permeabilities. The investment costs for the increase in membrane can be overcompensated by smaller vacuum pumps and compressors.

The advantage of these rubbery membranes besides high flux and acceptable selectivities is the preferential permeability of organic vapors. The preferred permeation of the condensable organic vapors is desirable in order to avoid condensation on the membrane surface.

1.3.2
Selectivity

The crucial parameter for gas transport through a dense polymer film is the interaction of the diffusion and the solubility coefficients of the feed gas in the polymer. The diffusion coefficient of a molecule generally decreases with the increase in molecular size. Glassy polymers with stiff polymer backbones act like a molecular sieve. Small gas molecules like hydrogen or helium permeate much faster through the rigid polymer backbone than the bigger molecules of hydrocarbon vapors. An elastomeric polymer acts more like a liquid. The gas transport through an elastomer is determined more by its solubility than by its diffusion coefficient. The high solubility of organic vapors in some elastomers is the reason for their high permeability. This item has been discussed in the chapter "Materials and transport mechanisms in the part gas separation with membranes."

The selectivities of various organic vapors over nitrogen of a rubbery thin-film composite membrane are shown in Fig. 1.1.

The length of the bar shows the highest average selectivity obtained by single gas measurements at ambient temperatures. The selectivity that can be achieved in a technical process is dependent on the membrane structure, module configuration and process parameters. The temperature and the partial pressure of the organic compound have a direct impact on the membrane selectivity.

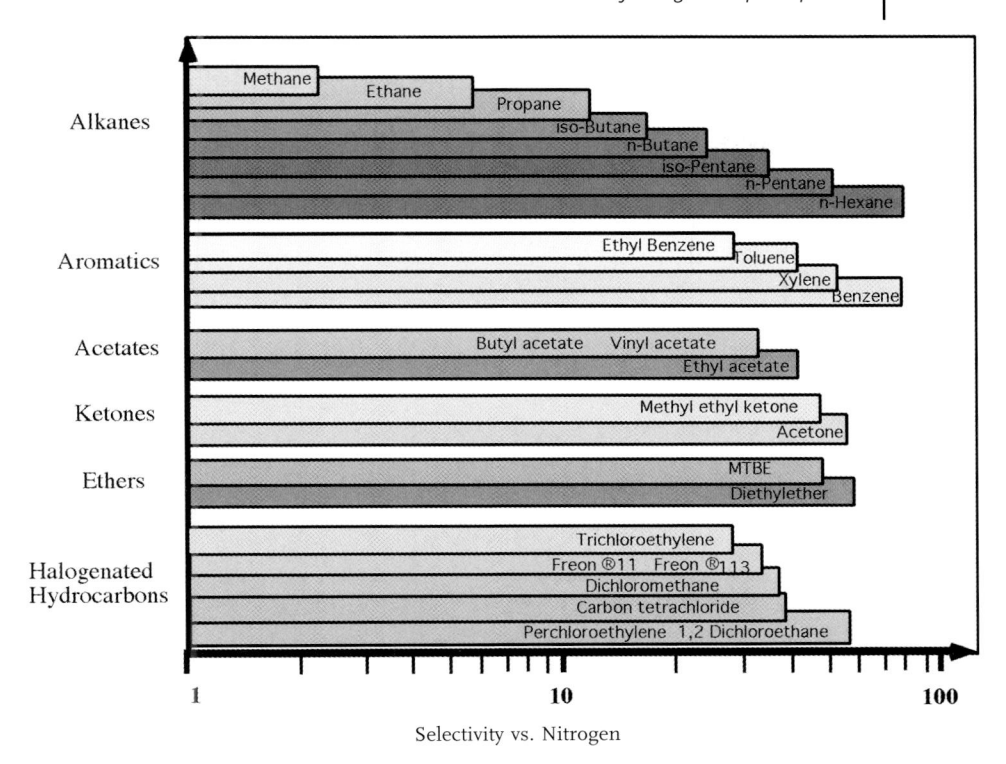

Fig. 1.1 Selectivities of various organic vapors over nitrogen.

1.3.3
Temperature and Pressure

The permeation of permanent gases increases with the increase in temperature. Because the permeation of organic vapors depends on its solubility, the flux increases with a decrease in temperature. It is advantageous to operate a membrane-based organic vapor separation process in temperatures as low as possible in order to achieve the highest possible selectivity. The second important parameter is the concentration or partial pressure of the organic vapor. Whereas the flux of permanent gases through a rubbery membrane is practically independent, the flux of the organic vapor is highly dependent on the vapor pressure. The higher sorption at high organic vapor pressure plasticizes the membrane material and causes an increase of the solvent diffusion coefficient. The flux density dependence of various organic vapors on vapor pressure is shown in Fig. 1.2.

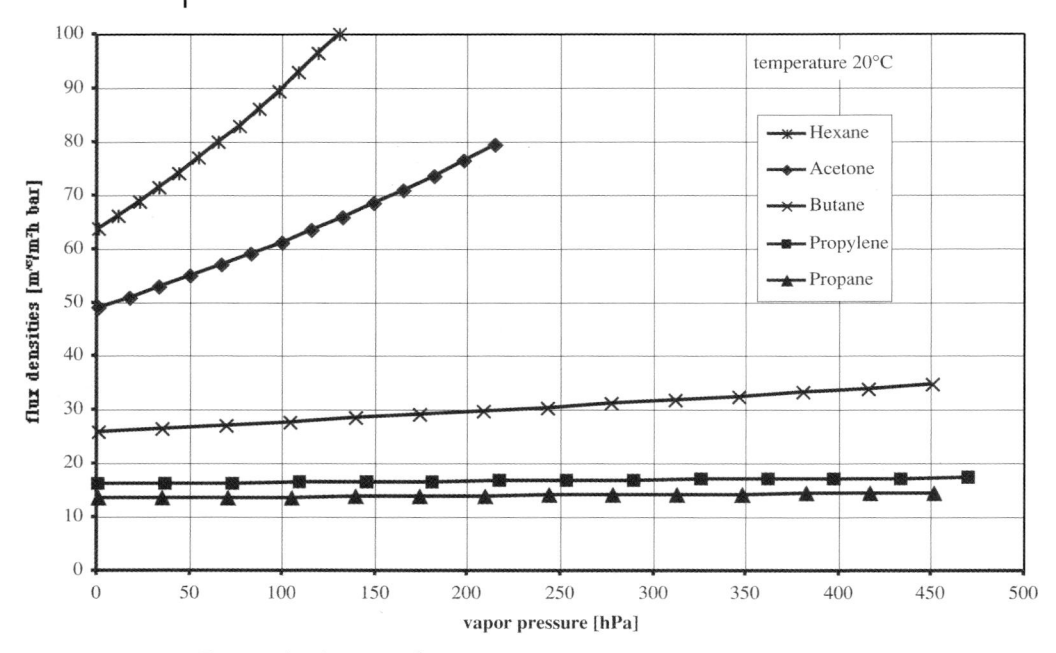

Fig. 1.2 Flux densities of various organic vapors over vapor pressure.

1.3.4
Membrane Modules

The membrane modules, which are commonly used in organic vapor separation, are spiral-wound modules or the envelope-type GKSS GS modules. Capillary or hollow-fiber modules are only used in small-scale laboratory applications. The spiral-wound module and the envelope module are based on flat-sheet membranes. Spiral-wound modules are compact and cheaper in comparison to installed membrane area, but there are limitations in mass transfer on both sides of the membrane. The packing density – the ratio of installed membrane area over pressure vessel housing volume – of a spiral-wound module varies from approx. 300 to 1000 m^2/m^3 (Fig. 1.3).

The envelope-type GS module offers advantages in flow distribution and a minimized pressure drop at the permeate side.

The membrane envelope consists of two membranes; fleeces and spacers between the membranes to provide an open space for an unrestrained permeate drainage. Thermal welding at the outer cutting edges seals the membrane sandwich (Fig. 1.4).

The membrane module consists of a pressure vessel, a central permeate tube and a stack of membrane envelopes. The stack of envelopes is divided into asymmetric arranged compartments by means of baffle plates. The design of a

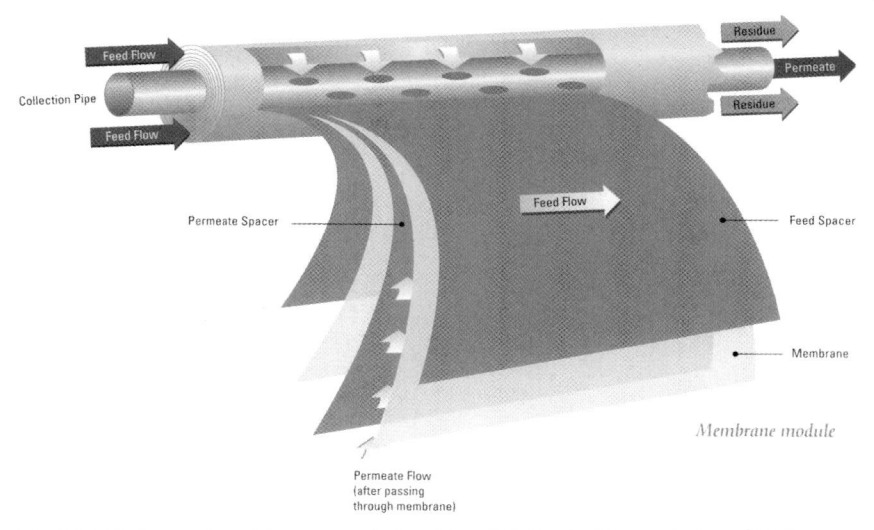

Fig. 1.3 Spiral-wound module as manufactured by MTR. Used with permission of MTR.

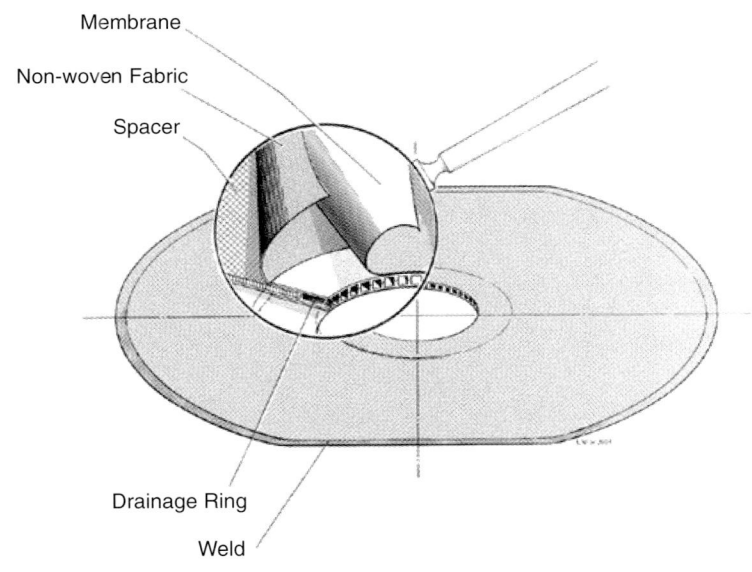

Fig. 1.4 Membrane envelope.

compartment is calculated according to a uniform flow velocity over the membrane surface. In correlation with the feed volume reduction caused by permeation the number of envelopes between two baffle plates are reduced. This special design feature allows a design velocity over the membrane surface in order to reduce boundary layer effects and to achieve the highest possible membrane

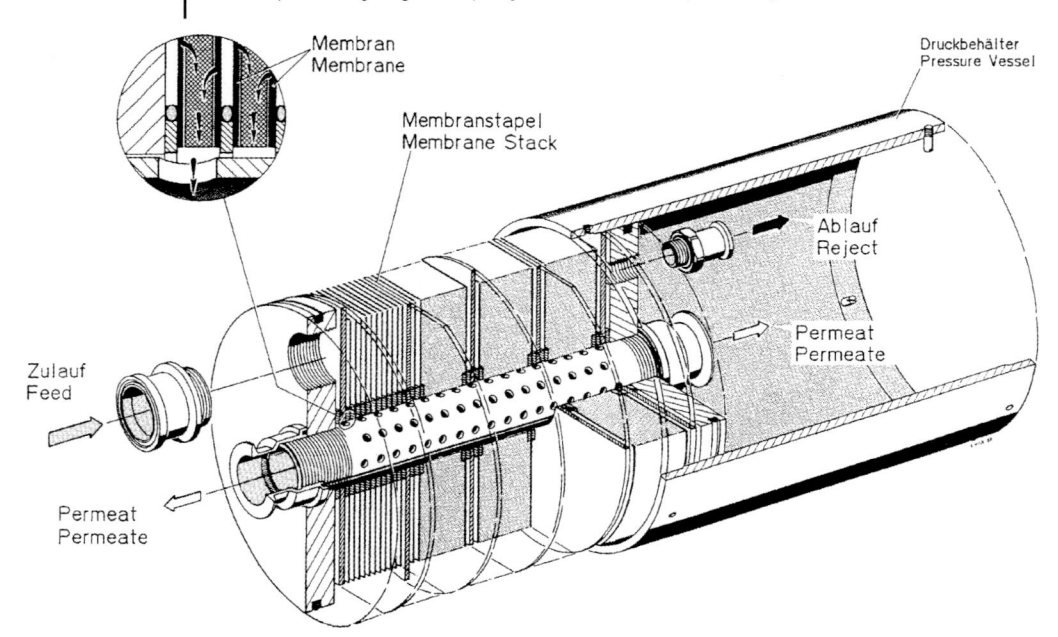

Fig. 1.5 GKSS GS-module.

selectivity. The packing density variation is dependent on the number of baffle plates from approx. 270 to 450 m^2/m^3 (Fig. 1.5).

1.4
Applications

1.4.1
Design Criteria

Most of the organic vapor separation applications are unique and the systems are tailored according to customer- or site-specific requirements. Figure 1.6 shows the basic input data, which are necessary to evaluate the membrane-separation process. The physical constants of the feed compounds, the operating conditions and the design concentration of the product have to be known in order to calculate the basic lay out. The feed pressure and the degree of saturation of the condensable feed compounds leads to the most effective location of the recovery by condensation or absorption. In the case of pressure increase by means of a compressor and a moderate to high organic vapor concentration of the feed the recovery unit should be placed in front of the membrane stage. If the separation process is operated without feed compression the recovery unit should be placed in the enriched permeate stream at the feed site of the recycle

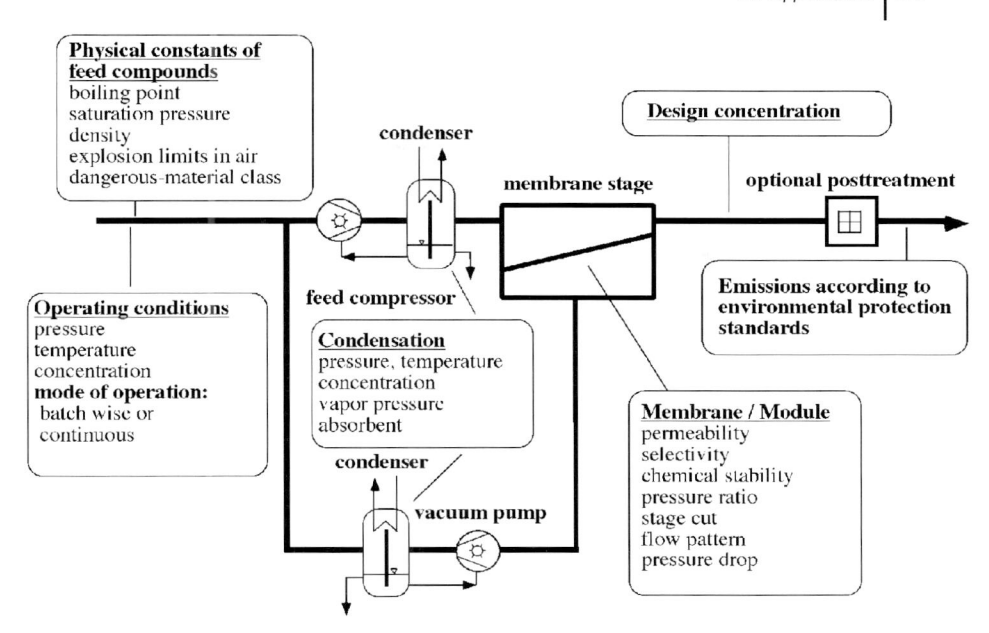

Fig. 1.6 Simplified flow scheme of a membrane-separation process.

pump. In the case of multicompound streams and very low retentate concentrations the combination of a membrane stage and a post treatment unit could provide the most economic solution. The polishing of the retentate according to stringent clean air regulations can be performed by adsorption, absorption, thermal combustion or catalytic conversion.

The design of single stage membrane units depends on the specific application
- efficient recovery of a valuable product
- achievement of the stipulated clean air requirements
- a combination of both
- separation.

The efficiency of a membrane stage is dependent on
- intrinsic membrane selectivity
- flow distribution in the membrane module to achieve the highest possible selectivity under operating conditions
- feed pressure
- pressure drop
- pressure ratio (feed pressure over permeate pressure)
- operating temperature.

There is a strong interaction between the demanded retentate concentration and the investment and operating costs of compressors, vacuum pumps and membrane modules. In any case a very low organic vapor concentration of the retentate leads to a high stage cut that requires higher suction capacities of vacuum pumps and compressors.

1.4.2
Off-gas and Process Gas Treatment

Typical applications are the treatment of small volume flows of solvent-contaminated off-gas streams in the chemical and pharmaceutical industry. Figure 1.7 shows the off-gas purification of a stripping column to treat edible vegetable oil. Nitrogen is used to strip the solvent from oil. The off-gas has to be treated according to an outlet concentration < 150 mg hexane/m^3 inert gas.

The process was designed according to the following data:
- Process vacuum and permeate pressure: 30 mbar
- Volume flow: 4 m^3/h inert gas with hexane vapors
- Feed concentration: 300 g/m^3
- Outlet concentration: < 150 mg/m^3
- Service liquid: hexane
- Temperature service liquid: –15 °C
- Recovery rate: 99.95%

This application shows that it is possible to meet the stringent German TI Air standards with a single-stage membrane unit. But this is also an example for the potential of the improvement of the process economics. If it is possible to recycle the retentate the following savings are possible:
- Reduction of stripping gas cost
- Reduction of vacuum pump capacity
- Reduction of required membrane area.

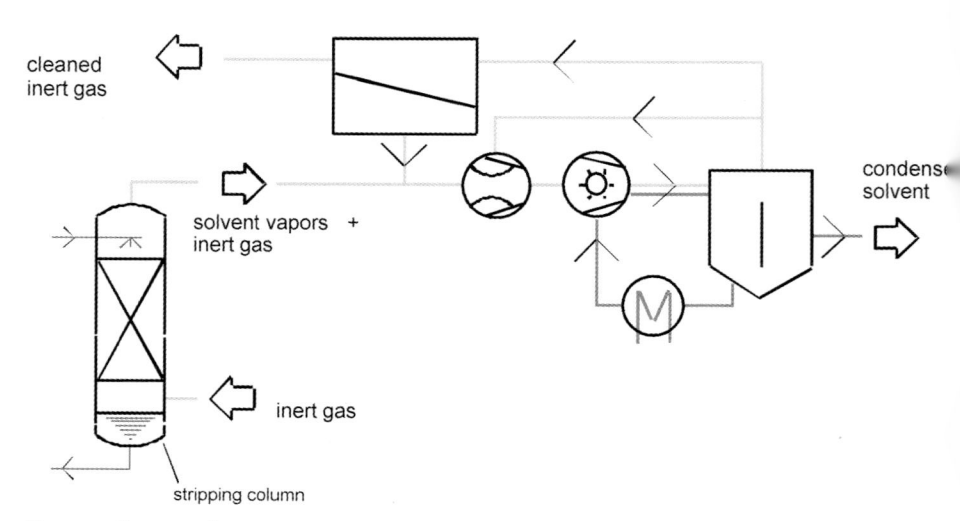

Fig. 1.7 Off-gas purification.

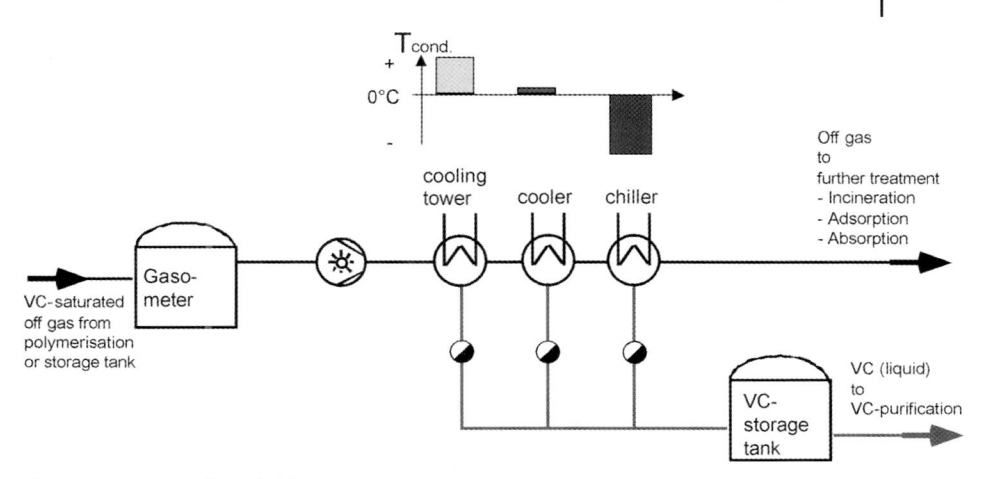

Fig. 1.8 Conventional vinyl chloride recovery.

An excellent example for the economic use of membrane technology is the vinyl chloride recovery from off-gases of the PVC production. The vapor pressure of vinyl chloride at 30 °C is 4.51 bar, the residual concentration of a compressed off-gas stream at 5.5 bar is approx. 2 kg/m^3 (STP). Vinyl chloride is a high-value product with a market price of approx. 500 €/t. The common technology to recover the vinyl chloride is multistage pressure condensation. The production off-gas is collected in a buffer. It is compressed up to approx. 5.5 bar and the condensation of the vinyl chloride takes place in condensers operated with cooling tower water, the second stage condensation at approx. –10 to –15 °C and the third stage at chilling temperatures of approx. –40 °C. The recovery rate at condensation at –40 °C is 92% and the residual concentration is 0.18 kg/m^3 (STP).

Vinyl chloride is considered as a carcinogenic solvent and the allowed emission concentration is <5 mg/m^3. All kinds of recovery units have a post-treatment system by adsorption and/or combustion (Fig. 1.8).

The replacement of the second and third condenser by a membrane stage enables a high separation of vinyl chloride at condensation temperatures of –70 °C or lower (Fig. 1.10).

Another advantage of the use of membranes is that it enables a continuous operation without defreezing the condenser surfaces in the case of the presence of water vapor in the off-gas.

1.4.2.1 Gasoline Vapor Recovery

The principle of emission reduction in the chain of gasoline distribution is vapor balancing. The volume of vapor, which is displaced when a volume of liquid is filled into a tank, is collected and returned to a tank from which a liquid is drawn of. But it is expected that a surplus of volume will be created during filling and transfer procedures. This is because of the possible change of temperatures and pressures as

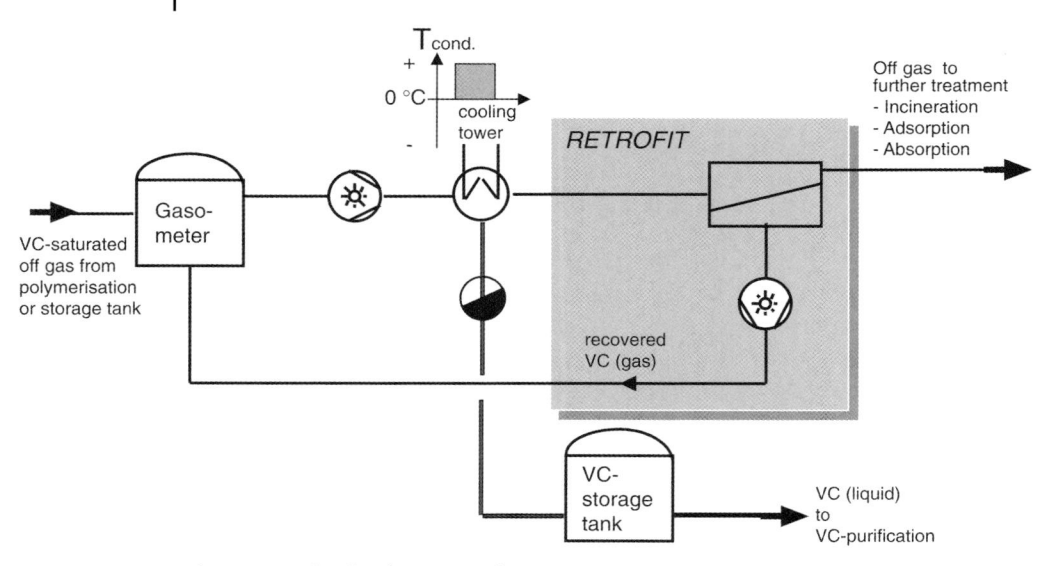

Fig. 1.9 Retrofit of multistage condensation unit.

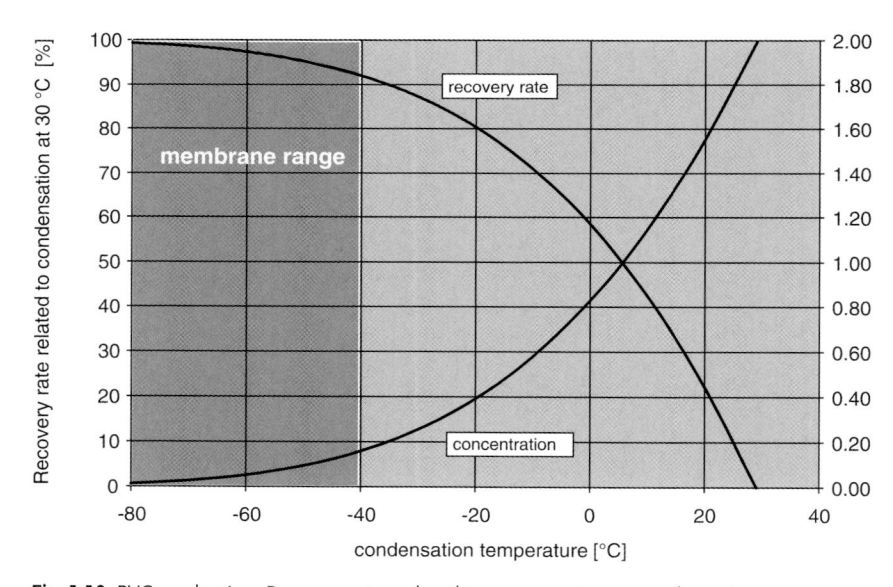

Fig. 1.10 PVC production: Recovery rate and outlet concentration vs. condensation temperature.

well as evaporation generated by turbulences of the liquid phase during filling procedures. The average gasoline vapor concentration in off-gases is approx. 20 to 40 vol%, which corresponds to approx. 600 to 1200 g hydrocarbon/m^3 air. The allowed vent gas concentration is governed by established clean air regulations. The European stage II directive 94/63/EC allows 35 g HC/m^2 air, whereas the German TI-Air is restricted to 150 mg/m^3, a value that is approx. 230 times lower.

In the case of gasoline vapor loaded off-gases the recovery of the organic vapors is favored over destruction.

Common used recovery techniques are:
- Adsorption
- Absorption
- Condensation
- Membrane separation.

In order to achieve high recovery rates and low investment and operating costs the combination of different recovery techniques has been realized.

A membrane-based gasoline vapor-recovery unit designed to meet the stringent TI-Air standards consists typically of
- a recovery stage by absorption
- a membrane separation stage and
- a post-treatment by means of a pressure swing adsorption unit
  or a gas motor where the retentate of the membrane stage
  is used as fuel gas to supply the combustion engine (Fig. 1.11).

The inlet vapor coming from a gasholder to balance volume peaks or coming directly from the loading facility is fed to a feed compressor. The recovery takes

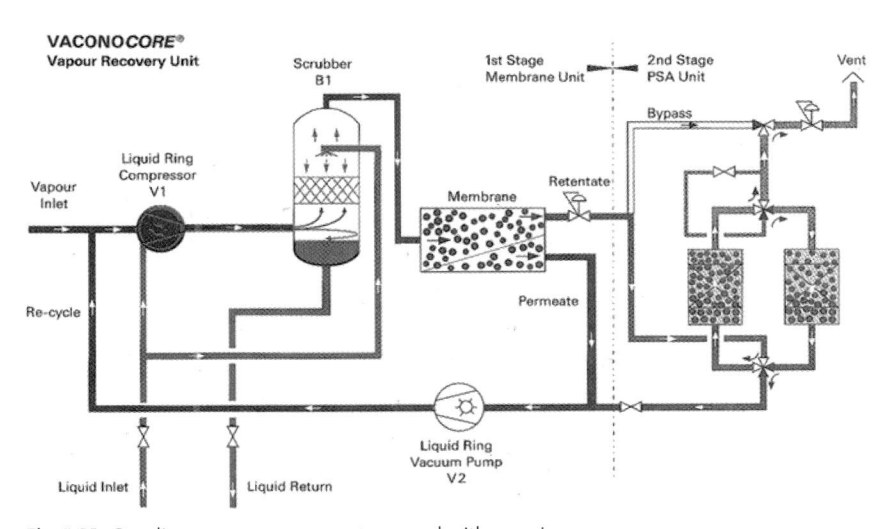

Fig. 1.11 Gasoline vapor recovery system used with permission of Borsig Membrane Gas Processing.

place in a scrubber. The system utilizes lean liquid product as service liquid for the liquid ring compressor and as scrubbing fluid for the recovery unit. The recovered product is absorbed by the scrubbing action and returned to storage as an enriched stream. The gas flow leaves the top of the scrubber with a residual concentration in accordance with temperature and pressure and is introduced into the membrane stage. The hydrocarbon-selective membrane separates the stream into a lean retentate stream and an enriched permeate stream. A vacuum is applied at the permeate side to support the permeation process. The permeate is recycled and mixed with the inlet vapor at the suction side of the feed compressor.

The retentate contains only a residual organic vapor content, which consists mostly of light hydrocarbons. The stream is introduced into the integrated pressure swing adsorption unit (PSA). The unit consists of adsorption beds arranged in parallel and a service of control valves. The control valves are cycled in a predetermined sequence. This allows the operation of one bed whilst the other bed is regenerating and provides a continuous process. Part of the clean vent stream is bypassed and recycled to be used as a purge gas stream for regeneration. The pressure of the membrane process is used for an enhanced adsorption whereas the regeneration is supported by the vacuum of the membrane process. The purge gas stream is returned and mixed with the inlet gas.

In the case of gasoline tank farm applications in Germany membrane-based recovery plants have the highest market share of more than 55%. The hybrid process of membrane technology combined with pressure swing adsorption was the best selling system in the case of TI-Air requirements.

In the case of gasoline vapor recovery two types of membranes are used: the PDMS (polydimethylsiloxane) membrane and the POMS (polyoctylmethylsiloxane) membrane.

The PDMS membrane is a membrane that is characterized by high flux densities and moderate hydrocarbon over air selectivities, whereas the POMS membrane has a higher selectivity and moderate gas fluxes.

The intake concentration of a unit to treat the off-gases from a tank farm varies from approx. 600 to 1200 g/m^3. The variation is caused by the vapor pressure of the gasoline (e.g. summer or winter quality), temperature and loading procedure (e.g. top loading or bottom loading).

A case study compares the use of PDMS or POMS membranes at given design criteria.

Capacity of feed compressor:	1000 m^3/h
Feed pressure:	3.5 bar
Permeate pressure:	150 mbar
Feed concentration:	20 vol% HC
Absorption medium:	lean gasoline
Temperature:	25 °C
Isothermal compressor efficiency:	0.38
Isothermal vacuum pump efficiency:	0.30
Inlet concentration PSA:	10 g/m^3

PSA purge gas ratio of retentate
Volume flow: 15
Vent gas purity: 150 mg/m³ (TI Air requirement)
                                = 99.99% recovery rate
                                10 g/m³ = 98.3% recovery rate
                                20 g/m³ = 96.7% recovery rate
                                35 g/m³ (EU stage 1 requirement)
                                = 94.2% recovery rate

Feed-gas composition: Methane: 0.017 vol%   Ethane: 0.049 vol%
                        Propane: 0.642 vol%   i-Butane: 3.581 vol%
                        n-Butane: 7.592 vol%   i-Pentane: 4.117 vol%
                        n-Pentane: 2.275 vol%   Hexane: 1.323 vol%
                        Heptane: 0.220 vol%   Benzene: 0.183 vol%
                        Nitrogen: 63.280 vol%   Oxygen: 16.720 vol%

	PDMS	POMS
	Retentate concentration [g/m³]	
	0.15 (TI Air)	
VRU capacity [m³(STP)/h]	524	611
Membrane area [m²]	60	186
Stage cut [%]	53.2	44.3
Specific energy consumption [kWh/m³(STP)]	0.334	0.262
	10 g/m³	
	612	709
	58	178
	44.2	33.9
	0.261	0.201
	20 g/m³	
	663	745
	47	139
	38.8	30
	0.227	0.183
	35 g/m³	
	(European stage I Ordinance)	
	705	775
	38.2	108.5
	34.3	26.6
	0.203	0.169

It is clearly shown that "slower" membranes require approx. 3 times more membrane area to achieve the same vent gas purity. The higher selectivity, on the other hand, causes a reduction of the stage cut, which leads to a reduced recycled permeate stream.

The suction volume of the compressor consists of the flow from the loading terminal and the recycled permeate flow. The reduced stage cut affects the relation regarding the flows share of both and causes an increase of VRU capacity, which is in direct relation to the specific energy consumption. The second influence on the stage cut is the required vent gas purity. Any increase of the allowed hydrocarbon concentration in the vent gas leads to a decrease of the stage cut.

In summary: Retentate concentration and membrane selectivity have a direct influence on the stage cut. The membrane permeability has an effect on the required membrane area. Investment costs for compressor, vacuum pump and membrane area have to be balanced with the operating costs. As a result of the stringent clean air requirements the POMS membrane was the first choice of all realized plants.

Retrofitting of Adsorption Units

Old vapor-recovery units often do not meet the new and more stringent clean air requirements. One possibility of retrofitting of an adsorption unit is the installation of a membrane unit in the feed line before the entrance into the adsorption unit. The membrane stage could be used to shave the peaks at fluctuating hydrocarbon concentrations and to reduce the amount of higher hydrocarbons. Figure 1.12 shows a simplified flow scheme of retrofitting an adsorption unit by means of the installation of a membrane stage.

Table 1.1 shows the effect of three steps to improve a vapor-recovery unit to meet the new TI Air standards and to operate the unit at the highest efficiency.

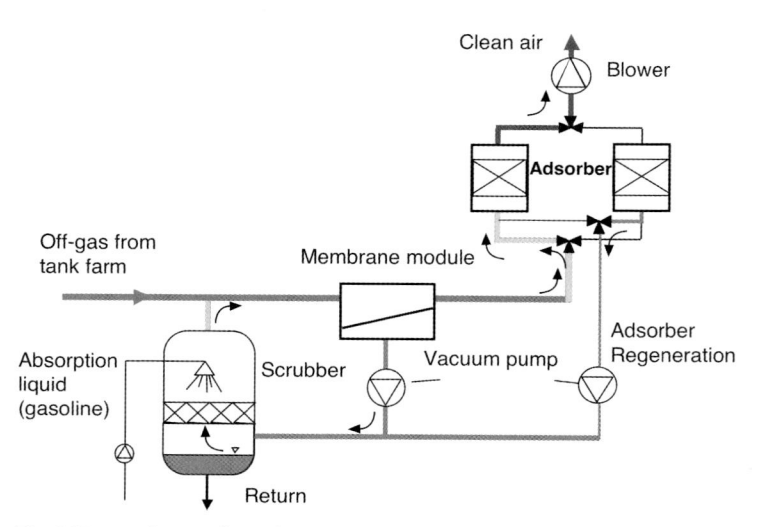

Fig. 1.12 Retrofitting of an adsorption unit to meet TI Air standards.

Table 1.1

Starting conditions	1 Retrofitting Enhancement of vacuum pump capacity	2 Retrofitting Installation of a membrane stage	3 Retrofitting Enhancement of blower capacity
Emissions >3 kg/h	3 kg/h	<150 mg/m³ goal 20 mg/m³ actual	<150 mg/m³ goal 70 mg/m³ actual
Electrical power 64.5 kW installed	94.5	131.5	131.5
Electrical power 57 kW actual	86	118	120
VRU capacity 280 m³/h	280	280	350
Specific power Requirement 0.2 kW/h	0.31	0.42	0.34

It is confirmed by realized installations, that it is possible to retrofit existing plants with acceptable investment costs to meet new clean air standards and to enhance the plant capacity.

1.4.2.2 Polyolefin Production Processes

The potential use of membrane separation in polyolefin production processes will be in the areas of raw material purification, chemical reaction and product purification and finishing [17–19]. Hydrocarbon vapors including propylene and ethylene can be separated and recovered from nitrogen and light gases such as the methane and hydrogen in polyolefin production vent streams. On the other hand, nitrogen can be purified to be reused as purge gas (Fig. 1.13).

In order to meet quality requirements the raw material has to be purified. In the case of independently separated isolated plants the purification takes place in a splitter column. The build up of nitrogen and light gases has to be removed from the overhead of the splitter column. Because of the significant amount of unreacted monomers, which represents a high commercial value, the separation of the organic compounds offers the opportunity to reduce production costs. The operation of the membrane process is supported by the available pressure of the column overhead. The monomer-depleted residual stream of the membrane stage, which contains nitrogen, hydrogen and methane is either fed for further treatment to a flare or used as fuel. The monomer-enriched stream is recompressed and recycled to the splitter. The olefin polymerization takes place in the presence of the monomer plus catalyst, various comonomers, solvents and stabilizers, which are contacted at high pressure in a polymerization reactor. The polymerization reaction is performed in the gas phase or in a slurry phase. Butane or hexane are used as organic solvents. The raw polymer product has to be treated because of the amount of nonreacted sorbed monomers, comonomers and processing sol-

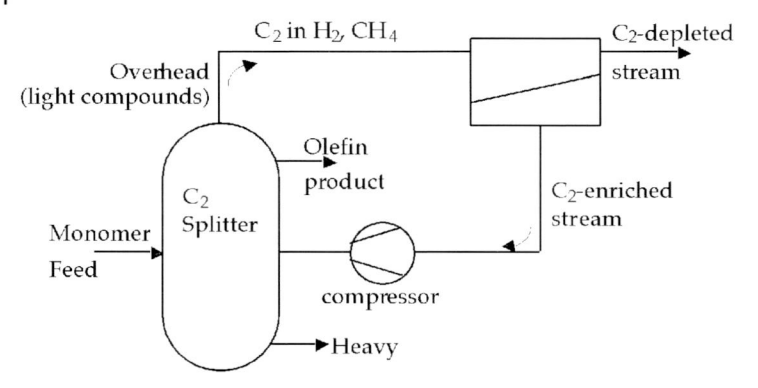

Fig. 1.13 Column overhead recovery.

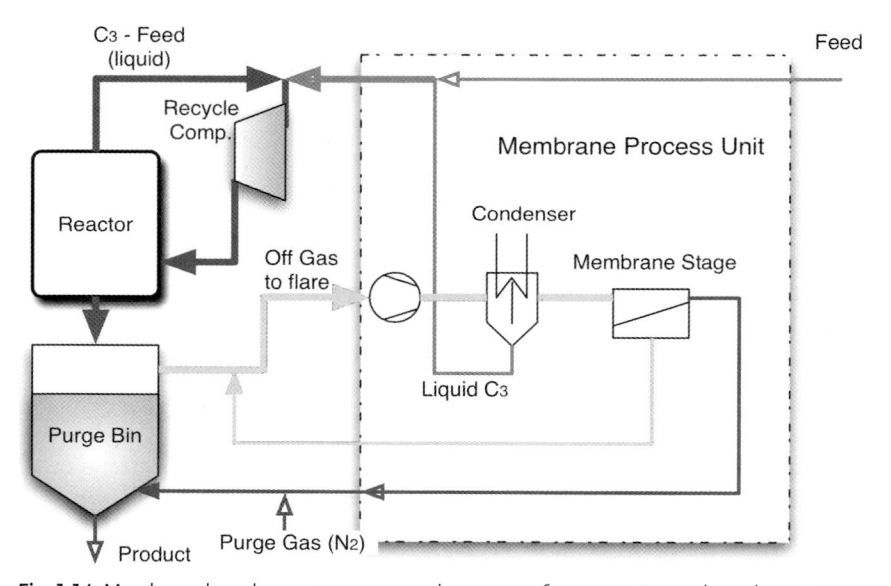

Fig. 1.14 Membrane-based monomer recovery by means of compression and condensation.

vents. The resin is introduced into a purge bin where nitrogen is used to remove absorbed monomer and process solvents. The composition of the vent stream leaving the purge bin depends on the degree of polymer purifying, the polymer product and the kind of polymerization process. The typical membrane-based monomer recovery process consists of compression of the purge bin vent stream, low-temperature condensation and residual monomer separation by means of selective membranes. The vent stream can be purified up to 99.9% nitrogen when the stream is recycled to the purge bin. High nitrogen purity requires an increase in membrane area and compressor capacity because of the increased stage cut that leads to lower permeate concentration and higher permeate volume (Fig. 1.14).

1.5
Applications at the Threshold of Commercialization

1.5.1
Emission Control at Petrol Stations

A spin-off of the activities of gasoline vapor recovery at gasoline tank farms is the development of a system to reduce emissions generated by the operation of petrol stations. In the case of car refueling, the connection between the dispenser nozzle and the tank filler pipe is the only open area to the atmosphere. To reduce the emissions during refueling vacuum-assisted vapor return systems have been introduced in many countries. In order to avoid the emission transfer from the tank-filling point of the car to the vent pipes of the storage tanks an air to liquid ratio of 1:1 of the vapor return system has been stipulated. An investigation of the TÜV Rheinland has shown that the efficiency of catching emissions by means of the 1:1 vapor return ratio is limited to an average of approx. 75% [20]. The difference of a minimum value of 50% and 90% maximum value of vapor return is caused by the different construction of the car filling pipes. In order to enhance the vapor return rates a surplus of vapor volume has to be returned. Tests have shown that the increase of the air to liquid ratio of 1.5:1 leads to an improvement of the efficiency of 95 to 99%, depending on the type of car [21]. The enhancement of the vapor return rate is only allowed if no additional emissions are generated. A membrane-based vapor separation system to treat the breather pipe vent gases of storage tanks enables the emission reduction during car refueling without creating any additional emissions. The essential requirement is a leak-proof installation of tanks, pipes and dispensers. Furthermore, the installation of over-/underpressure safety valves at breather pipes and check valves at the filling and vapor-balancing couplings of the storage tanks is also necessary. Because of the surplus of returned vapor volume a pressure build-up occurs in the storage tanks. At a given set point of a pressure gauge, which measures the differential pressure between tank pressure and atmospheric pressure, the vacuum pump of a membrane separation system is activated. This system is installed parallel to the vent stack of the storage tanks. A pneumatic valve in the retentate line of the membrane module is opened by the applied vacuum. The overpressure of the storage tanks causes a volume flow, which is released by passing the membrane stack. The gasoline vapors are separated from the off-gas and clean air leaves to the atmosphere. After the lower set point of the pressure gauge is reached, the system is deactivated (Fig. 1.15). Because of the experience that often small leakages in the gas-phase installation can occur a system was designed that can operate at set points around atmospheric pressure. In order to provide a reliable operation a retentate pump was installed at the outlet of the membrane module. This pump sucks the generated vapor from the headspace of the storage tanks. Another advantage of this design is the vapor flow with a constant flow velocity over the gasoline selection membrane.

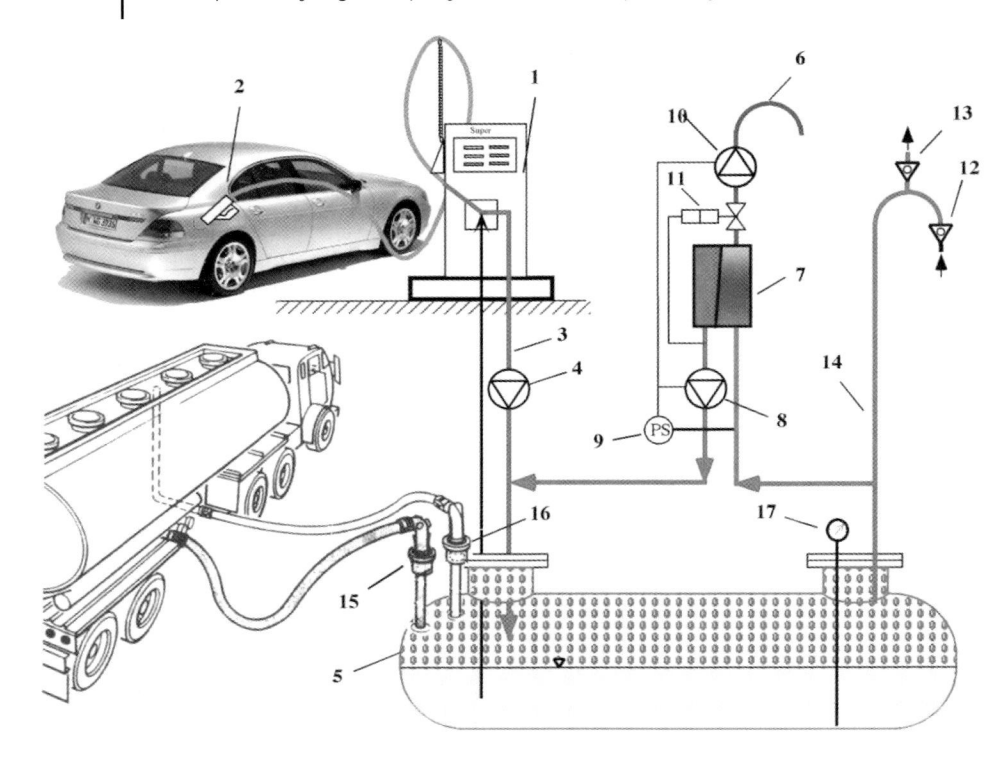

1	Petrol pump with activ gas return line	9	Pressure switch
		10	Retentate pump
2	Fuel dispenser	11	Pneumatic retentate valve
3	Gas return line		
4	Gas return pump	12/13	Pressure / vacuum valve
5	Underground storage tank	14	Breather pipe
6	Retentate	15/16	Product fill connectors
7	Membrane module	17	Electronic level gauge
8	Vacuum pump		

Fig. 1.15 System to reduce emission at petrol stations.

Besides the advantage of emission reduction the wet stock losses of gasoline storage can be reduced because diffusive emissions are avoided and most of the generated gasoline vapor is returned to the storage tank. It has been proven that the installation of a vapor-recovery unit at petrol stations causes a reduction of inventory losses of about 0.3% of the gasoline sales. A petrol station in California was equipped with a vapor-recovery unit. Monthly savings of approx. 3800 liter have been observed at gasoline sales of approx. 1 200 000 liter a month. Because of the simplicity and the nearly maintenance-free operation the system is particularly suitable for petrol station applications.

1.5.2
Natural Gas Treatment

Natural gas produced at the well head has to be treated in several processing steps especially dehydration and hydrocarbon dew pointing to meet the required pipeline and quality specifications. Water and higher hydrocarbon have to be removed in order to avoid the build-up of gas hydrates. The commonly used state-of-the-art processes, such as absorption and cryogenic condensation, have shortcomings with respect to environmental aspects, energy consumption, weight and space requirements. A reliable and proven membrane process could offer a serious alternative in comparison to established techniques.

Basic process data are available from the development of organic vapor separation. The real challenge is the transformation of the available knowledge into high-pressure applications. Several drawbacks such as the compaction of the substructure of composite membranes and the influence of the boundary layer on the membrane selectivity have to be overcome. Pour structure and polymer compositions have to be suited to the high operating pressure in the presence of higher hydrocarbons [22, 23].

Test plants to evaluate membrane performance and process options have been commissioned. Potential first applications are the treatment of smaller natural gas streams. Hydrocarbon dew pointing and fuel-gas conditioning for gas engines and turbines are very similar in terms of the basic engineering. A combustion engine needs a certain fuel-gas specification, which is defined by the methane number. Higher hydrocarbons have a severe impact on the antiknocking property. In many cases gas associated with oil production is flared. The need for a self-sufficient energy supply or up-coming environmental regulations lead to the use of associated gas as fuel gas. The composition of this gas is often highly dependent on the ambient temperature of the oil well. During hot summer times an increase of higher hydrocarbons occurs due to evaporation that causes an unstable operation of the gas engine. Membrane separation offers a simple technique to remove the higher hydrocarbon and so increase the methane number to a suitable value. The technology is proven by pilot tests. The methane number has been enhanced from approx. 35 to the target of higher than 50 under the given site-specific conditions. Hydrocarbon dew pointing of natural gas has been tested in parallel to conventional gas treatment systems. The arrangement is shown in Fig. 1.16. The long-term performance of the membrane module and the achieved dew-point reduction under given test conditions are illustrated by the phase envelope diagram. Based on the experience of the field tests the first commercial membrane-based hydrocarbon dew-pointing units have been commissioned.

Some chemical production processes use the municipal gas supply as chemical feed stock. In some cases the gas has to be treated to meet the production specifications. Pressure swing adsorption is often used to remove higher hydrocarbons. One drawback of this application is the limited lifetime of the adsorber in the presence of higher hydrocarbons and the costs for replacement and trans-

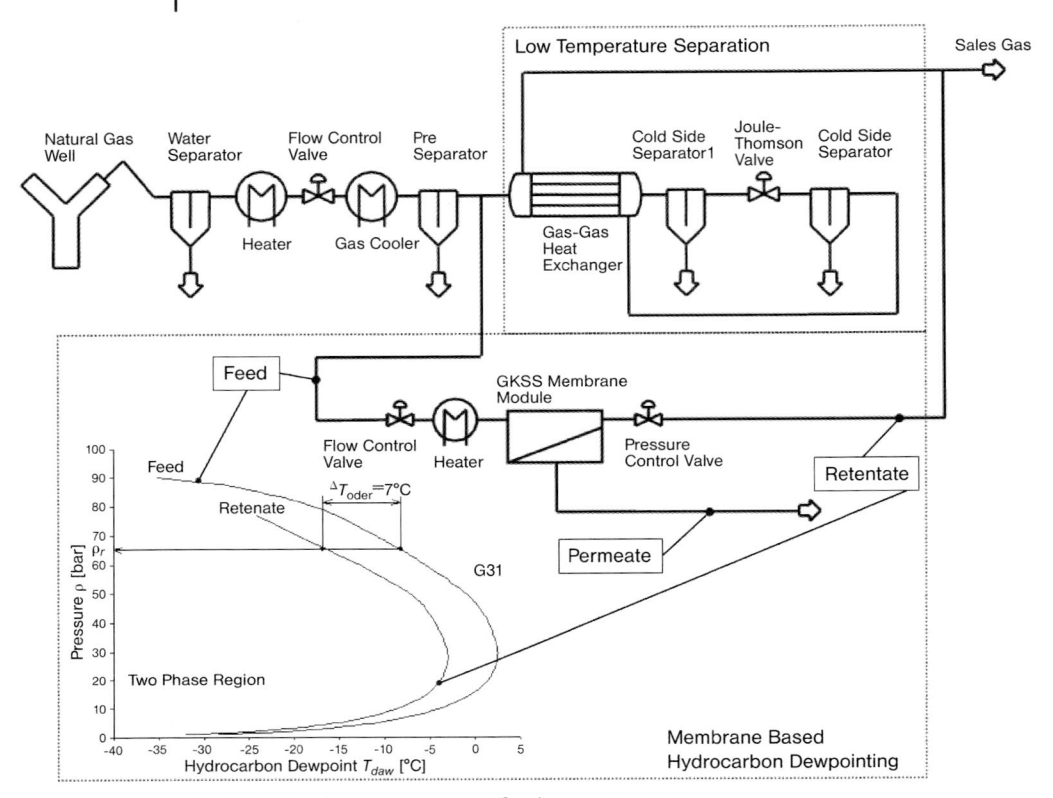

Fig. 1.16 Membrane arrangement for dew pointing tests.

portation of the used adsorber material. The separation of higher hydrocarbons by means of membranes provides a simple and reliable solution to overcome the drawbacks. Two design options can be offered:
- recycling of the permeate and remixing with the gas supply
- the use of the permeate as boiler fuel.

In the case of recycling no methane losses occur but the permeate has to be recompressed and a condensation unit has to be installed to separate the higher hydrocarbon from the enriched permeate stream. The realized system consists of a membrane stage and the permeate was used as boiler fuel (Fig. 1.17).

1.5.3
Hydrogen/Hydrocarbon Separation

Process and off-gas streams in refineries and the petrochemical industry often contain hydrogen and hydrocarbons. The hydrogen stream has to be purified before it can be reused. Large volume streams are treated by cryogenic condensation and fractional distillation and in the case of smaller streams by pressure

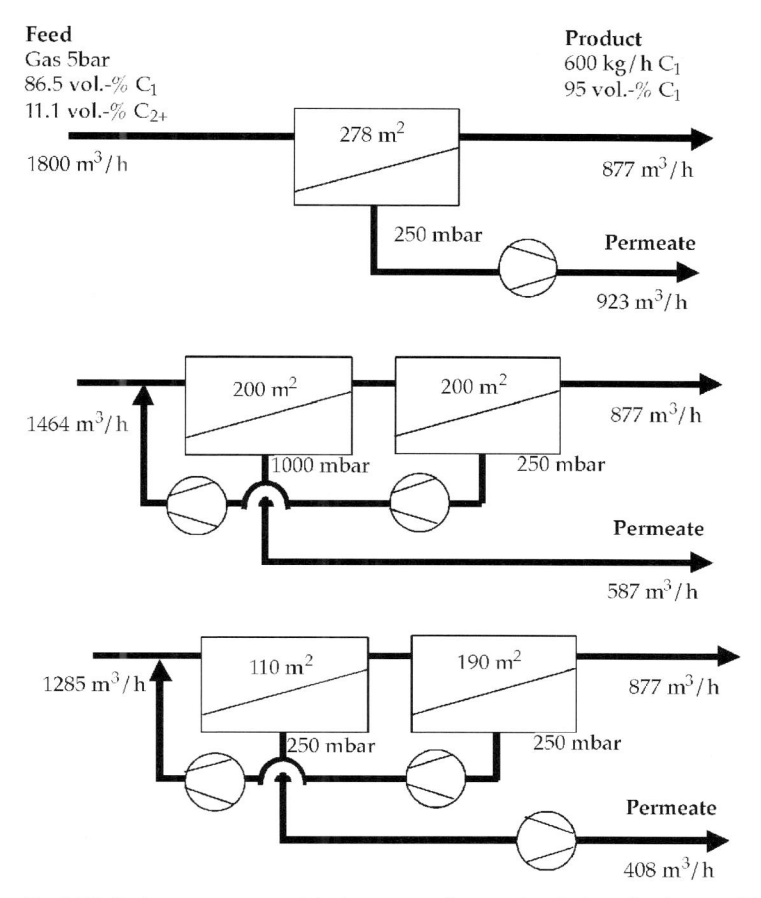

Fig. 1.17 Options to treat municipal gas according to chemical production specifications.

swing adsorption. Elastomeric membranes have certain hydrogen/hydrocarbon selectivities. The increase of selectivity is, in terms of hierarchy, similar to the depiction in Fig. 1.1 "Selectivities of various organic vapors over nitrogen". The real value is lower because of the higher permeation rate of hydrogen in comparison to nitrogen. Hydrogen-selective membranes, which are based on glassy polymers, are often sensitive with regard to condensed hydrocarbon. Condensation on the membrane surface can occur if the feed stream is depleted of nitrogen and the hydrocarbon dew point is achieved. Liquid hydrocarbons can cause unwanted swelling or destruction of the glassy membranes. The combination of a membrane separation stage consisting of an organophilic elastomeric membrane as the pretreatment step and a hydrogen-selective membrane based on a glassy polymer offers a technology to treat small or moderate volume flows. If very high purities are required a post-treatment system, e.g. pressure swing adsorption, can be combined with the membrane stage.

1.6
Conclusions and Outlook

The separation of organic vapors by means of membranes is, as yet, still a niche application. The growing acceptance brought above by references proving performance and reliability supports the growth of market share. Environmental regulations encourage recovery techniques rather than destruction. Membrane technology provides the opportunity to enhance the recovery conditions by means of condensation or absorption. Condensable compounds can be separated from noncondensable compounds. The simplicity of the process, the ease of operation, the long lifetime of membranes and investment and operating costs are demonstrated. The experience of more than 20 years leads to a higher degree of knowledge, popularity and confidence in the technology. The introduction of the membrane technology was supported by developing simulation tools for vapor permeation, gas permeation and pervaporation. These tools are compatible to commercially available process simulators. End-of-pipe installations have at present the higher market share. This trend will presumably switch in the future to process integrated membrane systems. If the use of membranes for hydrocarbon dew pointing of natural gas is widely accepted by natural gas producers, the technology will become a push concerning dissemination and total amount of produced membrane area.

References

1 R.D. PAUL, Y. YAMPOLSKII (eds.) Polymeric Gas-separation Membranes, Chapter 8, pp. 353–397. 1994, CRC Press Inc.
2 KATOH, M., INONE, N., BITOO, T., HASHIMOTO, K., TSUNEIZUMI, H., FURONO, N.: Hydrocarbon Vapor Recovery with Membrane Technology, *NKK Technical Review* **56**, 1993.
3 Technische Anleitung zur Reinhaltung der Luft (TA Luft) vom 27. Februar 1986, Gemeinsames Ministerialblatt, S. 95–202.
4 JOEL HILL, Controlling Emissions from Marine Loading Operations. *Chemical Engineering* May 1990, pp. 133–143.
5 Clean Air Act Amendments, 55 FR 25454–25519, June 21, 1990.
6 MITCHELL, J.K.: J. Med. Sci. (London) 13, 36, 1831; *J. Roy. Inst.* **2**, 101, 307, 1831.
7 T. GRAHAM: On the Adsorption and Dialytic Separation of Gases by Colloid Septa. *Philos. Mag.* **32**, 401 (1866).
8 G.J. AMERONGEN: Influence of Structure of Elastomeres on their Permeability to Gases. *J. Appl. Polym. Sci.* **5** (1950).
9 R.M. BARRER: Diffusion in and through Solids. Cambridge University Press, London 1951.
10 R.W. BAKER, I. BLUME, V. HELM, A. KHAN, J. MAGUIRE, N. YOSHIOKA: Membrane Research in Energy and Solvent Recovery from Industrial Effluent Streams, Energy Conservation. DOE/ID/12379-TI (DE 840 168 19).
11 Patent DE 3940855 A1, Process-integrated waste air purification system for vacuum pumps (11. 12. 1989).
12 H. WIJMANS: Recovery of Valuable Chemical from Process Streams. The 1999 Seventeenth Annual Membrane Technology/Separations Planning Conference, December 6–7, 1999, Newton, Massachusetts.
13 M. HERBST: Industrielle Gastrenntechnik mittels selektiver Membranen. Vak. Prax. S. 103–108 (1993).
14 K. OHLROGGE, J. WIND, K. STÜRKEN, E. KYBURZ: Membrane applications to sepa-

rate VOCs. npt process technologie 1, January–February 2000, pp. 25–30.
15 V. NITSCHE, K. OHLROGGE, K. STÜRKEN: Separation of Organic Vapors by Means of Membranes. *Chem. Eng. Technol.* 21 (1998) 12, pp. 925–935.
16 K. OHLROGGE, E. KYBURZ, K. STÜRKEN: Membrane Applications to Separate Organic Vapors from Off-Gas Streams. Seminar on the Ecological Applications of Innovative Membrane Technology in the Chemical Industry. CHEM/SEM. 21/R.1, United Nations. Economic Commission for Europe, Cetraro, Calabria, 1996.
17 H. IWASAKI: The Membrane Gas Separator for VOC – Especially GKSS Type Membrane Technology, Aromatics, Vol. 15 (issue 5&6) pp. 12–19.
18 R. W. BAKER, M. JACOBS: Improved monomer recovery from polyolefin resin degassing. *Hydrocarbon Processing*, March 1996.
19 M. JACOBS, D. GOTTSCHLICH, F. BUCHNER: Monomer Recovery in Polyolefin Plants Using Membranes – An Update. 1999 Petrochemical World Review, D. Witt & Company Inc., Houston, Texas, March 23–25, 1999.
20 D. HASSEL: Mess- und Überwachungsverfahren bei Gasrückführungssystemen. Praxis-Forum Umweltmanagement 23/94, S. 183–203.
21 K. OHLROGGE: Möglichkeiten der Anwendung von Membrantrennanlagen an Tankstellen. Praxis-Forum Umweltmanagement 23/94, S. 207–217.
22 K. OHLROGGE, B. KEIL, J. WIND: Membranverfahren zur Abtrennung höherer Kohlenwasserstoffe. *Chemie Ingenieur Technik* **9**, 2000, S. 1024–1025.
23 A. ALPERS, B. KEIL, O. LÜDTKE, K. OHLROGGE: Organic Vapor Separation: Process Design with Regards to High-Flux Membranes and the Dependence on Real Gas Behavior at High Pressure Applications. Industrial & Engineering Chemistry Research 1999, Volume 38, Number 10, pp. 3754–3760.

2
Gas-separation Membrane Applications

D.J. Stookey

2.1
Introduction

The previous chapter outlined the phenomena and theory associated with gas-separation membranes. The fundamentals of mass transfer and the process design equations that model membranes were also addressed. In this chapter, our attention turns to the industrial application of gas-separation membranes, specifically separations with polymeric membranes.

Gas permeation is a scale-independent phenomenon. Thus, it is not surprising that membranes applications are represented over a very wide range of membrane sizes. The smaller sizes are typically found in laboratories, but gas-separation membranes are now finding utility in commercial analytical products and instruments. One of the smallest is the square millimeter scale membranes incorporated in electronic chips by I-Stat for blood gas assays [1]. This chapter, however, will focus on applications where the gases separated have commercial value and industrial utility. Presently typical membrane installations of these types range in size from one to 10 000 square meters. However, the range is ever-increasing as membrane technology gains ever-wider acceptance. Proposals for membrane systems approaching the square kilometer (one million square meters) scale are now under study for gases-to-liquids projects [2, 3]. It is fair to say that gas-separation membranes have become an accepted technology.

There exists a variety of applications for which gas-separation membranes have been applied along with their commercial membrane suppliers. The process engineer facing the task of designing a gas-separation process is well advised to consult with these suppliers to assist in his design. If the task involves common gas species, it is highly probable that the application has already been commercialized or is under consideration. This chapter provides an overview of some of these commercial applications. Also provided is an overview of considerations, limitations, and hurdles to commercialization of new applications for gas-separation membranes. Development of a new membrane for a new application is not for the faint of heart.

Each successful application of gas-separation membranes is the result of a whole series of successful technical and commercial activities. Steps included in this series include the following:
1. Membrane material selection – Does a material exist with suitable permeability and selectivity to separate the components involved?
2. Membrane form – Can the membrane material selected be formed or applied into a film or hollow-fiber form suited to the application?
3. Membrane module geometry – Can the membrane formed be incorporated into a module geometry that accommodates conduits for feed and product gases, optimum driving force for the separation, efficient membrane area density, and with minimal pressure head loss (energy)?
4. Compatible sealing materials – Are sealing and tubesheet materials available that are compatible with the gases and process streams involved in the intended application and with the module manufacturing process?
5. Module manufacture – Can a reliable membrane module be manufactured in a cost-effective manner?
6. Pilot or field demonstration – Will the membrane module perform to expectations in the intended environment?
7. Process design – Can the membrane module be incorporated into a flowsheet with suitable controls and safeguards for optimal operation and accommodate non-routine events such as start-up and shut-down?
8. Membrane system – Can the membrane assembly be packaged into a deliverable system that will operate in concert with peripheral equipment in the intended environment?
9. Beta site – Can a customer or partner be identified who will accept the risks associated with new technology and initial installation at a meaningful and acceptable scale of operation?
10. Cost/performance – Does the membrane application perform against alternatives and meet competitive challenges?

2.2
Membrane Application Development

2.2.1
Membrane Selection

The selection of materials for gas-separation membranes requires the matching of performance characteristics of the materials available with the application. Much of the selection process has historically been a trial and error process involving many of the steps in the series outlined above. Not surprising, the selec-

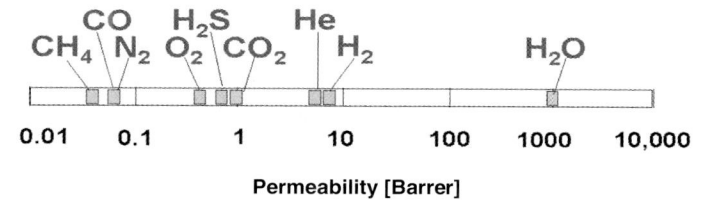

Fig. 2.1 Permeability of common gases through a polyetherimide film.

tion process is typically guided by a team of polymer chemists and physicists having insights into polymer mechanics and physics. It is not within the scope of this chapter to describe this selection process, but rather to provide an overview of some of the key elements and considerations.

The gas-separation characteristics have been measured and reported for the common gases for many polymers. Indeed, in the preceding chapter, materials have been identified and commercialized by membrane suppliers for many of the common separations. Having identified a gas-separation of interest, a thorough review of the available literature is in order. This must include patent literature as many researchers have sought protection for membranes formed of proprietary polymers and their application in many specific gas-separations if not for membrane separations or devices as a whole.

The gas permeabilities for some common polymers used in gas-separation membranes are reported in the preceeding chapter. Figure 2.1 shows the relative flux rates for a variety of gases for a polyetherimide film.

Inspection of this diagram shows that gas-permeability is not determined by the size of the gas molecule. One immediately sees a wide, nearly six-decade, range in transport rates of different gases. Notice also that the permeability of helium, the smallest molecule, is slower than the larger hydrogen and much slower than water in this polymer. Similarly, carbon monoxide transports much slower than its larger cousin, carbon dioxide, does. Furthermore, the relative fluxes vary between different polymers. Thus, it can be seen that the researcher must be largely guided by intuition and statistical analysis of data to guide his selection.

The membrane thickness, the application temperature, and other species present further complicate the selection of a membrane polymer for a specific gas-separation. Equations show that the gas fluxes through the membrane are inversely proportional to the membrane thickness, l. Most of the published gas-permeability data arises from pure gas measurements on flat, thick, dense films supported on a porous backing medium to accommodate the applied pressure differential. Thus while a specific polymer may be chosen for its favorable selectivity, it may very well be impractical if it cannot be fabricated into a very thin membrane-separating layer that is stable in the intended application.

Permeability and selectivity, the relative permeabilities of the species being separated from a mixture, are very strongly influenced by the membrane's operating temperature [4, 5]. This arises from the dependence of permeation in polymers on the solubility and diffusivity of gases in the polymer. Permeability, P_i, is typically correlated with absolute temperature, T, by the empirical relationship

$$P_i = K_i \exp(-E_i/RT) \tag{1}$$

where K_i and E_i are empirical correlation coefficients determined from laboratory and field test data. K_i and E_i are unique to each gas specie and can differ greatly between gases and polymers. The membrane selectivity, $a_{i,j}$, for a gas pair i-j is given by the relationship

$$a_{i,j} = P_i/P_j = (K_i/K_j)\exp(-(E_i - E_j)/RT) \tag{2}$$

Thus, it can be seen that the membrane researcher must consider operating temperature in making his recommendation for the membrane material.

The presence of condensible, polar, and associating components in the mixture being processed can also influence the permeabilities and selectivities of gas species being separated. Polymer free-volume, the void space between polymer molecules, plays a major role in permeation of gases through membranes [6, 7]. In addition to forming sub-micron membrane thicknesses, membrane chemists have employed a variety of techniques for increasing the free-volume locked within the membrane skin during its formation [8, 9]. Others have sought to alter the free-volume and modify the surface by various chemical treatments [10, 11]. In altering the polymer free-volume in the membrane skin, variations in transport rate and selectivity can be accomplished. Sorption of components in the mixture being separated within the free-volume and on the membrane surface can substantially alter gas permeabilities. The sorbed species effectively occupy space and hinder the passage of other components. This phenomenon is employed in many vapor-recovery membranes discussed elsewhere in this book. Sorption can be so complete as to block the transport of otherwise highly permeable species [12, 13]. An example is the separation of hydrogen from hydrocarbons with PTMSP. Auvil and others have shown that the free-volume in PTMSP can under proper conditions so completely fill with sorbed and condensed heavy species such as hydrocarbons as to render the membrane nearly impermeable to light species such as helium or hydrogen [14]. To the extent that all polymers have some free-volume, sorption and free-volume filling can become a factor in performance of all membranes and in the selection of the polymer for the gas-separation membrane.

Commercial membrane suppliers have each selected and developed membranes with these factors plus a host of other considerations in mind, usually with a specific target application in mind. Sizeable investments in development and manufacturing capital are now in place. Thus as new gas-separation applications surface, membrane manufacturers first screen the separation on available products. Frequently, the new separation if fitted to the available products rather than the product to the application. Polymers more suited to the separation in terms of selectivity may indeed exist, however, the choices may be restricted to less than optimum performance due to economics and availability. Some manufacturers are developing composite structures as a means for increasing the flexibility and choice of polymers employed in the separation. This is discussed in more detail in the next section.

2.2.2
Membrane Form

The transmembrane flux is inversely proportional to the membrane thickness and is directly proportional to the membrane area and to the applied pressure differential across the membrane. Thus, a membrane manufacturer's primary objectives typically revolve around means for producing the thinnest possible membrane in a structural form that will accommodate the applied pressure while maximizing the membrane surface area. Thus, knowledge of the structural and mechanical properties of the membrane material is of paramount importance. Unfortunately, many polymers of interest as gas-permeable membranes are rubbery materials with poor mechanical strength. Hence, many membranes require an underlying support material that can accommodate the applied pressure load.

Membrane properties, as discussed in the previous section are typically measured in dense polymer film approaching one millimeter in thickness, less than 100 cm^2 area, and supported on a porous ceramic or metal backing plate. Scale up of flat-film membranes to commercial scale has been quite limited, typically to applications involving small quantities of gas, low-pressure differentials, and to high flux, low selectivity membranes.

Other flat-film membrane designs employ textiles, non-woven fabrics, or porous polymeric sheets as backing material to the membrane film. Composite membranes of this construction enable the separating layer thickness to be reduced to a few microns. When high flux polymers are employed, the transport resistance of the non-selective support layer can become a significant resistance. The effect of resistances in series with the separating layer resistance has the effect of reducing the overall membrane selectivity [15]. Hence, much attention is given to minimizing the supporting layer's resistance.

A number of researchers have produced capillary tubing and hollow-fibers from materials of sufficient strength to avoid the need for a porous support. These materials are typically melt spun into hair-size fibers having dense walls of sufficient strength to obviate the need for a supporting layer.. While the dense layer offers substantial resistance and limits the permeation flux, the hair-size of the hollow-fibers enables designs that accommodate high membrane surface area densities (area per unit volume) [32].

Loeb and Sourirajan [16] introduced the unitary asymmetric membrane. These membranes consist of a microscopically thin skin on a porous support formed of the same material as the skin. By judicious selection of solvents, coagulants, and processing conditions, these researchers were able to precipitate polymer solutions to form both skin and porous support in a single processing step. The asymmetric membrane is known to produce the thinnest of membranes. Kesting [17] has shown membrane thicknesses below 50 nanometers (500 Angstroms) are possible in the unitary asymmetric membrane. Unitary asymmetric membranes are formed in both flat-sheet and hollow-fiber forms. Polymers selected for the asymmetric form must obviously be of sufficient strength to perform the role of the porous support, while also serving as the separating medium. Unitary asymmetric hollow-fibers are produced with collapse pressures in excess of 100 atmospheres. In this case, the economics of the membrane product is controlled by the availability of low-cost polymers that also have permselective properties. The unitary asymmetric membranes are thus limited to selections from the cellulose acetate, polysulfone, polycarbonate, and polyimide families.

Many polymers have been tailored for specific permselective properties, unfortunately often not without penalty. Many of these materials have poor mechanical properties or are costly due to the exotic materials involved or their relatively small production volumes. Many researchers pursue composite membrane forms as a means of overcoming these limitations. These involve creating a composite thin film or an asymmetric layer of the desired separating material on a porous support [18, 19]. While this is a seemingly obvious approach, it is not without great difficulty. First, the porous material must be compatible with the processing conditions used in applying the composite film or the asymmetric structure on the support surface. Aggressive solvents are frequently required to solution the specialty polymer employed in the separating layer. Thus, it may become necessary that the porous support be formed of a specialty material as well lest it be attacked during the composite-layer deposition. Second, adhesion of the asymmetric skin to the porous support can be a major limitation. Any differences between the two composite layers with respect to thermal, mechanical, shrinkage, and swelling properties during either the membrane processing or in the membrane gas-separation application can lead to cracking, delaminating, and failure akin to peeling paint. Third, the additional materials, solvents, and processing steps needed to form the composite membranes all contribute to increased product costs. Thus, a substantial performance improvement is usually necessary to make the composite membranes a viable product.

Innovations by DuPont [20, 21] enable formation of composite hollow-fibers that overcome many of the problems enumerated above. DuPont co-spun hollow-fibers from two different dopes, one forming a thin separating layer, the other forming the porous support structure. They applied a specialty polyimide polymer on a polyimide support made of a commercial low-cost polyimide polymer. By comparison to the unitary structure, this co-spinning forms a composite asymmetric hollow-fiber composed of less than 10% specialty polymer, with the other 90% being a commercial polymer. Considering that as little as 1% specialty polymer might be required in applying the separating layer to a preformed porous support, the co-spun asymmetric hollow-fiber overcomes many of the other limitations noted above. First, co-spinning allows the porous support to be made of materials of the same polymer class, thereby enabling the same solvent to be employed for both polymer dopes. Second, co-spinning insures intimate polymer contacting and phase mixing at the composite interface during the polymer precipitation step. The likelihood that the specialty and commercial polymers can be from the same polymer class also increases the chances for comparable properties and thereby improves adhesion between the composite layers. Finally, co-spinning involves little additional processing facilities; save the additional dope supply system.

2.2.3
Membrane Module Geometry

The function of the gas-separation membrane module is fourfold. First, it must contain the membrane within a pressure housing rated for the application. Second, it must have fittings to introduce feeds and to collect and distribute products leaving the device. Third, it must have internal means for gas-tight sealing between the feed and permeate sides of the membrane and with the containment vessel. Fourth, it must direct the gases in a prescribed manner uniformly over the membrane surfaces.

Module geometries generally revolve around the form of the membrane, the membrane flux, and the volumetric flows of the feed and product streams. Module geometries have also been patterned after earlier dialysis, reverse osmosis, and filtration module designs. Early flat-film designs were constructed of plates and membrane films sandwiched together in a filter-press-type assembly. The plate and frame design has a relatively low membrane packing density (membrane area per unit of volume). It can be used effectively in low-pressure and vacuum applications, particularly for high-flux membrane applications. Such applications frequently require large flow channels to accommodate the volumetric flow with minimal hydraulic resistance. These assemblies have found application in oxygen enrichment for respiratory care patients. Another flat-film membrane device is shown in Fig. 2.2.

GKSS presently employs a flat-film supported on a stack of hollow ceramic discs in their vapor-recovery devices. Baffling between the discs directs the feed-gas flow across the membrane-covered discs in the stack. The interior of the discs communicates with a central permeate-collection conduit. The short permeate flow path of this design is particularly suited to operating the permeate side under vacuum conditions.

The more common design employing the flat-sheet is the spiral-wound membrane design shown in Fig. 2.3.

In this design, a spacer containing permeate flow channels is enclosed in a bag or envelope made of two pieces of flat membrane sheet that are sealed on three edges. The open edge attached and sealed to a slotted mandrel that acts as the permeate conduit and the support for the membrane assembly. Other spacers are placed between a number of leaves attached to the mandrel. The leaf and spacer assembly is then rolled around the mandrel to form the spiral arrangement from which it derives it name. The assembly is completed by applying a covering and insertion in a pressure vessel. Feed gases pass axially

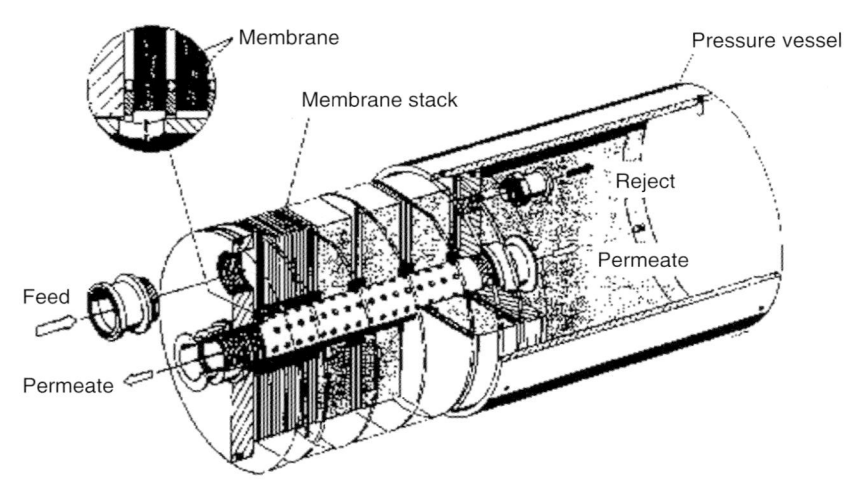

Fig. 2.2 Flat-film membrane stack design by GKSS (used with permission of GKSS).

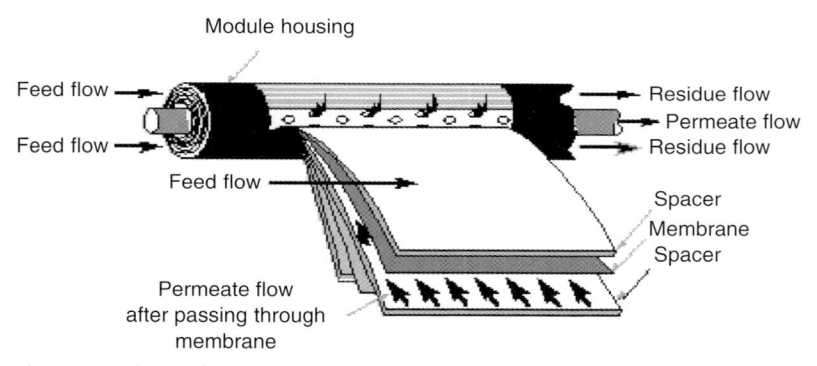

Fig. 2.3 Spiral-wound module (used with permission of Membrane Technology & Research, Inc.).

through flow channels in the spacer-filled gaps between the spiral leaves. Gases that permeate through the membrane pass along the spiral permeate channels to the central collection pipe. Manufacturers utilizing the spiral-wound design have some flexibility in the number and the length of the leaves they employ, spacer thickness and design, and the axial length and package diameter. Particular care is taken in the spacer design to ensure that loading arising from high flows and corresponding high pressure differentials between the axial ends do not cause deformation and damage to the spiral assembly. Some manufacturers add a perforated backing plate mounted on the downstream side of the spiral assembly to minimize distortion and damage to the spiral membrane assembly.

Hollow fibers enable substantially higher membrane area densities than are possible with either of the flat-sheet designs. They also enable operation at substantially higher operating pressure and at higher transmembrane pressure differentials. Hollow fibers also afford flexibility in module design and in alternative feed and product flow geometries. Feed and permeate flows can easily be aligned in cocurrent, countercurrent, or crossflow orientations as may be desired for the specific application. Module designs employing hollow-fibers are often tailored to application-specific requirements by adapting the fiber dimensions, the fiber lay-up orientation, and the fiber packing density. These variations enable the pressure losses in the respective flowing streams to be minimized thereby maximizing the driving forces for the separation. Figure 2.4 describes an axially oriented fiber bundle with two tubesheets separating feed and product streams.

This diagram shows the pressurized feed and product on the shell-side while the sweep enters and permeate leaves from the hollow fiber bores that pass through the tubesheets on either end of the device. The sweep gas shown in this diagram is not necessary for many applications and a three-ported geometry is employed. In this case, the second tubesheet is often eliminated by simply sealing or plugging the bores of the hollow-fibers.

Another three-ported hollow-fiber arrangement applied in a number of applications places the pressurized feed to the bore side of the module in Fig. 2.4 and collects the permeate on the shell side. This arrangement avoids the need for the shell to be a pressure-containment device since only the end caps distributing the feed and product to and from the respective tubesheets and the hollow fibers themselves are pressurized. Hence, the shell can be made of plastic pipe or light gage metal tubing. The bore-fed arrangement can be particularly attractive when high degrees of removals of permeating species are desired in a single countercurrent arrangement. The bore-fed arrangement ensures that each element of the feed must necessarily be exposed to the membrane and subjected to the partial pressure driving force before exiting at the opposite end.

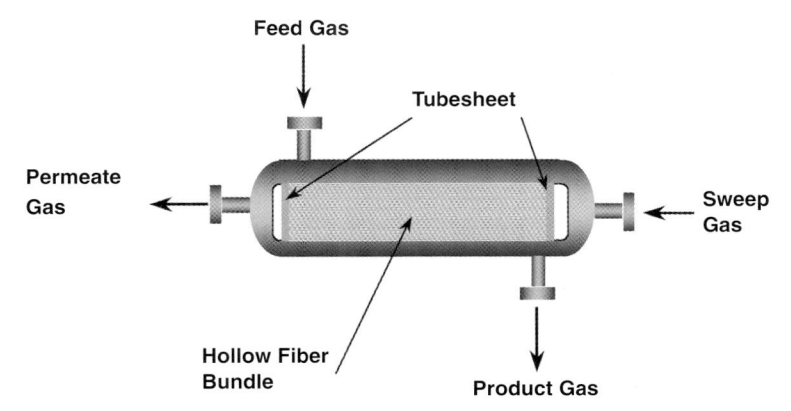

Fig. 2.4 Axially oriented hollow fiber module with bore and shell feeds (used with permission of Air Products and Chemicals, Inc.).

Thus, there can be no bypassing or channeling of flow as is sometimes possible in shell-fed axial fiber bundles [22]. However, the bore-fed module can also suffer from flow non-idealities arising from fiber-to-fiber variation in bore dimensions and in membrane transport properties. The fourth-order dependence of flow on the bore diameter makes it essential that the fiber-bore dimensions be precisely controlled in order to maintain uniform residence times between the multiplicity of parallel bore flow channels in the bore-fed design [23].

Uniformity and control of flow pattern on the shell side of the hollow fiber membrane module is also important and is addressed in a variety of manners. The flexibility of the hollow fibers frequently allows movement, redistribution, and settling of fibers within the unconstrained axial bundle geometry. Variety of means have been applied to minimize the deleterious effects on membrane module performance arising from the non-idealities in flow patterns arising from fiber movement. Monsanto introduced crimped fibers that maintained a degree of bulkiness and preserved fiber–fiber spacing within the axial bundle [24]. Toray has wound a textile fiber around the hollow-fiber in a spiral fashion to provide fiber–fiber spacing [25]. Air Products recently introduced an axial bundle containing an axial textile fibers comingled with the axial hollow fibers to aid in preservation of flow patterns within the bundle under varying operating conditions [26].

Others have introduced helically wound or patterned fiber lay-up in the bundle geometries as a means for controlling and preserving the shell-side flow patterns within the module [27]. The use of hollow-fiber fabrics has also been proposed as a means for precision spacing of the hollow-fibers within the membrane module [28–31]. Many of these are assembled around a central mandrel that also serves as a fluid conduit for feed introduction or product or permeate withdrawal. Modules assembled in this fashion also accommodate distributors and baffles inserted within the fiber pack to aid in the direction and control of the flow shell-side flow pattern.

As can be seen with the wide variation in module geometries already discussed, the options available for gas-separation modules are limited only by the ingenuity of the designers. Current innovations stem largely from refinements on the basic designs employed in earlier liquid-phase membrane devices to deal with parameters and variables unique to gas phase separations and applications. Major distinctions between the liquid- and gas-phase designs arise from the phase density differences. Film-boundary resistances and concentration-polarization concerns that are significant in liquid-phase separations are usually negligible and can be ignored in gas separators. That the gas-separator designs can be radically different and independent from the traditional liquid-separator geometries is reflected in a recent proposal for a carpet-like module [32]. This design proposes that very high area densities can be attained by constructing a carpet from microhollow fibers. Short lengths of fibers or fiber loops pass through a sealed backing material. By cutting the fibers on the underside of the backing, bores are opened on the underside of the carpet. It is expected that refinements and step-out innovations such as this will continue to drive new membrane module designs.

2.2.4
Compatible Sealing Materials

A number of gas-tight sealing means are typically required in the assembly, manufacture, and application of the membrane device. First, provision must be made to incorporate the membrane module into the process piping system with related manifolds and related equipment up and downstream of the membrane. This is usually accomplished via standard piping systems and is guided by piping design codes. Consideration need only be given to the wide variety of flanges, screwed thread fittings and geometries, connectors and disconnects, etc. that are available and must be either accommodated or adapted to in the design and application of membrane devices.

Seals required on the membrane elements vary with the membrane type and the module geometry. Many flat-film designs employ thermal and ultrasonic welding techniques and thereby avoid the introduction or minimize the use of dissimilar sealing materials.

Most manufacturers though form at least one tubesheet around the membrane or hollow fiber in order to separate the permeate product from the feed and non-permeate product streams. Selection of the tubesheet material is typically guided by the intended application conditions, the quality of the sealing surfaces, and the manufacturing process. The tubesheet material is typically a liquid, curable resin that solidifies upon cooling or crosslinking reaction. Common materials are from the epoxy, silicone, and urethane resin families. Particular attention must be given to adhesion to and compatibility with the membrane surface. Without good bonding to the tubesheet, pressurized gases can leak through a gap between the membrane and the tubesheet. Many of the materials of choice have a solvency or plasticization effect on the membrane polymer. Thus, preserving the integrity of the microscopically thin membrane skin on the asymmetric membrane can be particularly troublesome both during assembly and in separation service. In the extreme, the asymmetric structure may be attacked or crack during the tubesheet-formation step resulting in gas leakage past the membrane. Since the tubesheet typically must form a rigid structure capable of also forming a gas-tight seal to the pressure-containment vessel, any shrinkage and dimension changes during curing or set-up of the sealant or swelling during exposure to the process streams can also be a concern. In some applications, chemical degradation, temperature limits of the materials, and swelling or temperature-induced stress within the membrane assembly can control the material selection or even the feasibility of the application.

Sealing between the membrane element and the pressure containment vessel is typically done with O-rings and gaskets. Primary consideration in the module design and manufacture is the provision of good sealing surfaces to accommodate the selected sealing means. Most of these seals involve compression loading of the O-ring or gasket against the sealing surfaces. Maintenance of the compressive load throughout the device's lifetime becomes the controlling parameter in the selection of these components.

Most of the gas-separations involving membranes involve the handling or formation of flammable, asphyxiants, or poisonous gases. Thus, reliable sealing of the membrane elements in the system is frequently a major environmental and safety consideration. As the size of the gas-separation installations grows, there is a proportional increase in the number of elements employed and a corresponding increase in the potential for leaks. Many manufacturers have reduced the number of connections by incorporating larger and multiple membrane elements within a pressure containment vessel. Others have resorted to completely welded closures and connections to piping to eliminate they need for gaskets and seals to ambient.

Membrane manufacturers frequently have as much technology and art invested in the selection and processing of the tubesheet and sealants as in the membrane itself. Understandably, much of the details about tubesheet and gasket material selection and processing is considered proprietary and trade, or closely guarded, secret information by membrane manufacturers.

2.2.5
Module Manufacture

Manufacture of gas-separation membrane modules is largely a machine-assisted, labor-intensive operation. Polymer dopes are typically prepared batchwise with sufficient hold time to insure uniformity. The membrane performance is largely controlled by the polymer precipitation step and very dependent upon phase behavior and precipitation kinetics. Thus, it is essential that processing conditions be maintained as uniformly and as constant as possible if product quality and uniformity is to be preserved. For this reason, membrane-film formation and hollow-fiber spinning processes are usually operated continuously or for extended run times. Since the intermediate film or fiber must eventually be converted into discrete items, the continuous process is typically interrupted by collection of the membrane formed on spools or fiber skeins where it may be inventoried briefly before batch processing into the final assembly resumes.

Gas-separation-module assembly is typically an operator-machine interactive process, because the scale of operation cannot justify the type of automation necessary for the high-volume dialysis and filter-module assembly lines. This also has some advantages in enabling products to be sized and tailored to the application.

2.2.6
Pilot or Field Demonstration

Commercial process streams can rarely be accurately replicated in a laboratory setting. Minor and trace compounds are frequently unknown. Lubricants and corrosion inhibitors added to the process seldom appear in a feed stream composition analysis, but indeed they can make their presence known after a membrane is put into service. Other compounds may arise intermittently or occur on infrequent process upsets or under dynamic process conditions. Hence, there is no substitute for measuring the membrane performance in the intended operating setting for an extended time.

Many of the commercial membrane applications were developed by first testing performance of small membrane modules installed in pilot and field slipstreams. In recent years, commercial-scale modules have been used increasingly in new application field trials. Valuable development time can be saved when representative process conditions can be tested at the larger scale. When operating conditions permit, online, real-time data acquired from an instrumented pilot unit can provide valuable data with which to quickly map membrane performance over a wide range of operating condition. Such units also enable quick diagnosis of the membrane's response and ability to recover from upset conditions and perturbations over the operating dynamics.

The instrumented pilot unit is not always necessary, or even justified, though. Valuable data can be measured from field samples carried to a laboratory for analysis, however, particular attention must be given to the sampling and analytical process, particularly when temperatures vary greatly from the operating conditions or when reactive species are present. Accurate measurement of flow rates and compositions on all streams involved, namely feed, permeate, residue, and, if present, sweep, are essential if meaningful membrane performance data is expected. Regression of permeance and selectivity coefficients is strongly dependent on having accurate material, and especially, component balances around the membrane.

Membrane performance often changes with time. Some of this change arises from creep, plastic deformation under pressure differential loading, even in benign streams. Other changes occur as minor or trace moieties accumulate on or react with the polymer membrane over extended service time. Thus, it is recommended that membrane sample testing on the field stream should be for an extended period, at least one month, preferably for six months. Performance should be tracked over the period until the magnitude of the creep or performance change can be accurately represented since these data must be used to project the useful lifetime of the membrane elements.

It is essential that field-tested samples be returned and retested on a well-calibrated laboratory test stand to make comparison of performance changes against other experiences. Finally, after all calibration tests are completed, the membrane element should be autopsied and inspected for any observable changes in dimensions, color, effects on components, corrosion, and the like. This important step may be easily overlooked, particularly after successful permeation tests. When large modules have been tested, removal of small portions of the membrane from various locations in the geometry and re-testing on laboratory bench-scale equipment can also be instructive. Frequently, subtle changes may be occurring locally that are not detected on the larger-scale device due to inaccuracies, magnitude of change, etc.

The output of the pilot and field-testing program is a characterization of the membrane's performance in the intended service. Sufficient information needs to be assembled to evaluate the economics and minimize the risks associated with a first-time commercial membrane installation.

2.2.7
Process Design

The process design of a membrane system involves the determination of the system size and the configuration necessary to meet the project scope and specifications. Presently this exercise is performed by the membrane supplier and application developer. Membrane suppliers have developed a proprietary membrane-simulation model related to their specific membrane material, module geometry [33, 34], field and application experience, and performance parameters. Suppliers typically provide performance predictions and guarantees with their equipment and therefore often take responsibility for providing a process design package relating to their component. Thus, the process design of a membrane system typically requires close working relations and trust between the gas-membrane supplier and the customer to insure that the process design package provided can be flawlessly integrated into the customer's flowsheet. Successful design and application typically requires exchange of information about operations immediately up or down stream of the membrane unit as these often interact with the control and operation of the membrane unit, particularly during start-up and shut-down.

Process controls of gas-separation units typically revolve around four membrane operating parameters; temperature, pressure, flow rate, and membrane area. Since temperature adjustment frequently involves heat exchange and energy requirements for the system, it is seldom the manipulated variable. Pressure is the most commonly manipulated variable, however, it is frequently limited by the available feed pressure or product pressure requirements imposed by down-stream processing. Frequently, manipulation of the permeate pressure is the only control parameter at the process designer's disposal, however, its impact on product purity and recovery can impose bounds on its range. When product purities are limiting, the process designer often resorts to provision for adjusting the membrane area in service to accommodate variations in feed rates, etc. Frequently this can be done by simply blocking the permeate flow from selected membrane elements while leaving the feed pass through the remaining idled elements. This enables nearly instantaneous response to the changing performance demand.

Membrane systems are unique among gas-processing technologies in that they can be easily expanded to meet changing demand. Where other technologies may require the installation of full-capacity towers and vessels to be operated at fractional capacity during a product or market growth period, membranes afford the customer the option of deferring the membrane capital investment until demand is realized. Consideration only need be given for provision for tie-ins and space in the plot plan for deferred additions to the membrane unit.

2.2.8
Membrane System

Membrane units are usually supplied as a complete assembly requiring minimal field erection, save connections to process and utility piping and instrumentation systems. Fabricators and OEMs working closely with the membrane suppliers have developed membrane-system packages that address a wide variety of application and industry code requirements. The nearly 20 years of experience for most membrane suppliers has afforded many membrane-module, system-design and cost-reduction innovations to be incorporated into these packages. New applications will most certainly rely heavily on these experiences. The photographs in Figs. 2.5, 2.6, and 2.7 provide examples of the diversity of these designs. Each system shown generates nitrogen, but each is operated in a very different environment and is constructed to meet the rigors unique to it.

Nitrogen generators of the type pictured in Fig. 2.5 supply gaseous nitrogen to industrial gas customers. Systems of this type can be installed in a plant warehouse or utility area to provide local nitrogen requirements. Modem connections enable the units to be monitored from remote locations.

Mobile membrane nitrogen generators of the type shown in Fig. 2.6 provide nitrogen for aircraft tire and strut inflation on flight lines. A diesel engine

Fig. 2.5 Onsite industrial nitrogen generator. (Used with permission of Air Products and Chemicals, Inc.).

Fig. 2.6 Mobile nitrogen generator. (Used with permission of CvB Company).

Fig. 2.7 Nitrogen generator for an offshore platform. (Used with permission of Air Products and Chemicals, Inc.).

powers air and nitrogen compressors to deliver the 99.5% nitrogen to a high-pressure receiver for supplying high volumes of nitrogen for aircraft servicing. Membrane elements used in such systems must be capable of handling shock and vibration associated with the movable nature of the unit and the close proximity to engine and compressors.

The system shown in Fig. 2.7 produces over 2000 N m3/h of nitrogen for inerting equipment on a North Sea production platform. Such systems must be constructed for operation under severe environment conditions, particularly ambient temperature extremes, ice loading, and sea-water corrosion.

2.2.9
Beta Site

For each successful application of new technology, there must be a first-time user. The beta-site or beta customer is where the new membrane application grows it wings. The beta-site differs from the field test in that the installation's function shifts from one of gathering technical information and demonstrating performance to the confirmation of commercial viability and utility of the membrane in the application. The membrane supplier or application developer is dependent on identifying a customer or partner who finds the risks associated with the new technology to be minimal and acceptable for the projected rewards or benefits. The testimonials of the beta-site customer largely determine the successful penetration of the technology into the new market and application. Thus, the selection and the development of a strong partnering relationship becomes a critical step in the successful introduction of a new membrane application.

Early beta-site customers were frequently the parent companies or petrochemical affiliates of the membrane developers. Thus, there were convenient relationships and an internal willingness to take the risk of being a beta-site. Realignment of the membrane suppliers and changes within the petrochemical industry has severed many of these relationships. Thus, two organizations must become convinced of the risk/reward merits of the new undertaking. While this provides two independent audits of the merits, the identification and negotiations for a beta-site customer have become doubly more involved.

The difficulty in identifying beta-sites for new applications is also shifting with the acceptance of membrane technology by the process industry. Many of the initial applications were installed with back-up provisions or in non-critical services. Industrial gas nitrogen generators were typically supported by liquid nitrogen supplies to cover the risks associated with outage or downtime. Recovery of hydrogen from a fuel-gas header only bore the risk of associated capital investment with minimal operating cost penalty should there be problems with the application. The acceptance of membrane technology as a viable gas-separation technique has now been largely proven thanks to over 20 years of operating experience, much of it in non-critical services [35]. With their acceptance however, membranes are being proposed increasingly for applications within the revenue stream where there is little opportunity for back-up or parallel technologies to absorb the risk of failure. Carbon dioxide removal from natural gas is an example of a membrane application that is within the main revenue stream. This application was developed to the large-scale membrane systems in operation today because of a number of small beta-site installations. However, incremental and evolutionary growth is likely to be difficult, if not impossible, in some of the new membrane applications now under consideration, making the identification of the beta-site or beta customer ever more critical. The Starchem methanol process cited earlier is a case in point. In this case, the membrane systems are an integral part of a flowsheet intended for megascale operation. Its development will likely depend heavily on the costly pilot demonstration of the entire flowsheet, the risks and costs of which can only be borne by a consortium of partners.

2.2.10
Cost/Performance

The ultimate test of a successful membrane application is whether it is cost effective relative to other alternatives. The membrane supplier and application developer generally have good insights into the merit of the applications they pursue to develop. They would not pursue development if they did not believe there were potential rewards for their effort. Project cost principals can be applied to provide good insight into capital and operating costs and their sensitivity to cost parameters. Product value pricing demands that the supplier have good insights into the merits and potential for their products.

However, the economics of many applications are site specific; hence, the relative merits of a membrane approach can be unique to each installation. Often their true value can only be determined by the customers themselves. For example, the economics of many applications are strongly dependent upon the utility and energy pricing for the site, parameters that are likely to be known only by the customers. Frequently, industrial gas customers conduct make-versus-buy analyses in which case the equipment supplier may have few clues about the costs involved in an alternative supply chain. Thus, there is an intuitive concern among many suppliers that it is difficult to quantify value and price their products to the true customer value. For example, it is hard to capture the value for having onsite availability and independence from outages created by transportation or delivery interrupts. In some cases, particularly those involving portable or movable gas-separation units capable of remote operation, membranes offer capability where none existed before.

2.3
Commercial Gas-separation Membrane Applications

The key commercial membrane-based gas-separations and suppliers are listed in Table 9 in Part I. The key features of these applications are discussed below.

2.3.1
Hydrogen Separations

Hydrogen is one of the more readily membrane-separated gases. It has reasonably high selectivity relative to the other gas species in the mixtures with which it is commonly associated. It is also usually present in pressurized mixtures arising from the high hydrogen partial pressures employed to promote hydrogen's reactivity. Fortunately, high partial pressures are also essential for driving membrane separations. The cost of hydrogen manufacture is closely connected to energy prices since most hydrogen is produced by reforming of natural gas and hydrocarbon compounds. Since the introduction of gas-separation membrane in the late-1970s there has been the ever-increasing demand for hydrogen owing largely to the higher severity in processing of heavier sour crude oils while simultaneously reducing the sulfur content of fuels to meet environmental regulations. Thus, the combination of technical feasibility plus the environmental and economic drivers has made hydrogen separations a much sought and highly successful application of membranes.

Synthesis gases formed by steam reforming typically have hydrogen:carbon monoxide ratios on the order of 3:1. Carbonylation processes, however, typically require a 1:1 ratio. Adjustment of the hydrogen:carbon monoxide ratio to satisfy the stoichiometric requirements of a carbonylation process [36] was the first commercial application of Monsanto's Prism® Membrane technology. The application simply involves passing the 3:1 feed mixture across the membrane while withdrawing the hydrogen-rich permeate at reduced pressure. Control of the 1:1 hydrogen:carbon monoxide ratio in the carbonylation synthesis gas is accomplished under automatic process control by simply adjusting the membrane area, the feed rate, and the permeate pressure.

Hydrogenation of unsaturated hydrocarbons and hydrotreating to remove sulfur from fuels are two major consumers of hydrogen. Hydrogen supplied to reactors in these processes typically contains low concentrations of methane. Methane and other inerts formed during the hydrogenation and hydrotreating are inerts that accumulate in the reactor and reduce the hydrogen partial pressure and correspondingly, the rate of the hydrogenation reaction. Thus, it is necessary to bleed off a quantity of the reactant hydrogen to purge the reactor of these inerts. Directing these pressurized purge gases to a membrane system enables a large fraction of hydrogen lost in the purge stream to be recovered as permeate and recycled to the reactor system. Inerts are thus rejected from the process at a higher concentration. The membrane system may also enable the hydrogenation to be optimized to operate at higher hydrogen partial pressures by increasing the reactor purge rate without a severe penalty on hydrogen yield. Indeed the scheme frequently enables improved throughput or hydrogenated product quality.

Ammonia-synthesis reactions are also frequently limited by the accumulation of inerts in their reactor systems. Argon, arriving with the nitrogen reactant from air, and methane, arriving with the hydrogen, accumulate as inerts within the ammonia reactor loop. The purge from the ammonia loop is typically available in excess of 100 atm. pressure and contains in excess of 60% hydrogen. Membrane systems processing this high-pressure purge typically recover over 90% of the hydrogen from these purge gas. These membrane systems typically employ two membrane units operated in series. The first recovers and recycles a large fraction of the hydrogen permeate to the final compression stage in the ammonia plant feed compressor, typically about 70 atm. In the second membrane unit, the permeate is recovered at lower pressure and is recycled to the intermediate compressor stage typically around 25 atm. The rejected gas containing the concentrated inerts is typically burned as fuel in the primary reformer. The membrane systems enable ammonia plants to be optimized around higher productivity or energy savings [37–39]. The concentrated inerts can also be further processed by cryogenic or adsorption techniques for argon recovery [40, 41].

Methanol synthesis also deals with the accumulation of inerts in the reactor loop. In this case, the reformed gases supplied to the methanol reactor contain hydrogen, carbon monoxide, carbon dioxide, and small amounts of methane. The first three components react when the pressured gases are circulated over the synthesis catalyst to form methanol and water. By cooling the reactor effluent gases, methanol and water are condensed and removed, leaving the methane to concentrate in the reactor circulation loop. By removing a small purge from the loop, the concentration of methane is controlled at the expense of the reactants also vented in this stream. These purged gases are available at the reactor loop pressure, typically 60 to 100 atm., thereby having sufficient partial pressures of hydrogen and carbon dioxide to drive a membrane recovery unit. The purge gas is also saturated with methanol, a slow-permeating specie for most membrane materials. Thus, it must be removed before the membrane system to avoid condensation upon concentration by removal of the fast-permeating species. This is accomplished in a water-scrubbing tower, the effluent of which can be combined with the crude methanol for product distillation. Hydrogen and carbon dioxide are then recovered from the scrubbed gases in a membrane unit and recycled to the synthesis gas compressor. Figure 2.8 is a photograph of a methanol purge gas recovery unit. Carbon monoxide unfortunately is a slow-permeating specie and therefore concentrates with the methane in the membrane reject stream that becomes fuel. Thus, as with the ammonia purge recovery, the membrane unit provides efficient recovery of valuable reactants from the purge stream and provides the methanol operator with increased productivity and flexibility in plant optimization [42]. Starchem has proposed methanol production from a synthesis gas that is rich in nitrogen and only about 90% of the stoichiometric amount of hydrogen needed to convert the carbon monoxide and carbon dioxide. They propose the use of a membrane unit to recover and recycle hydrogen to improve the reactant stoichiometry [43].

Hydrogen-recovery systems have found similar utility in oil refining. In addition to recovery of valuable hydrogen [44], the hydrogen separation system often enables the refiner to operate at higher purge rates to increase the hydrogen partial pressure in the hydroprocessing unit. This can have a significant impact

Fig. 2.8 Methanol purge-gas hydrogen recovery system showing water-scrubbing tower (right) and membrane elements (center). (Used with permission of Air Products and Chemicals, Inc.).

on catalyst lifetime and extend run times. It also can increase the unit's throughput and product quality. Hydrogen recovered from fuel streams can also be used to supplement hydrogen production, frequently delaying or deferring the need for added hydrogen production capacity.

Membrane processing of hydrogen-rich fuel gases can also improve the quality of the fuel gas since the heat content of hydrogen is about one third that of methane and hydrocarbons. Thus by removing hydrogen, the heating value of the fuel gas can be adjusted to permit use of the fuel in conventional burners and equipment sized for the heating content of natural gas or even sold as pipeline fuel.

The hydrogen-separation membranes are also frequently used in conjunction with other processing and separation technologies. Production of high-purity hydrogen via pressure swing adsorption (PSA) results in a waste stream containing over 40% hydrogen. Frequently, additional hydrogen and fuel values can be realized by compression of the PSA purge and processing in a hydrogen membrane system [45, 46]. Removal of the hydrogen from hydrocarbon-rich fuel streams also enables recovery of condensible hydrocarbon liquids by processing the hydrogen-lean fuel gases in conventional expander plants.

2.3.2
Helium Separations

Like hydrogen, helium has high selectivity against most other components present in its mixtures for a wide variety of polymers. The primary source of helium is natural gas reservoirs usually in concentrations below 1%. Considering that the low concentrations translate to low partial pressures, it can be seen that membranes find limited application in helium production since multiple stages would be required to reach even crude helium purity levels.

Membranes have found application in helium gas recovery. Deep-diving gases employ helium mixtures that become contaminated. A combination of membranes and adsorbents are employed in purification of these gases for reuse. Small membrane systems are also employed in the purification of helium used in blimps. With time, the helium gas that inflates these lighter-than-air craft becomes contaminated with atmospheric gases. These contaminants are readily rejected by operating a small compressor and membrane unit to reject the contaminants and return purified helium to the craft.

2.3.3
Nitrogen Generation

Generation of nitrogen and nitrogen-rich atmospheres has become one of the largest uses for gas-separation membranes. The process simply involves passing filtered compressed air across the membrane. Since oxygen permeates faster than does nitrogen, it is driven across the membrane by its partial pressure and concentrates in the permeate stream leaving reject product enriched in nitrogen. The degree of oxygen removal is simply controlled by the residence time the nitrogen product has to contact the membrane. By reducing the product draw rate from the membrane, there is a correspondingly higher degree of oxygen removal. Thus, the oxygen concentration in the nitrogen-rich gas can be simply controlled with a process controller adjusting the product delivery rate to meet the specified oxygen level. Oxygen control levels vary with the application but typically fall within the 0.5% to 2.0% range. These can typically be achieved in a single countercurrent membrane element. Concentrations below 0.1% typically employ two or more membrane elements in series [47]. Argon, carbon dioxide, and water vapor also transport readily across the membrane. Thus, the membrane-produced nitrogen is also very dry with dew points below $-80°C$ measured.

Most nitrogen generators process compressed air at 7 to 20 atm. pressure and deliver the nitrogen-rich product within one to two atmospheres of the feed pressure. The degree of pretreatment of the air supplied to the membrane varies with the membrane employed and manufacturer. The performance of some membranes is sensitive to the moisture content of the air. In these cases, the compressed air is dried before processing. Other membranes are sensitive to organic vapors and oils that are removed by carbon filters or absorption beds. Most generators are also supplied with filtration systems to remove condensate, mist, and any particulates that might plug the membrane flow channels or fiber bores. Since the permeability and selectivity of the membranes can be highly temperature dependent, most applications require the air feed to be temperature controlled in order to control the product nitrogen quality or the membrane's performance. The nitrogen product is typically delivered under pressure and flow regulation to a receiver from which it is dispensed to users on demand. The coproduced permeate is typically simply vented back to the atmosphere.

The primary utility for the nitrogen is as an inerting atmosphere, typically for fire suppression. The Critical Oxygen Concentration, the minimum level of oxygen required to sustain combustion, is typically between 9% and 12% oxygen for many hydrocarbons and organic compounds. Thus, maintaining an inerting supply below 3% oxygen provides ample safety margin for many inerting applications. These levels can readily be reached with membranes having oxygen-nitrogen selectivities greater than 4. Thus membrane-generated nitrogen finds ready application anywhere flammable materials are stored, processed, or handled. Many fruits, vegetables, and produce are also preserved by maintaining them in a low-oxygen atmosphere. In controlled atmosphere storage of fruits and vegetables, a low-oxygen atmosphere, typically in the 2 to 10% range, depending upon the produce, is required. This is readily accomplished by purging the storage atmospheres with nitrogen or a nitrogen-rich gas as supplied by a membrane generator.

Membrane-based nitrogen-generation equipment is supplied by a number of manufacturers (see Table 9, Part I). Most are industrial gas suppliers who also use the equipment to supply smaller gas sales accounts. Membrane units are also used to supplement nitrogen supply at larger sale of gas accounts. Frequently, the membrane units are sized to handle a continuous, base-load requirement with truck-supplied liquid nitrogen providing back up and covering peak demand requirements. Many of the membrane manufacturers also market membrane modules to original equipment manufacturers, OEMs. The OEMs incorporate the membrane modules and nitrogen-generation capability into their own packaged equipment offerings. Since compressed air is necessary for nitrogen generation, it is not surprising that many OEMs are also compressor manufacturers and air compressor distributors. Other OEMs are specialty equipment manufacturers whose equipment is dependent on a nitrogen supply. For example, a metal-processing furnace manufacturer often incorporates the inerting atmosphere generation as part of his offering. Similarly, an instrument manufacturer may incorporate a small nitrogen generation system in his package to free his product from dependence on laboratory or cylinder nitrogen supply. Their features of portability, lightweight, compact, and robustness have aided in their incorporation into nitrogen-generation packages that literally can operate on land, at sea, or in the air. Following are some of the applications being served by membrane nitrogen-generation equipment.

Shipboard inert gas – Nitrogen generated onboard ships is used to purge the vapor space in chemical and flammable hydrocarbon liquid vessels on chemical tankers, the seals on cryospheres on LNG tankers, and various electronic systems on navy ships.

Controlled atmosphere storage – Membranes are used to maintain controlled atmospheres in perishable warehouses, cargo containers, and the holds on banana reefer ships.

Tank and process equipment inerting – Many offshore platforms, marine terminals, refinery and chemical tank farms are supplied with membrane-generated nitrogen.

Industrial gas sales – Many membrane suppliers are owned by or affiliated with industrial gas companies. They use onsite nitrogen generators to supply nitrogen to meet customer demand from baseload to total site nitrogen requirements, frequently supplemented with liquid nitrogen supply or backup.

Laboratories, photoprocessors, etc. – Many facilities have replaced nitrogen cylinders with small wall mounted cabinets that generate nitrogen for laboratory use from the facility's compressed air.

Metals processing – Membranes generate nitrogen for inerting many sintered-metal process furnaces, soldering station blanketing and the like.

Tire inflation – Flight line ground-support carts generate and store high-pressure nitrogen for aircraft tire and strut inflation. Tire-molding operations and truck and automobile tire service centers are turning increasingly to generators to provide nitrogen as a replacement for compressed air for tire inflation.

Oil and gas field service – Membrane-produced nitrogen is replacing air in pneumatic drilling, used in oil-well servicing, and for pipeline purging.

Flare and seal purging – Small nitrogen generators are supplying seal gas to a variety of seals on rotating machinery. It is also supplying purge to flare piping systems.

On-board inert gas generation system – OBIGGS – Nitrogen membranes fly at supersonic speeds on the USA Air Force F-22. A small bleed of compressed air from the jet engine's turbine compressor is processed into nitrogen to purge the fuel tanks onboard this aircraft. OBIGGS is gaining increasing interest for military and commercial aircraft with the discovery of a fuel-tank explosion as the probable cause of the TWA Flight 800 disaster. Major aircraft manufacturers are beginning to incorporate fuel-tank inerting in the design of new commercial aircraft. Older commercial aircraft will also be retrofitted with on-board fuel-tank inerting systems under a USA Federal Aviation Authority mandate over the next five to seven years.

Hypobaric atmospheres – Athletes and racehorses are training and sleeping in reduced-oxygen atmospheres to simulate high-altitude training and hypoxic-induced physiological changes conductive to enhanced performance.

2.3.4
Acid Gas-Separations

Figure 2.1 shows that the acid gases carbon monoxide and hydrogen sulfide have permeabilities in polyetherimide film that are about 20 times those of gases such as methane, nitrogen, and carbon monoxide, that frequently accompany them in mixtures. Indeed, many commercial gas-separation membranes have sufficiently favorable selectivities and permeance to make their application viable in a variety of applications. Commercial acid gas removal applications include pipeline grade natural gas production [48, 49], carbon dioxide recovery and recycle in enhanced oil recovery (EOR) [50–52], methane recovery from landfill and biogas, [53, 54], and carbon dioxide recovery from flue gases [55, 56].

Figure 2.9 depicts the performance of a typical acid gas removal membrane. The graph provides operating lines for several feed composition levels and predicts combinations of the permeate and non-permeate product composition that can be expected when the membrane feed to permeate pressure ratio is greater than 20.

Inspection of this diagram reveals the dilemma associated with membrane applications for acid gas-separations, namely that it is difficult to produce high-purity products in either permeate or non-permeate at high recoveries with a single membrane stage. Thus, it is necessary to have an alternative application for the impure coproduct, or resort to a more complicated process involving compression of the permeate to feed successive membrane stages to achieve high degrees of separation [57].

Study of the diagram reveals that processing a typical raw natural gas to a pipeline grade natural gas containing less than 2% carbon dioxide, while rejecting carbon dioxide at greater than 90% purity would require at least three membrane stages with permeate compression between each stage. The economy for such makes it practical in only a limited number of applications.

Despite this limitation, membranes are finding increasing application in acid gas removal. The successful applications depend upon identifying utility for the low-purity coproducts in another service or application. For example, processing a 20% CO_2 feed to pipeline grade natural gas produces a medium-heating quality fuel that is used for power generation. In another case, a membrane unit produces a high CO_2 permeate gas suitable for injection into an oil reservoir for enhanced oil recovery while the co-product is processed further with a conventional amine sweetening process.

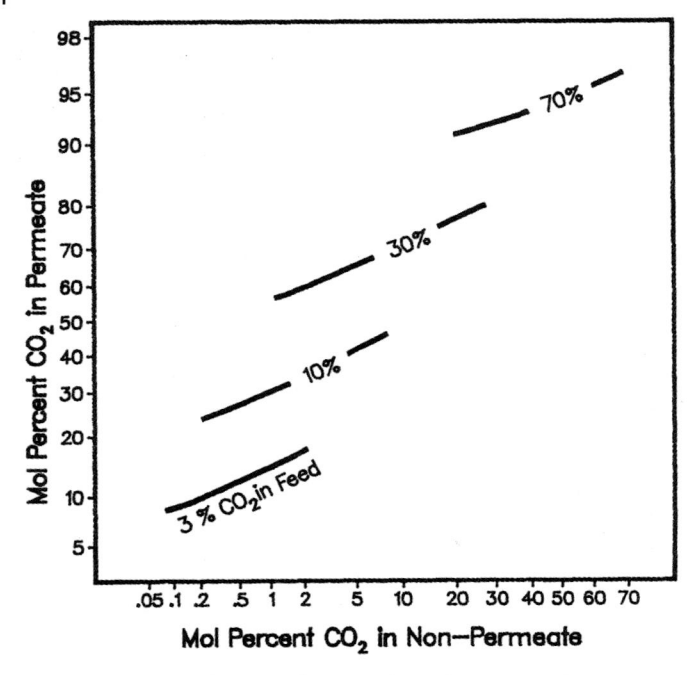

Fig. 2.9 Typical membrane performance for acid gas removal from mixtures with methane or n gen. (Used with permission of Air Products and Chemicals, Inc.).

2.3.5
Gas Dehydration

The capability of membranes to dry gases was recognized in the early hydrogen, acid gas, and nitrogen membrane applications. Indeed some nitrogen-generation systems were marketed for their ability to produce a very dry product suited for purging of cryogenic and LNG equipment. It was recognized early that a membrane's productivity of dry gas was dependent upon the creation of sufficient permeate volume to maintain the very low partial pressure driving force associated with a product containing only parts per million of water. Monsanto introduced the Cactus® membrane air dryer in 1987 that relied on a different, higher flux membrane on the dry product end of the separator [58]. This allowed for the internal generation of countercurrent sweep flow of dried product to create a good partial pressure driving force along the dryer length. Other dryers have since been introduced by a number of suppliers that rely on a four-ported membrane design as shown in Fig. 2.4 to provide a controlled amount of sweep gas.

Most membrane dryers such as the Cactus® Membrane Air Dryer line shown in Fig. 2.10 are marketed to OEMs and compressor and instrument distributors.

Fig. 2.10 Cactus® membrane compressed air dryers. (Used with permission of Air Products and Chemicals, Inc.).

These devices process filtered compressed air to deliver compressed air having dew points as low as −40°C at 10 atm. pressure.

Using membranes for drying natural gas was suggested as early as 1984 [59], but it was 1996 before the concept became reality when Air Product's Permea organization introduced a swept dryer design that was capable of drying gases to pipeline specifications at pressures as high as 70 bar [60]. Since then several demonstration units with capacities as high as 200 KN m3/day have been installed.

The high-pressure dryer has also found application in high-pressure compressed air systems onboard US Navy vessels [61]. These dryers are typically operated between compressor stages at 30 to 70 bar pressures and the dry product returned to the next compression stage for final boost to the 200 to 330 bar receiver pressure.

2.4
Developing Membrane Applications

2.4.1
Oxygen and Oxygen-enriched Air

All nitrogen generators necessarily produce an oxygen-enriched air, OEA, coproduct, however, it is generally vented to the atmosphere. This is probably because the OEA is at low pressure, varies in composition due to process controls on the nitrogen generator, and volumes are small in comparison to meaningful OEA application requirements.

One commercial application generates OEA for use in intermittent catalyst regeneration. This application is unique though as there is also an onsite application for nitrogen generation. The installation serves as a beta-site for the OEA application.

Membranes have also been employed to produce OEA for ozone generators. Most attempts have involved the OEA coproduct from compressed air-driven nitrogen generators. Since ozone generation is proportional to oxygen concentration and is favored by atmospheric-pressure operation, feeding the OEA to an ozone generator would appear to be a logical application. However, the economics are highly dependent on recovering value from the nitrogen coproduct stream. The application involves feeding the membrane with dry compressed air since any moisture present in the feed copermeates with the OEA and interferes with the ozone generation.

Only a small amount of the oxygen supplied is converted in an ozone generator. In a typical application, the ozone-rich stream is processed in an absorber or contactor tower to decompose organics in waste streams. Vent gases leaving such equipment are rich in oxygen when the ozone generator has been fed by OEA or oxygen. Thus, there has been interest in recovery and recycle of these vent gases to the ozone generator. Air separation membranes have been operated in such a recycle arrangement to enrich the OEA to as high as 60 to 70% oxygen. At these levels, the ozone generator becomes very effective and more than doubles in productivity over an air feed. Such is the good news. Unfortunately, any residual ozone in the vent-gas recycle stream also attacks the polymeric membranes. With time, even ppm levels of ozone have a cumulative, irreversible, degrading effect on the membranes and shortening of their useful life. Design of guard beds, ozone-destruct catalysts, and the like to protect the membranes have been successful during normal operation, however, control systems designed for low ozone levels were not sufficiently responsive in detecting and responding to overload and upset conditions before damaging the membranes in the demonstration unit.

OEA production from air is favored by vacuum-driven membranes operating at low oxygen recovery. In this case, the OEA purity is simply a function of the membrane's selectivity and the vacuum level at which the OEA permeate is withdrawn. Unfortunately, the high-selectivity membranes employed in nitrogen generators have low flux and would require huge membrane areas at subatmospheric pressure. Presently, high-flux materials are being used but OEA purities are limited to less than 30% oxygen. There is much research activity in pursuit of higher flux, higher selective membranes that will bear watching for this application.

2.4.2
Nitrogen Rejection from Natural Gas

A number of membrane developers and suppliers are pursuing nitrogen rejection from natural gas with high-flux, low-selectivity membranes. Modest nitrogen reduction (30–50% nitrogen removal) in the pressurized natural gas has been demonstrated at a beta-site with plasma-treated polydimethlysiloxane membranes supplied by Neomecs. [62]. Permeate from the nitrogen-reduction membrane unit is nitrogen enriched. As the acid gas removal cases discussed above, this application also depends upon the disposition of this low-energy gas.

2.4.3
Nitrogen-enriched Air (NEA)

Compact Membrane Systems is actively pursuing NEA as a replacement for Exhaust Gas Recycle, EGR, as a NOx abatement measure on diesel engines. The CMS membrane is high flux, low selectivity capable of enriching the turbocharged engine air to 82 to 84% nitrogen. Such levels have been demonstrated as being effective in substantially reducing NOx components [63, 64].

References

1 G. Davis, I. R. Lauks, R. J. Pierce, and C. A. Widrig, Method of measuring gas concentrations and microfabricated sensing device for practicing same. US Patent No. 5514253, Assigned to I-Stat Corporation, May 7, 1996.

2 C. P. Van Dijk and L. D. Fraley, Process for producing and utilizing an oxygen enriched gas. US Patent No. 5245110, Assigned to Starchem, Inc., Sept. 14, 1993.

3 C. P. van Dijk, Methanol production process using a high nitrogen content synthesis gas with a hydrogen recycle. US Patent No. 5472986, Assigned to Starchem, Inc., Dec. 5, 1995.

4 S.-T. Hwang and K. Kammermeyer, Membranes in Separations, John Wiley, (1975).

5 MacLean, D. L., D. J. Stookey, and T. R. Metzger, Fundamentals of Gas Permeation., Hydrocarbon Processing, (August, 1983).

6 A. Y. Alentiev and Y. P. Yampolskii, Free volume model and trade off relations of gas permeability and selectivity in glassy polymers. J. of Membrane Science, 165 (2000) 201–216.

7 D. R. Paul and Yuri P. Yampol'skii, Polymeric Gas-separation Membranes, Chapter 4, N. Plate and Y. Yampol'skii, Relationship Between Structure and Transport Properties for High Free Volume Polymeric Materials., CRC Press, 1994, p. 155–207.

8 R. E. Kesting, A. K. Fritzsche, M. K. Murphy, A. C. Handerman, C. A. Cruse, and R. F. Malon, Process for

forming asymmetric gas-separation membranes having graded density skins. US Patent 4871494, Oct. 2, 1989.

9 R. E. Kesting, A. K. Fritzsche, M. K. Murphy, A. C. Handerman, C. A. Cruse, and R. F. Malon, Asymmetric gas-separation membranes having graded density skins. US Patent 4,880,441, Nov. 14, 1989.

10 M. Langsam, Fluorinated polymeric membranes for gas-separation processes., US Patent 4657564, Assigned to Air Products and Chemicals, Apr. 14, 1987.

11 P. W. Kramer, M. K. Murphy, D. J. Stookey, J. M. S. Henis, and E. R. Stedronsky, Membranes Having Enhanced Selectivity and Method of Producing Such Membranes. US Patent No. 5215554, June 1, 1993.

12 R. M. Conforti, T. A. Barbari, P. Vimalchand, and M. D. Donohue, A Lattice-Based Activity Coefficient Model for Gas Sorption in Glassy Polymers. *Macromolecules*, 1991, **24**, 3388–3394.

13 G. G. Lipscomb, Unified Thermodynamic Analysis of Sorption in Rubbery and Glassy Materials. *AIChE Journal*, Oct. 1990, **36(10)**, 1505–1516.

14 R. Srinivasan, S. R. Auvil, and P. M. Burban, Elucidating the mechanism(s) of gas transport in poly[1-(trimethylsilyl)-1-propyne] (PTMSP) membranes. *J. of Membrane Science*, **86** (1994) 67–86.

15 D. R. Paul and Yuri P. Yampol'skii, Polymeric Gas-separation Membranes, Chapter 8, R. W. Baker and J. G. Wijmans, Membrane Separation of Organic Vapors from Gas Systems., CRC Press, 1994, p. 353–397.

16 S. Loeb and S. Sourirajan, Sea water demineralization by means of an osmotic membrane. *Advanced Chem. Ser.* **38** (1962), 117.

17 R. E. Kesting, Asymmetric gas-separation membranes having graded density skins., US Patent 4,880,441, Nov. 14, 1989, Assigned to Pumea.

18 M. Haubs and W. Prass, Composite membrane, process for its production and its use. US Patent No. 5342432, Assigned to Hoechst Aktiengesellschaft, August 30, 1994.

19 P. S. Puri, Process for making highly permeable coated composite hollow-fiber membranes. US Patent No. 4756932, Assigned to Air Products and Chemicals, Inc., July 12, 1988.

20 C. R. Gochanour, Gas-separation membrane with ultrathin layer. US Patent No. 5160353, Assigned to E. I. Du Pont de Nemours & Company, November 3, 1992.

21 S. E. Moore, Sr., Apparatus for spinning multicomponent hollow-fibers. US Patent No. 5,320,512, , Assigned to E. I. Du Pont de Nemours & Company, Jun. 14, 1994.

22 A. Frank, G. G. Lipscomb, and M. Dennis, Visualization of Concentration Fields in Hemodialyzers by Computed Tomography. *Journal of Membrane Science*, **175** (2000) 239–251.

23 J. Lemanski and G. G. Lipscomb, Effect of Fiber Variation on the Performance of Counter-current Hollow Fiber Gas-separation Modules. *Journal of Membrane Science*, **167** (2000) 241–252.

24 R. Leonard, Crimped Hollow Fibers for Fluid Separations and Bundles Containing the Hollow Fibers., Canadian Patent No. 1,114,307.

25 Fujii et al, Fluid separation apparatus., US Patent 4293418, Assigned to Toray Industries, Inc, Oct. 6, 1981.

26 D. G. Kalthod and D. J. Stookey, Hollow-Fiber Membrane Device with Inert Filaments Randomly Distributed in the Inter-Fiber Voids., US Patent No. 5779897, July 14, 1998.

27 S. R. Wickramasinghe, M. J. Semmens, E. L. Cussler, Hollow-fiber modules made with hollow-fiber fabric. *J. Membrane Science* **84** (1993) 1–14.

28 E. L. Cussler, Hollow Fiber Contactors., In: J. E. Crespe, K. W. Böddeker (Ed.), Membrane Processes in Separation and Purification, Kluwer Academic Publishers, Netherlands, 1994, pp. 375–394.

29 A. Gableman and S-T. Hwang, Hollow-fiber membrane contactors. *J. of Membrane Science*, **159** (1999), 61–106.

30 J. Rogut, Fiber membrane elements and modules and methods of fabrication for improved fluid separations. US Patent 5,238,562, Aug. 24, 1993.

31. P. E. ALEI, ET AL, Loom processing of hollow-fiber membranes. US Patent 5,598,874, Assigned to MG Generon, Inc., Feb. 4, 1997.

32. J. ROGUT, Hollow-fiber membrane carpet manufacturing method and an elementary carpet member and carpet. US Patent 5716689, Feb, 10, 1998.

33. P. J. HICKEY and C. H. GOODING, Modeling spiral wound membrane modules for pervaporative removal of volatile organic compounds from water. *J. of Membrane Science*, **88** (1994) 47–68.

34. D. T. COKER, B. D. FREEMAN, and G. K. FLEMING, Modeling Multicomponent Gas-separations Using Hollow-Fiber Membrane Contactors. *AIChE Journal*, 1998.

35. D. J. STOOKEY, Gas-separation Membranes – 20 Years Old, Still Growing. presentation at 1997 Membrane Technology/Planning Conference sponsored by Business Communications Co., Inc., October 26-28, 1997, Newton, MA.

36. E. C. MAKIN, J. L. PRICE, and Y. W. WEI, Carbonylation Process. US Patent 4255591, Mar. 10, 1981.

37. E. PERRY, Process for Hydrogen Recovery from Ammonia Purge Gases. US Patent 4172885, Assigned to Monsanto Co., Oct. 30, 1979.

38. H. R. NULL and E. PERRY, Process for hydrogen recovery from ammonia purge gases. US Patent No. 4180553, Assigned to Monsanto Co., Dec. 25, 1979.

39. T. E. GRAHAM, D. L. MACLEAN, Process for Hydrogen Recovery from Ammonia Purge Gases. US Patent 4180552, Assigned to Monsanto Co., Dec. 25, 1979.

40. D. L. MACLEAN, R. KRISHNAMURTHY, and S. L. LERNER, Argon recovery from hydrogen depleted ammonia plant purge gas utilizing a combination of cryogenic and non-cryogenic separating means. US Patent No. 4750925, Assigned to BOC Group, Inc., June 14, 1988

41. D. L. MACLEAN, R. KRISHNAMURTHY, and S. L. LERNER, Argon recovery from ammonia plant purge gas utilizing a combination of cryogenic and non-cryogenic separating means. US Patent No. 4752311, Assigned to BOC Group, Inc., June 21, 1988.

42. E. C. MAKIN and K. K. OKAMOTO, Process for Methanol Production. US Patent No. 4181675, Assigned to Monsanto Co., Jan. 1, 1980.

43. C. P. VAN DIJK, Methanol production process using a high nitrogen content synthesis gas with a hydrogen recycle. US Patent No. 5472986, Assigned to Starchem, Inc., Dec. 5, 1995.

44. L. G. POSEY, JR., Processes. US Patent No. 4367135, Assigned to Monsanto Co., Jan. 4, 1983.

45. G. M. INTILLE, Selective Adsorption Process. US Patent No. 4229188, Assigned to Monsanto Co., Oct. 21, 1980.

46. E. PERRY, Selective Adsorption Process. US Patent No. 4238204, Assigned to Monsanto Co., Dec. 9, 1980.

47. A. W. RICE, Process for capturing nitrogen from air using gas-separation membranes. US Patent No. 4894068, Assigned to Permea, Inc., Jan. 16, 1990.

48. T. E. COOLEY and W. L. DETHLOFF, Field Tests Show Membrane Processing Attractive. Chemical Engineering Progress, October 1985, 45–50.

49. T. E. COOLEY and A. B. COADY, Removal of H2S and/or CO_2 from a light hydrocarbon stream by use of gas-permeable membrane. US Patent No. 4130403, Dec. 19, 1978

50. C. S. GODDIN, Pick treatment for high CO_2 removal. Hydrocarbon Processing, May, 1982, 125–130.

51. W. J. SCHELL, and C. D. HOUSTON, Process gas with selective membranes. Hydrocarbon Processing, Sept. 1982, 249–252.

52. W. A. BOLLINGER, D. L. MACLEAN, and R. S. NARAYAN, Separation Systems for Oil Refining and Production. Chemical Engineering Progress, Oct. 1982, 27-32.

53. Recovering Methane At A Small Landfill. Waste Age, Nov. 1983, 28–30.

54. D. J. STOOKEY, K. BOUSTANY, R. L. KILGOUR, Recovery of Methane from Biogas with Monsanto's Prism Separators. BioEnergy 84 World Conference and Exhibition, Gothenburg, Sweden, June 18–21, 1984.

55. R. A. CALLAHAN, Process and apparatus for producing liquid carbon dioxide. US Patent No. 5233837, Aug. 10, 1993.

56 P. A. Daus, C. R. Pauley, J. W. Koenst, and F. Coan, Membrane process for producting carbon dioxide. US Patent No. 6085549, Jul. 11, 2000.
57 D. J. Stookey, T. E. Graham, and W. M. Pope, Natural gas processing with Prism® separators. Environmental Progress, August 1984, 212–214.
58 D. J. Stookey, Fluid Separation Membranes. US Patent No. 4,687,578, Aug. 18, 1987.
59 F. Fournie and J. P. Agostini, Permeation: A New Competitive Process for Offshore Gas Dehydration. OTC '84 Proceeddings, 1984 Offshore Technology Conference, Houston, TX, May 7–9, 1984.
60 D. J. Stookey, K. Jones, D. G. Kalthod, and T. Johannessen, Membrane Dehydrators – A New Alternative for Drying High Pressure Gases. Presention at The 1996 Membrane Technology/Planning Conference sponsored by Business Communications Co., Inc., October 29, 1996, Newton, MA.
61 T. Theis and S. Titus, The Development of Permeable Membrane Air Dehydrators for the U.S. Navy. *Naval Engineers Journal,* May, 1996, 243–265.
62 R. A. Callahan, Enerfex Inc., private communications to author, October, 2000. See http://www.enerfex.com/index.htm
63 Compact Membrane Systems website, http://www.compactmembrane.com/
64 S. Nemser, D. Stookey, J. Nelson, Diesel Engine NO_x Reduction via Nitrogen-Enriched Air. *Fluid/Particle Separation Journal,* Vol 15, No. 1, 2004, p. 69–80.

3
State-of-the-Art of Pervaporation Processes in the Chemical Industry

H. E. A. Brüschke

3.1
Introduction

In 1917 P. A. Kober published a paper [1] in which he described his observation that "a liquid in a collodium bag, which was suspended in the air, evaporated, although the bag was tightly closed". Kober was not the first researcher to observe this phenomenon that a liquid can evaporate through a tightly closed "membrane", but he was the first to realize its potential for the separation of liquid mixtures that otherwise are difficult to separate, and to separate them under moderate conditions by means of a membrane. He introduced the terms "pervaporation", and "perstillation", and the first term is today in use to describe in general a process in which one component out of a fluid mixture is selectively permeating through a dense membrane, driven by a gradient in partial vapor pressure, leaving the membrane as a vapor, and being recovered in a condensed form as a liquid.

In the years following Kober's publication a few papers were published describing membranes and processes for pervaporation. The related phenomena were mainly investigated in research laboratories but without looking intensely for any practical applications. This was mainly due to the lack in understanding of membrane processes in general and the absence of suitable membranes in detail. Later, during the 1950s, the picture changed and a considerable effort was devoted in industrial research for effective membranes in order to introduce pervaporation as an additional industrial separation process. The interest focused on membranes and processes for the separation of different classes of hydrocarbons and of isomers [2–4] and a number of patents were granted [5, 6]. Membrane materials disclosed were natural and synthetic rubbers, cellulose esters and ethers, and several treated and untreated polyolefines. Due to insufficient flux and selectivity, however, none of these early membranes could be applied in any industrial process.

With the development of the integral asymmetric cellulose acetate membranes for desalination of saline water a few researchers turned again to pervaporation [7–9]. But it was not before 1982 that a first pervaporation membrane useful for the removal of water from organic liquids on an industrial scale was developed and introduced into the market by a small German company, GFT (Gesellschaft für Trennverfahren) [10–12]. In 1983 a first plant started its operation for the dehydration of azeotropous ethanol in Brazil, with a capacity of 1200 l/d of anhydrous ethanol. This plant was followed by others [13, 14], first for the production of anhydrous ethanol only.

Following the experience gained in ethanol dehydration, in 1988 the first plant started its operation in the chemical industry for the dehydration of an ester. Soon other applications for dewatering followed, covering today a broad range of solvents and solvent mixtures, especially those forming azeotropes with water. In 1994 a first plant started its operation in which water was continuously removed from a reaction mixture, in order to shift the reaction equilibrium towards the wanted product, in this case a diester [15], and, by nearly totally converting one of the educts, to increase the yield and to facilitate the downstream purification of the product.

Removal of water from organic mixtures by pervaporation or vapor permeation is now a widely accepted state-of-the-art technology. Meanwhile, new membranes have been developed that allow separating simple alcohols like methanol or ethanol from their mixtures with organic solvents whereby these mixtures are virtually anhydrous. The first industrial plant of this kind started its operation in 1997 [16], separating methanol out of its azeotropic mixture with TMB (trimethylborate). Many commonly used organic solvents form azeotropes with methanol that cannot always easily be separated. The respective simple esters of methanol and ethanol also form azeotropes with the alcohol, and the often-used water wash for splitting of the azeotrope may lead to unwanted hydrolysis. It can be assumed that the number of applications of pervaporation processes will increase in the future for this type of separation.

Removal of low volatile organic components (VOCs) from aqueous streams by means of pervaporation through organophilic membranes has been tested [17–20], but not yet found an industrial application. If the aqueous stream is a waste water competing processes like air or steam stripping, or distillation, and biological treatment are introduced and usually cheaper, especially as the substances recovered from a mixed waste water stream are low in volume and not of a high value as they have to be further treated and purified. When the substance to be separated from the aqueous mixture has a high value and is otherwise difficult to be recovered this application of pervaporation [21, 22] may gain new interest in the future. The separation and recovery of aroma components from natural products or from microbiological production is one such example that is widely investigated in research laboratories, although still no industrial application exists.

Separation of low volatile organic components (VOCs) from gas streams through organophilic membranes, however, has become an accepted and widely used technology. Monomers like ethene, propene, or vinylchloride [23] are recovered from strip gas or waste gas streams and recycled to the upstream process, and gasoline vapors are separated from waste air streams in tank farms [24]. In these applications the recovered material is sufficiently pure for further use and has a high value as it is otherwise lost and wasted, e.g. by incineration. These applications of vapor permeation will be dealt with in a separate chapter.

Separation of different organic components from each other is still a matter of laboratory investigation. In the past 15 years considerable efforts have been devoted to develop polymeric membranes to separate, for example, aromatic hydrocarbons from aliphatic ones which resulted in several patents [25, 26], or olefins from paraffins or to separate isomers, e.g. para- and ortho-xylenes, from each other. In the last years additional membranes [27] have become available and the first industrial applications have been reported, e.g. the separation of sulfur-containing aromatics from gasoline [28] and of benzene from a stream of saturated hydrocarbons [29]. Further development of membranes, especially of the mixed-matrix type, may lead to improved selectivity and a broadening of these applications.

3.2
Principles and Calculations

3.2.1
Definitions

Pervaporation, vapor permeation and gas permeation are very closely related processes. In all three cases the driving force for the transport of matter through the membrane is a gradient in the chemical potential that can best be described by a gradient in partial vapor pressure of the components. The separation is governed by the physical-chemical affinity between the membrane material and the species to be passed through and thus by sorption and solubility phenomena. The transport through the membrane is affected by diffusion and the differences in the diffusivities of the different components in the membrane can play an important role for the separation efficiency, too. All three processes are best described by the "solution-diffusion mechanism", their main differences are determined by the phase state and the thermodynamic conditions of the feed mixture and the condensability of the permeate.

Pervaporation
A liquid feed mixture is in contact with one side of the membrane, all partial vapor pressures at the feed side are at saturation. The feed-side pressure is of the order of several bar, just sufficient to keep the feed mixture in the liquid state at the operation temperature. The gradient in partial vapor pressure between the feed and the permeate side of the membrane is maintained by a reduction of the permeate-side partial vapor pressure. The permeate leaves the membrane as a vapor and is usually condensed, and removed as a liquid. The heat necessary for the evaporation of the permeate has to be transported

through the membrane, and this transport of energy is coupled to the transport of matter. The evaporation enthalpy is taken from the sensible heat of the liquid feed mixture, which leads to a reduction in feed-side temperature.

Vapor Permeation
The feed mixture is in the vapor phase, the partial vapor pressure of at least the critical (better permeating) component in the feed mixture is at or close to saturation, which may require compression of the feed mixture. The gradient in partial vapor pressure is maintained by a reduction of the permeate-side partial vapor pressure, too. The permeate leaves the membrane as a vapor and at least the critical (better permeating) component in the permeate can be condensed and removed as a liquid. Due to changes of saturation conditions (temperature or pressure) with changing composition of the feed mixture along its passage over the membrane some of the feed vapor may condense on the membrane surface and will be separated by pervaporation.

Gas Permeation
The feed mixture is in the gas phase (the temperature for all components is above the critical temperature); the partial vapor pressures of all components in the feed mixture are far below saturation. The gradient in partial vapor pressure is usually maintained by an increase of total feed-side pressure that may reach more than one hundred bar. The permeate cannot be condensed and is removed as a gas.

Therefore all three processes are but different aspects of the same transport mechanism and the same membranes are used at least for pervaporation and vapor permeation, sometimes even for gas separation. Today, wherever the term "pervaporation" is used it should be well understood that it includes at least "vapor permeation" as well.

Transport through the membrane can best be described by a so-called "solution-diffusion mechanism". In this mechanism it is assumed that a component of the feed having a high affinity to the membrane is easily and preferentially adsorbed and dissolved in the membrane substance (Fig. 3.1). The more soluble a component is the more matter is dissolved in the membrane and the more the membrane will swell and change its composition. Swelling effects are highest in pervaporation, as a high-density fluid is contacting the feed side of the membrane. They are somewhat lower in vapor permeation and of lesser importance in gas separation due to the much lower density of the feed mixture, but may become important at high pressure for real gases.

Following a concentration gradient the components migrate through the membrane by a diffusion process and are desorbed at the downstream side of the membrane into a vapor phase. In vapor permeation and gas separation the phases on both sides of the membrane are identically gaseous (or vaporous). In pervaporation the components passing through the membrane are absorbed out of a liquid phase but desorbed into a vapor phase, as the permeate-side partial vapor pressures are maintained below the respective saturation values existing

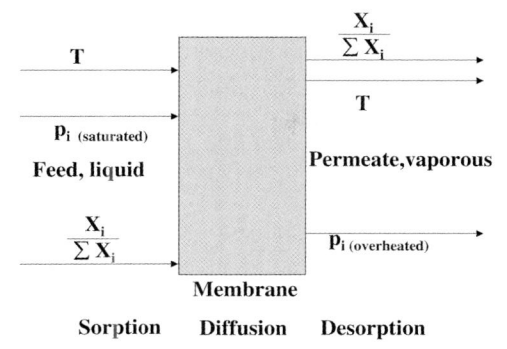

Fig. 3.1 Pervaporation, solution-diffusion mechanism.

on the feed side. The energy for the phase transition, the evaporation enthalpy has to be transported through the membrane as well, which makes pervaporation unique compared to all other transport processes involving membranes.

Substances with lower or no solubility in the membrane material cannot be dissolved or reach only low concentrations and thus low transport rates. As the diffusion coefficients of small molecules in a polymeric matrix do not differ too much, the separation characteristics of the membrane are primarily governed by the different solubilities of the components in the membrane material and to a lesser extent by their diffusion rates. When a smaller molecule is better dissolved in the membrane substance solubility and diffusion enhance each other. This is at least the case in dehydration processes where water is both the better soluble and faster diffusing component. In the removal of VOCs from gases where large molecules are removed and the larger molecule is the better soluble one, the diffusion step may counteract solubility and reduce the overall selectivity towards smaller molecules.

3.2.2
Calculation

Like any other kinetic process the transport through a dense membrane is directly proportional to the driving force of the process and inversely proportional to the transport resistance.

transport rate = driving force/resistance

In a pervaporation process the driving force is the gradient in the chemical potential which is defined as

$$\mu_i(T, p, x) = \mu_i^0(T, p^0) + R \cdot T \cdot \ln a_i(T, p^0, x_i) + \tilde{V}_i \cdot (p - p^0) \tag{1}$$

with $\mu_i(T, p, x_i)$ the standard chemical potential of the pure component, whereas the other terms describe the dependence on concentration and pressure.

In a pervaporation process the pressure is sufficiently low (of the order of a few bar), thus the last term in Eq. (1) can be neglected and thus

$$\Delta\mu_{i/PV} = R \cdot T \cdot \ln \frac{a_{i,\text{Feed}} \cdot p_{i,\text{sat.}}}{p_{i,\text{Permeate}}} = R \cdot T \cdot \ln \frac{x_{i,\text{Feed}} \cdot \gamma_i \cdot p_{i,\text{sat.}}}{x_{i,\text{Permeate}} \cdot p_{\text{Permeate}}} \tag{2}$$

$$= \tilde{V}_i \cdot \left(\frac{R \cdot T}{\tilde{V}_i} \cdot \ln \frac{p_{i,\text{sat.}}}{x_{i,\text{Permeate}} \cdot p_{\text{Permeate}}} - \pi_{i,\text{Feed}} \right) \tag{3}$$

The partial pressure of a component at the feed side of a pervaporation membrane has always to be higher than that on the permeate side, otherwise no transport will occur. For a dehydration process one obtains

$$x_{H_2O,\text{Feed}} \cdot \gamma_{H_2O} \cdot p_{H_2O,\text{sat.}} \geq x_{H_2O,\text{Permeate}} \cdot p_{\text{Permeate}} \tag{4}$$

When a high-selective membrane ($x_{H_2O,\text{ Permeate}}=1$) is used for the dehydration of ethanol at a temperature of 100 °C (373 K) with an activity coefficient γ_{H_2O} of 2.75 and a permeate-side pressure of 20 mbar, a retentate with a molar concentration of water of $x_{H_2O,\text{ Retentate}}=0.00694$ can be obtained, or an ethanol with a purity of 99.973% by weight. This example demonstrates the potential of the pervaporation process to purify in general and especially to dehydrate organic liquids to very high purity and low final water content.

For any calculation a few simplifications have to be introduced. The "solution-diffusion mechanism" assumes that the dense membrane layer behaves like a liquid and no coupling does exist for the transport of the different species. It is assumed that the respective sorption-desorption equilibria are established for all components at both sides of the membrane and that they follow Henry's law.

$$x_i^f = p_i^0 \cdot S_i \tag{5}$$

and

$$x_i^p = p_i^p \cdot S_i \tag{6}$$

where x_i^f and x_i^p denote the respective concentration of the component at the feed and the permeate side in the membrane, p_i^f the saturation partial vapor pressure in the feed and p_i^p the respective partial vapor pressure at the permeate side, and S_i the sorption coefficient, depending on the temperature and nature of the system.

The transport rate of a component through the membrane will then be determined by its concentration and mobility and the respective driving force

$$\dot{n}_i'' = -x_{i,\text{membrane}} \cdot b_{i,\text{membrane}} \cdot \frac{\partial \mu_{i,\text{membrane}}}{\partial z} \tag{7}$$

The concentration in the membrane depends on the outside activity and the sorption or partition coefficient of the species, the mobility on the nature of the membrane. The driving force for a component is a function of the process parameters, e.g. temperature, pressure, and concentration. In a pervaporation process usually the minor component is removed from a mixture. For the retained major component the driving force will always be higher than for the transported one. The selectivity of the membrane is then determined by the differences in the product of mobility and concentration and not by a difference in the driving force.

Following the Nernst-Einstein equation the relation between the mobility $b_{i,\text{membrane}}$ and the thermodynamic diffusion coefficient is given by

$$D_{i0} = R \cdot T \cdot b_i \tag{8}$$

Introduction of this relation into Eq. (7) leads to

$$\dot{n}_i'' = -x_{i,\text{membrane}} \cdot \frac{D_{i0,\text{membrane}}}{R \cdot T} \cdot \frac{\partial \mu_{i,\text{membrane}}}{\partial z} \tag{9}$$

As sorption equilibria exist on both sides of the membrane, the overall rate of the transmembrane transport is determined by the diffusion step only. Fick's law can be used to describe the diffusional transport of a component i through the membrane

$$\dot{n}'' = -x_{i,\text{sat.membrane}} \cdot D_{i0,\text{memb.}} \cdot \frac{\delta x_i}{\delta z} = -x_{i,\text{sat.memb.}} \cdot D_{i0,\text{memb.}} \cdot \frac{dx_i}{dz} \tag{10}$$

where $\dot{n}_i''$ is the partial flux or permeation rate of the component, D_i its diffusivity or diffusion coefficient and dx_i/dz is the concentration gradient across the membrane.

The diffusivity of a component dissolved in a liquid or in a polymeric film depends strongly on its concentration. As the concentration of the dissolved species change from the feed to the permeate side of the membrane, concentration-dependent diffusion coefficient have to be introduced into Eq. (10). Different expressions have been proposed to relate diffusivity to concentration. One of the more commonly used relations is

$$D_{i,\text{memb.}} = D_{i0,\text{memb.}} \cdot \exp\left(\tau_{(T,p,x_i)} \cdot x_i\right) \tag{11}$$

with D_{i0} the diffusion coefficient at zero concentration, x_i the respective concentration in the membrane and τ a plastification coefficient that describes the dependence of the free volume of a polymer on the process parameters temperature, pressure, and concentration.

In the literature other transport equations can be found, some of them following the terms of the theory of irreversible thermodynamics. Here, the flux through a membrane can be calculated by

$$J_i = \frac{Q_j}{l} \cdot (p_{\text{feed}} - p_{\text{permeate}}) \tag{12}$$

with Q_j the so-called permeability, l the thickness of the membrane, and p_{feed} and p_{permeate} the respective partial pressures at the feed and permeate side of the membrane. Permeability is a rather complex function of solubility and diffusivity, but often it is assumed to be a material constant of the respective polymer from which the membrane is made. Thus, permeability Q_j for a single component could be measured by pervaporation tests, knowing the thickness of the membrane, with some correction factors for the influence of temperature and pressure. With the knowledge of the respective permeability of the single components in the membrane and their respective temperature dependencies, fluxes for a component through a given membrane could be calculated and a process for any mixture engineered.

Unfortunately, such an approach is practical for the separation of inert gases in polymeric films only, where no interaction occurs between the membrane and the migrating molecules. As soon as one of the components of the feed mixture interacts with the membrane material and is dissolved in the membrane to any larger extent the effective fluxes and selectivity for a mixture cannot be calculated from single-component data [30]. This is due to the change of the membrane material by the dissolution of at least one of the components in the polymer. When a first small portion of a substance is dissolved in the membrane material the latter starts to swell and to change its properties. Thus, the next portion of the same substance is dissolved into a membrane different from the first one. When different substances get into contact with the membrane material out of a feed mixture strong coupling effects can be observed for both solubility and diffusion [31].

One substance A may have a high affinity with and be highly soluble in the membrane polymer. This will lead to high swelling and eventually to a high permeability, whereas another substance B may not be soluble at all and thus its permeability may be close to zero. By measuring the single-component data and calculating the selectivity as the ratio of the permeabilities a high, so-called "ideal" selectivity will result. When bringing the binary mixture of A and B in contact with the membrane, however, no selectivity at all may be measured, as component B will now be easily passing through the membrane highly swollen by component A. This coupling of solubility, swelling and flux between the two components is usually dependent on the concentration of A and B in the mixture, and may even occur at fairly low concentrations of substances with a high swelling potential. It may be shown for polyvinyl alcohol membranes [32] that simple alcohols like methanol, ethanol, or 2-propanol are nearly insoluble in this membrane material, whereas water is easily dissolved and swells the mem-

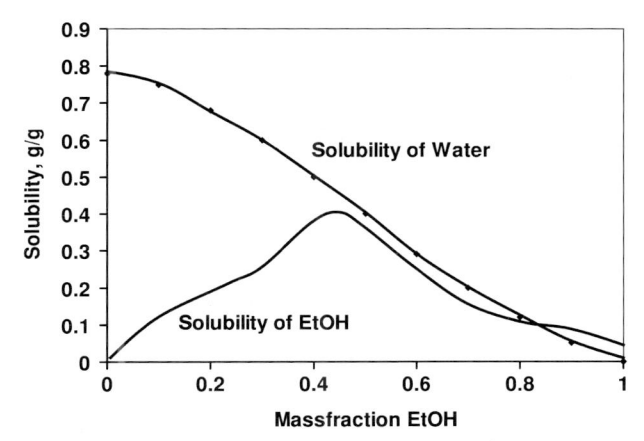

Fig. 3.2 Solubility of water and EtOH in a PVA-membrane [32].

brane. With increasing water content in the feed mixture, more and more of the alcohol is dissolved in the water swollen membrane. The solubility of the alcohol goes through a maximum and drops with decreasing alcohol content in the feed (Fig. 3.2).

Similar behavior is observed for the diffusion coefficient. Calculation of flux and selectivity for a membrane even for a simple binary mixture from single-component data therefore requires measurements of solubility and diffusion for both components over the whole range of composition and of temperature of the mixture with high accuracy. For any practical application and engineering design of a pervaporation plant such an approach is not realistic.

The design of any real pervaporation and vapor-permeation installation has thus to be based on experimental data measured in the laboratory under conditions as similar as possible to those of the subsequent full-size plant. These conditions include the flow regime of the feed mixture, the temperature and the geometry of the feed side, the composition and nature of the feed mixture, the permeate side geometry and partial vapor pressure. From the experimental data the partial transmembrane fluxes of all components of a mixture and thus the selectivity can be determined as a function of composition, temperature and permeate-side conditions for the respective mixture and geometry. In practice the permeate-side conditions (total pressure, condensation temperature) are kept as close as possible to those expected in the final plant, thus changes of these parameters do not need to be considered. Figure 3.3 depicts the partial fluxes of EtOH and water measured for a PVA-membrane.

Any suitable equation, which still may have a resemblance to a transport equation but does not need to refer to any physical model, can then be used to describe the experimental results with sufficient accuracy, preferably with a minimum of adjustable coefficients. Simple binomial functions are commonly used to calculate the flux of a component as a function of its concentration and of the operation temperature. In many applications it is even possible to reduce

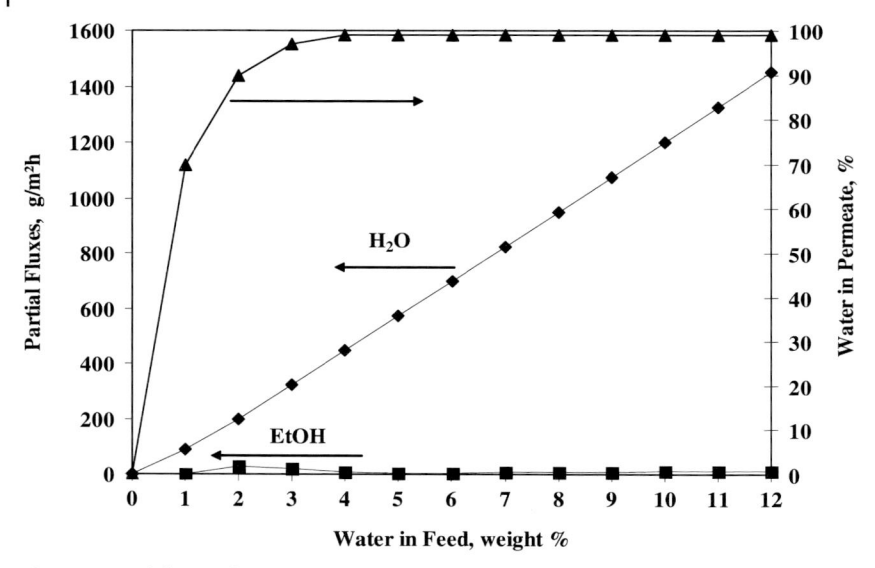

Fig. 3.3 Partial fluxes of EtOH and water, PVA-membrane.

multicomponent mixtures to binary systems by calculating the partial flux of the more-permeable component on one side and summarizing the less permeable components as one retained component on the other side. Equations of the form

$$\dot{n}''_{transported} = A \cdot x + B \cdot x^C \tag{13}$$

and

$$\dot{n}''_{retained} = D \cdot (1-x) + E \cdot (1-x) \cdot x \tag{14}$$

have proven to be quite useful to describe the dependencies of the partial fluxes of the transported and retained component for dehydration applications, respectively; where x is the mass (or molar) fraction of the more-permeable component, $\dot{n}''_{transported}$ is the partial flux of the more-permeable component (e.g. water in hydrophilic pervaporation), $\dot{n}''_{retained}$ the flux of the non- or lesser permeating component (one or several organic substances in dehydration), and A, B, C, D, E are adjustable coefficients that have to be determined from tests with each individual mixture.

The dependence of the fluxes on temperature can be described with a simple Arrhenius-type equation.

$$\dot{n}''_T = \dot{n}''_0 \cdot \exp-\left(\frac{E_A}{R} \cdot \left(\frac{1}{T} - \frac{1}{T_0}\right)\right) \tag{15}$$

or

$$\dot{n}''_T = \dot{n}''_0 \cdot \exp-\left(T_A \cdot \left(\frac{1}{T} - \frac{1}{T_0}\right)\right) \qquad (16)$$

with $\dot{n}''_T$ the flux at temperature T and $\dot{n}''_0$ that at the reference temperature T_0, R the gas constant and E_A and T_A the apparent activation enthalpy or activation temperature. In some cases, however, the apparent activation enthalpy (E_A) or apparent activation temperature (T_A) seem to be not a constant, but will depend on the concentration x as well and will have to be described by a two-parameter linear or exponential function.

All constants in these equations have to be determined experimentally. From the total flux

$$\dot{n}''_{total} = \dot{n}''_{transported} + \dot{n}''_{retained} \qquad (17)$$

and the concentration $x_{i,\,Permeate}$ of the more-permeable component in the permeate the partial flux of the more-permeable component and that of the retained component can be calculated.

$$\dot{n}''_{transported} = x_{i,transported,Permeate} \cdot \dot{n}''_{total} \qquad (18)$$

$$\dot{n}''_{retained} = \dot{n}''_{total} - \dot{n}''_{transported} = \left(1 - x_{transported,Permeate}\right) \cdot \dot{n}''_{total} \qquad (19)$$

For the determination of the four constants A, B, C, D of the flux equations (13) and (14), at least four experiments at different concentrations, but constant temperature, have to be performed. Another measurement, at an already fixed constant concentration, but a second temperature will yield the (then constant) activation enthalpy or activation temperature. When the activation temperature additionally depends on the feed concentration, two more measurements are necessary. In all these experiments other test parameters like feed velocity or permeate-side conditions should be as close as possible to those of the later real plant. By using Eqs. (13) to (15) the performance of the membrane can then be calculated with sufficient accuracy and even large industrial plants can be designed within the range of concentrations and temperatures of the experiments.

Such calculations are generally performed stepwise by separating the total membrane area into sufficiently small membrane increments, assuming constant composition and temperature for each small membrane increment. The amount of the permeate passing through the membrane and its composition are calculated as well as the loss in temperature caused by the evaporation of the permeate. The temperature and composition of the residual feed stream leaving the first increment now gives the respective values for the second increment. By means of the above Arrhenius equation the reduction of flux caused by the temperature drop for each step or membrane increment is calculated, too. The total membrane area required for a wanted separation is then obtained as the sum of all membrane increments.

A certain ratio of partial vapor pressures of the more-permeable component at the permeate and at the feed side is usually fixed and maintained in the laboratory experiment. When calculating the performance of a real plant in the above-described manner this ratio has to be kept even for the last increment of the membrane area, otherwise a transfer of the laboratory data to the full-scale plant will lead to large errors. By an additional efficiency factor corrections for any differences between the more ideal conditions in the laboratory experiment and the more realistic conditions in an industrial plant may be introduced.

More simple empirical relations may be used in the calculation of a vapor-permeation process. Due to the lower density of the vaporous feed compared to a liquid one swelling effects may be of lesser importance. Sometimes simple equations analogous to Eq. (12) may be sufficient

$$\dot{n}''_{i/VP} = R_i \cdot (p_{i,\text{Feed}} - p_{i,\text{Permeate}}) = R_i \cdot \Delta p_i \tag{20}$$

$$\dot{n}''_{k/VP} = R_k \cdot \Delta p_k \tag{21}$$

Here R_i and R_k are the permeabilities of the better permeable and the retained component and Δp_i and Δp_k differences in the respective partial vapor pressures. Both R values have to be determined experimentally and are assumed to be constants for a given feed mixture and membrane and a narrow concentration range. Otherwise the same equations (12) to (15) as for pervaporation can be used and the respective constants have to be determined by regression analysis. Calculation of any practical installation is performed analogous to the method as described above for pervaporation plants.

For small concentration changes between feed and retentate and for a first estimation of the membrane area necessary for a specified separation a simple but useful relation can be derived. It is assumed that the change in the concentration of a component removed from a certain amount of feed mixture is proportional to the applied membrane area, to the concentration of that component, and to the so-called pure component flux, and inverse proportional to the amount of the mixture

$$\frac{dx_i}{dt} = -x_i \cdot A \cdot J_0 \cdot \frac{1}{m} \tag{22}$$

Integration of this relation between the starting and final concentrations $x_{i,\text{start}}$ and $x_{i,\text{final}}$ yields for the membrane area

$$A = \frac{m}{t} \cdot \frac{1}{J_0} \cdot \ln\left(\frac{x_{i,\text{start}}}{x_{i,\text{final}}}\right) \tag{23}$$

Here,
A the membrane area,
m the amount of feed to be treated during time t,

J_0 is the pure component flux, the measured flux at a certain concentration divided by that concentration.

Equation (23) assumes constant temperature of the process, infinite selectivity of the membrane (only the removed component is passing into the permeate), no loss of matter from feed to product (constant volume of the feed, or small change from feed to product concentration) and a linear relation between the concentration of the more-permeable component in the feed and its flux through the membrane. As long as these limitations are kept in mind this relation is quite useful for a first estimation of membrane area for a given separation problem, especially in vapor permeation. For a given installation with fixed area of a membrane the influence of changes in one of the parameters like plant capacity (feed treated per unit of time), feed or product concentration on the other parameters of the plant can be estimated. Equation (23) allows, in addition, a direct scale up from a laboratory or pilot-plant test to a full-size plant, if the same membrane is used and all parameters are the same in the small and large plant. Furthermore, this equation reveals the relative small influence a small change in the starting concentration will have on the performance of a plant, and the relatively large influence a change in the final concentration will have. It can be seen that for the same ratio of starting to final concentration the same membrane area is required, and the membrane area increases exponentially with the wanted final concentration.

3.2.3
Permeate-side Conditions

In pervaporation and vapor-permeation processes the partial vapor pressures of the components at the feed side are fixed by the nature of the components, composition, and temperature of the feed, whereas the total pressure is of no influence, as long as the liquid mixture can be regarded as incompressible. Only by increasing the temperature of the liquid mixture the partial vapor pressure can be increased for a given feed mixture. Therefore the driving force for the transport of matter through the membrane is applied and maintained by reducing the partial vapor pressure at the permeate side.

The influence of the permeate side partial pressure can best be seen from a combination of Eq. (9) and Eq. (2).

$$\dot{n}_i'' = -x_{i,\text{membrane}} \cdot \frac{D_{i0,\text{membrane}}}{R \cdot T} \cdot \frac{\partial \mu_{i,\text{membrane}}}{\partial z} = -x_{i,\text{memb.}} \cdot \frac{D_{i0,\text{memb.}}}{R \cdot T} \cdot \Delta \mu_i \qquad (9)$$

$$\Delta \mu_{i/PV} = R \cdot T \cdot \ln \frac{a_{i,\text{Feed}} \cdot p_{i,\text{sat.}}}{p_{i,\text{Permeate}}} = R \cdot T \cdot \ln \frac{x_{i,\text{Feed}} \cdot \gamma_i \cdot p_{i,\text{sat.}}}{x_{i,\text{Permeate}} \cdot p_{\text{Permeate}}} \qquad (2)$$

This combination leads to:

$$\dot{n}_i'' = -x_i \cdot D_{io,\text{memb.}} \cdot \ln\left(\frac{x_{i,\text{Feed}} \cdot \gamma_i \cdot p_{i,\text{sat.}}}{x_{i,\text{Permeate}} \cdot p_{\text{Permeate}}}\right) = -x_i \cdot D_{i0,\text{memb.}} \cdot \ln\left(\frac{p_{i,\text{Feed}}}{p_{i,\text{Permeate}}}\right) \quad (24)$$

The argument under the logarithms in Eq. (24) is the ratio of the partial vapor pressure of the component in the feed over that on the permeate side. In most practical application $x_{i,\text{Feed}}$ is fairly small (the minor component has to be removed from the feed), but $x_{i,\text{Permeate}}$ will be close to unity for a high-selective membrane and the partial permeate-side pressure is thus close to the total pressure. The only means to increase the partial pressure on the feed side is an increase in the saturation pressure, hence an increase in the temperature at the feed side. At any ratio of the partial vapor pressures larger than unity the system will be operable, however, in order to assure sufficient efficacy in industrial plants, the permeate partial vapor pressure should be chosen such that this ratio is somewhere between values of seven and ten.

Different means have been proposed in order to reduce the permeate-side partial vapor pressure (Fig. 3.4):

a) All permeating vapor is removed by means of a vacuum pump. It is easily understood that such a system is applicable when the volume of permeating vapor is relatively small, or the permeate-side pressure is not too low. Otherwise vacuum pumps of extremely large capacities are required and the pumps will consume too much energy. After recompression the vapor may be condensed at the downstream side of the pump, this will always be necessary when emission control regulations have to be observed.

b) The permeated vapor is condensed at sufficiently low temperatures. This is the most cost-effective way to maintain the partial vapor pressure at the permeate side at the required low value. Condensation temperatures may be reached simply with cooling water, in some applications cooling media with

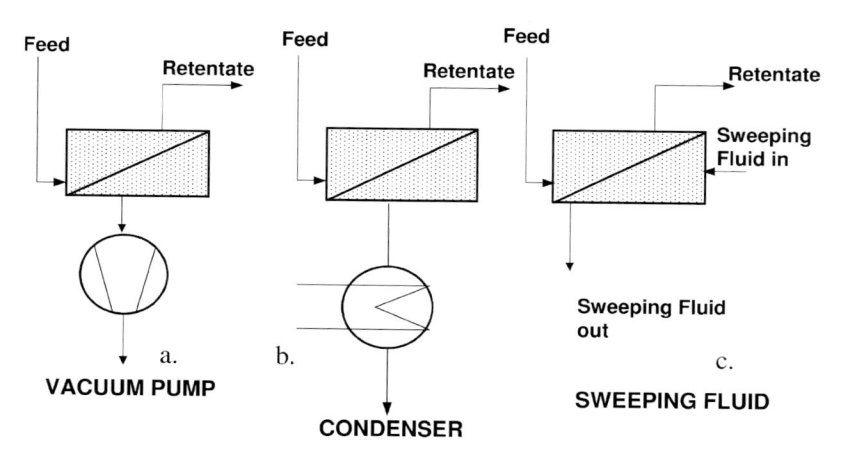

Fig. 3.4 Methods to reduce permeate-side partial vapor pressure.

temperatures as low as −20 °C are required. At these very low temperatures, however, the amount of permeate to be condensed is relatively small, and the cooling power required still economically acceptable. As the condenser surface will be installed at a certain distance from the permeate side of the membrane noncondensable gases have to be removed from the permeate compartment in order to minimize any resistance to the transport of the permeate vapor to the condenser surface and any pressure losses. However, especially when low final concentrations in the feed have to be reached not only will the more-permeable component pass through the membrane, but also an increasing amount of the originally retained component. Then, calculations for the design of the condenser have to be performed considering partial condensation along the dew line of an at least two-component mixture. Depending on the condensation temperature and the composition of the permeate freezing of the permeate may occur and has to be avoided. An additional condenser at the exhaust of the vacuum pump will control any emission of residual permeate vapors.

c) The permeate side of the membrane is swept with an inert gas in which the partial vapor pressure of the critical (preferential permeating) component is kept sufficiently lower than that on the feed side. This procedure is often discussed in the academic literature but not yet really introduced into practical application (with the exemption of air drying by means of membranes where part of the produced dry air is used as a sweeping gas). In all other applications the inert sweeping gas stream has to be preconditioned and, when laden with the permeated vapor, can usually not be wasted but has to be reconditioned and recycled. Reconditioning is generally done by condensing the permeated vapor out of the sweep gas stream at a sufficiently low temperature, followed by reheating in order to reduce the relative humidity and increase the capacity of the sweep gas stream. If a low partial vapor pressure has to be reached, the relative capacity of the inert gas stream will always be low, and large gas volumes have to be recycled and conditioned. This is uneconomically compared to direct condensation. When normal composite membranes with a porous substructure are used only diffusive transport from the permeate side of the separating layer through the pores will occur, forming an additional transport resistance and reducing the total flux through the membrane.

In nearly all industrial pervaporation and vapor-permeation installations the permeate is therefore directly condensed under vacuum. Depending on the nature of the organic components in the feed, which partially pass through the membrane, together with the preferential permeating component, depending on the final concentration to be reached in the product, and depending on the selectivity of the membrane condensation temperatures for large installations may vary between approximately +10 to −20 °C. When the required condensation temperature drops below the value of −20 °C recompression in a two-stage vacuum system with intermediate condensation at more acceptable pressures and temperatures can be an option.

Calculation of the required condenser surface is not trivial. In contrast to the common applications where saturated vapors are condensed the permeate is a superheated vapor mixture. For design calculations the selection of appropriate heat-transfer coefficients has to consider the cooling to saturation conditions, the presence of noncondensable gases, and the partial condensation of the components along the respective dew lines. Total condensation of the more volatile components of the permeate vapor will often not be possible, but any losses of permeate vapor through the vacuum pump have to cope with the respective emission control regulations. An important factor is the solubility of the components of the permeate in the liquid phase. An additional condenser at the high-pressure side of the vacuum pump is a feasible option.

From an analysis of Eq. (24) it follows that the operation parameters of an installation are closely related to the selectivity of the membrane. The latter has to be high at high concentrations, and thus at high partial pressure of the component to be removed from the feed because then only moderate permeate side partial pressures and condensation temperatures are required. When very low final concentrations (e.g. below 1000 ppm) have to be reached in the retentate high selectivity of the membrane is no longer optimal. At lower selectivity the partial vapor pressure of the critical component in the permeate vapor is reduced even at higher total pressures when a portion of the otherwise retained component passes through the membrane. Even if the permeate then contains more of the retained component than of the transported one the absolute amount of the retained component lost in the permeate is sufficiently small and high recovery ratios of the wanted, highly purified component will be obtained. With membranes that allow for an increase of the concentration of the retained component in the permeate with decreasing concentration of that component in the feed any final purity of the product can be obtained, e.g. water concentrations in ethanol as low as 10 ppm.

3.2.4
Transport Resistances

In the above calculations and considerations diffusion through the nonporous layer of the membrane was assumed to be the rate-determining process and thus the only transport resistance. In every membrane process, however, additional transport steps at the feed occur, usually summarized as "polarization". By the preferential transport of one component out of a mixture through the membrane the fluid layer directly adjacent to the membrane surface will be depleted of that component, and its concentration will be lower than that in the bulk of the feed mixture. This (unknown) lower concentration determines the sorption and thus the effective activity and partial vapor pressure of the component directly at the feed side of the membrane. The flux reduction caused by the additional resistance for the transport of matter by diffusion through the liquid layer adjacent to the feed side of the membrane is known as "concentration polarization", and effective in all membrane processes. Due to the phase change

Fig. 3.5 Polarization effects.

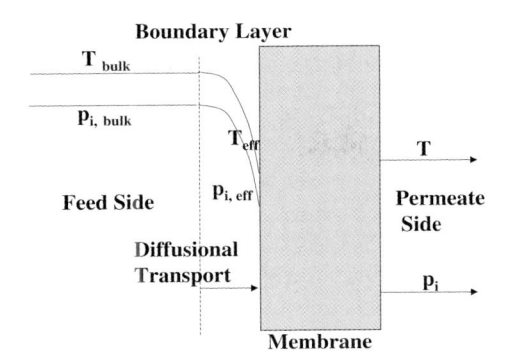

Diagram description (labels only): Boundary Layer; T_{bulk}; $p_{i,bulk}$; T_{eff}; $p_{i,eff}$; Feed Side; Diffusional Transport; Membrane; Permeate Side; T; p_i.

from the feed to permeate side a pervaporation membrane is a heat sink and a temperature difference exists between the bulk of the feed and the feed side of the membrane (Fig. 3.5). The resulting reduction in driving force and hence in performance is referred to as "temperature polarization" and this effect is unique for pervaporation processes. The saturation pressure of a component at the feed side of the membrane is determined not by the temperature of the bulk of the liquid but by the temperature directly at the feed side of the membrane, and the heat for the evaporation of the permeate has to be transported through the boundary layer and the membrane by heat conductivity.

Both flux reductions are coupled via the heat of evaporation. This is indicated in Fig. 3.7 that flux changes linearly on a change in concentration, but exponentially when the temperature is changed. This means that temperature polarization is the more important one and the development of high flux pervaporation membranes requires the development of modules in which temperature polarization is effectively reduced. The effect of temperature polarization is more important at high fluxes, thus for a given process it needs more attention at high concentrations of the component to be removed. At very low concentrations and thus fluxes the effect of temperature polarization is of less importance, whereas concentration polarization does not change significantly with changes of the composition of the feed. Like in other membrane processes polarization effects can be reduced in pervaporation also by a turbulent flow of the liquid on the feed side of the membrane. In industrial application costs for such a flow regime have to be balanced against loss in performance and higher costs for larger membrane areas. Polarization effects are the more pronounced the higher the flux through the membrane, and reductions of the obtainable flux to values of below 50% of the theoretical value can occur. Development of high-flux, high-cost membranes may thus eventually not be the most desirable option, but that of low-cost, low-flux membranes an overall better choice.

Another transport resistance does exist at the permeate side for the desorption step. For real membranes the pressure at the permeate side of the separation layer cannot be measured directly, as these membranes have a porous support and the pressure loss in the pores depends on pressure and volume flow of the permeate and thus on the operation parameters of the process. Knudsen diffusion is generally assumed for the flow in the pores of the substructure, and tortuosity factors are introduced in order to calculate the effective pore length from the thickness of the porous membrane. Knudsen flow may increase or reduce the effective selectivity of the separation layer, depending on the nature of the more-permeable component. Usually this effect is not measured separately but included in the overall experimental performance of the membrane. Any equation and calculation derived from a physical-chemical model that includes polarization effects and the effect of the porous substructure needs a large number of adjustable coefficients that have to be determined experimentally for the respective feed mixture, membrane, and operation conditions. Such a model would be very complicated, and still of insufficient accuracy.

Additional pressure losses caused by hydrodynamic resistances in the permeate pass from the permeate side of the membrane to the condenser or the vacuum pump will be even more detrimental to the performance of the pervaporation process. When an alcohol–water mixture has to be dehydrated to a final water content of 1000 ppm even at 100 °C the partial water vapor pressure at the feed side will be of the order of 10 mbar. Using a high-selective membrane the partial water vapor pressure at the permeate side of the membrane will have to be kept at a few millibar. As this pressure is determined by the temperature of the condensing liquid permeate there has to be an unobstructed flow of the permeate vapor from the membrane to the condenser. It is obvious from Eq. (24) that even a pressure drop of one or two millibar in the permeate channel of a module will have a severe effect on the ratio of the partial vapor pressure and thus on the performance of the system.

Fortunately another transport resistance, which is extremely important in the filtration processes and in reverse osmosis, namely fouling, is of no concern in pervaporation or vapor permeation with polymeric membranes. In these membranes no pores exist that can be blocked by any precipitation out of the liquid or vapor phase. Even if precipitation, e.g. of salts in dehydration processes, does occur the growth of the salts crystals may attack and eventually destroy the separating layer of the membrane, but will usually not influence the flux of water to the membrane.

3.2.5
Principles of Pervaporation

In Fig. 3.6 a principal scheme of a pervaporation process is shown. The liquid feed mixture is heated to the highest temperature compatible with its own stability, the stability of the membrane and all other parts (e.g. gaskets, module elements) in the system. All partial vapor pressures are at saturation and fixed by the temperature and composition of the liquid mixture, and by the nature of the components. On the permeate side all noncondensable gases are removed by means of a vacuum pump, and the permeated vapors are condensed at a sufficiently low temperature in order to maintain a sufficiently low vapor pressure

3.2 Principles and Calculations | 169

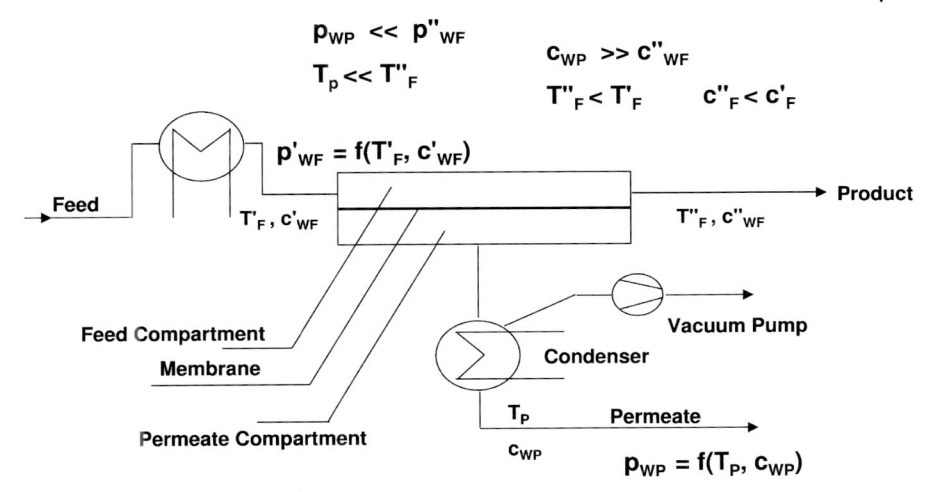

Fig. 3.6 Pervaporation, principle.

[Figure 3.6: Schematic of pervaporation process showing Feed entering a heat exchanger then the membrane module with Feed Compartment, Membrane, and Permeate Compartment. Labels: $p_{WP} \ll p''_{WF}$, $T_p \ll T''_F$, $p'_{WF} = f(T'_F, c'_{WF})$, T'_F, c'_{WF} at inlet; $c_{WP} \gg c''_{WF}$, $T''_F < T'_F$, $c''_F < c'_F$, T''_F, c''_{WF} at outlet going to Product. Permeate side shows Condenser at T_P, c_{WP}, Vacuum Pump, Permeate outlet, and $p_{WP} = f(T_P, c_{WP})$.]

at the permeate side. As the liquid feed mixture flows over the membrane and the more-permeable component is removed and its concentration lowered, the heat for the evaporation of the permeate passes through the membrane, too. The only source for this evaporation enthalpy is the sensible heat of the liquid. Thus a drop in concentration and in temperature occurs between the entrance of the feed on the membrane and its exit.

This drop in temperature has several consequences. The partial vapor pressure of the critical (more-permeating) component is decreased not only by the reduction of its concentration, but also by the reduction in temperature. Furthermore, as the diffusional transport through the membrane is temperature dependent an additional reduction of the transmembrane flux results. Whereas the flux drops approximately linearly with the reduction in concentration and the concentration drop is unavoidable (it is the goal of the process), the flux is reduced exponentially with the decrease in temperature (Fig. 3.7). When the lost heat is not replaced the flux will soon drop to unacceptable levels.

Different means have been proposed to overcome the effects of this heat loss: Direct heating of the membrane from the permeate side by steam, electrical heating of the membrane support, or direct heating of the liquid flowing over the membrane by an additional heat exchanger are some examples. For practical applications only an arrangement as shown in Fig. 3.8 has proven to be useful. The total membrane area required for a specific separation is split into several so-called stages that are arranged in series, with an intermediate heat exchanger between each two stages.

After passing over the membrane of the first stage, the lost heat is replaced in the intermediate heat exchanger (Fig. 3.8) before the feed gets in contact with the membrane area of the next stage. The total number of stages and size of each of the stages, and the tolerated temperature drop per stage are matters of

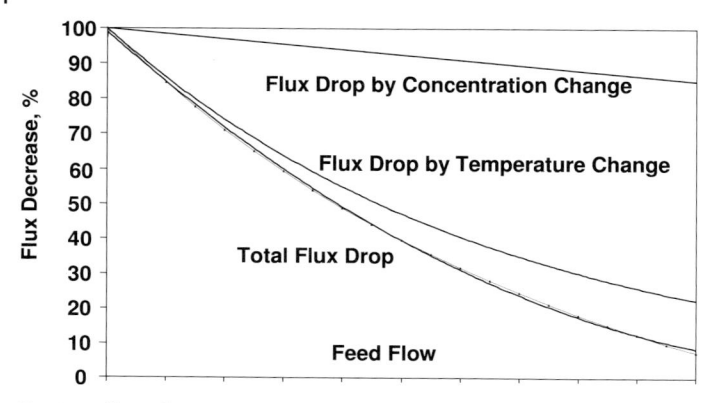

Fig. 3.7 Effect of concentration and temperature polarization.

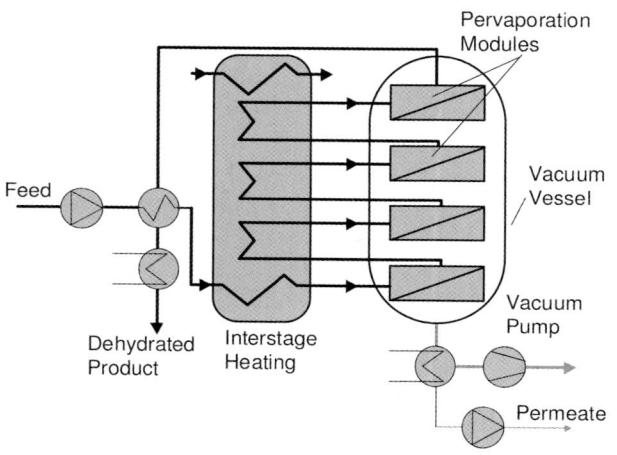

Fig. 3.8 Continuous pervaporation plant.

optimization for the respective application and plant. Usually all stages will have the same membrane area and the size of the intermediate heat exchangers are adapted to the temperature loss per stage, but constant sizes of heat exchangers and adapted sizes of the stages have been verified as well. By a pressure-control valve the liquid feed is kept under pressure when the operation temperature is higher than the atmospheric boiling temperature.

The real flux in an installation according to Fig. 3.8 will thus follow a complex pattern, as can be seen from Fig. 3.9. In each membrane stage the flux drops exponentially, due to the change in concentration and loss of heat. After reheating in the intermediate heat exchanger, and before entering into the next stage, the original temperature will be restored, but the flux has dropped to a lower value than that at the inlet to the first stage, as the concentration of the more-permeable substance has been reduced.

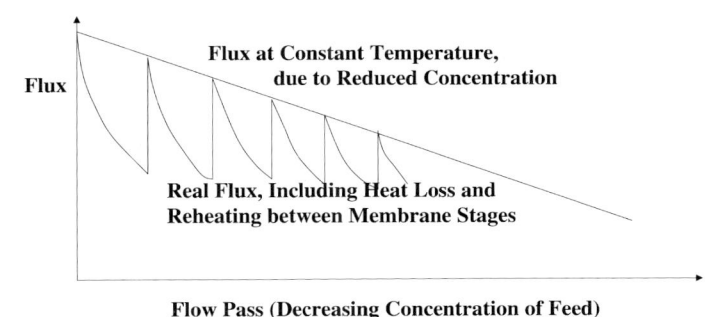

Fig. 3.9 Isothermal and real flux in pervaporation.

For large plants with large membrane areas optimization will lead to a larger number of stages in order to minimize the overall membrane area. Hydraulic pressure losses caused by too many stages and heat exchangers in series may be an important factor in the design of a large pervaporation plant. If, on the other hand the plant capacity is small one would rather waste some membrane area and allow a higher temperature drop per stage, and thus reduce the number of stages, heat exchangers and the respective piping.

It is evident that the arrangement as shown in Fig. 3.8 reduces the energy consumption of the pervaporation process to a minimum value. Only the heat required for the evaporation of the permeate has to be supplied and is lost in the process, the sensible heat of the product can be recovered to any extent, limited only by the costs of the interchanger.

3.2.6
Principles of Vapor Permeation

Vapor permeation differs from pervaporation, as stated above, insofar as the feed mixture to be separated is supplied as a vapor. At least the more-permeable component is kept as close to saturation conditions as possible. Thermodynamically there is no difference between a liquid and it's equilibrium vapor, the partial vapor pressure and thus the driving force for the transport through the membrane are identical and the same "solution-diffusion mechanism" is valid. However, the density of the vaporous feed and thus the concentration of molecules per volume is lower by two to three orders of magnitude than that of the liquid. As a consequence the membrane is usually less swollen than when in contact with a liquid feed. As the feed mixture getting in contact with the membrane is already in the vapor phase no phase change occurs across the membrane and thus no temperature polarization will be observed. Concentration polarization, however, is still an issue. Although the diffusion coefficient is much higher in a vapor than in a liquid, this is at least partially outbalanced by the lower density of the vapor, and therefore concentration polarization effects may be observed at all concentrations of the component to be removed. Minimum flow rates of the feed over the membrane have to be observed, especially when large volume changes from feed to retentate occur.

Today vapor-permeation processes are widely used in the dehydration of organic solvents, or in the removal of methanol from other organic components, or in the removal of VOCs from gas streams. In the literature the term "vapor permeation" is often related to the removal of organic vapors ("VOCs") from air or gas streams only. In these applications the more-permeable component is brought close to saturation by cooling, compression, or both pretreatment steps. Thus there is no real reason for such a narrow definition and the means by which the vapor has been produced has no influence either on the nature of the membrane or the mechanism of the separation process.

A principle scheme of a vapor-permeation plant is shown in Fig. 3.10. The liquid feed mixture to be separated is preheated and totally evaporated; the saturated vapor is fed to the membrane system. The whole membrane area is arranged in one stage, and will, in general, operate at nearly the same pressure. Intermediate heat exchanges with the respective temperature controls and the interconnecting piping are no longer required. By means of a pressure controller the vapor is kept under constant pressure. Recovery of the heat of evaporation from the product is possible in principle, but usually not economical, except that part required for preheating the liquid feed to the boiling temperature.

The evaporator may be part of the plant, in many applications the saturated vapor comes from the top of an upstream distillation column. Thus a vapor permeation step may be coupled with one or more distillation columns in a so-called hybrid system.

As can be seen from comparison of Figs. 3.8 and 3.10 the permeate-side arrangement remains unchanged and the same means for maintaining a sufficiently low partial pressure at the permeate side are used.

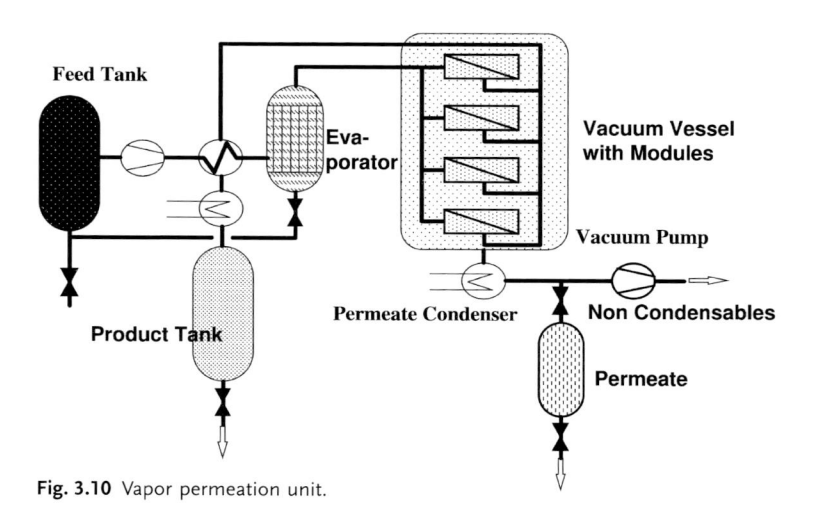

Fig. 3.10 Vapor permeation unit.

Superheating of the vapor should be strictly avoided. When a vapor is superheated at constant pressure the partial pressures of its components are not increased. At low degrees of superheating the vapor in contact with the membrane will behave as if it were supplied at the respective lower saturation temperature. Larger degrees of superheating will result in a drop of the performance of the membrane even below the values observed at the equivalent saturation temperature, as the density of the vapor and the activity coefficients (or fugacity coefficients) of the components drop.

When saturated vapor is fed to the membrane it is unavoidable that a small portion of the vapor will condense on the membrane. This will happen during start up of the installation when the first vapor reaches the still-cold membrane and module, and during shutdown, respectively. Additionally, some heat will always be lost to the outside of the modules during operation.

Besides these heat losses caused more by the physical arrangement of the installation, there are two more, caused by the laws of thermodynamics:

1. The vapor is expanded from the high pressure at the feed side to the low pressure at the permeate side. This will cause a Joule-Thompson effect, which in general will lead to a slight temperature drop from the feed to the permeate side and cool the membrane. Although this effect will reduce the temperature by only one to three degrees centigrade for most of the mixtures treated in practical application, it will lead to condensation of a small part of the vaporous feed.

2. In most applications the composition of the feed mixture will be at or close to a minimum boiling-point azeotrope. By removing one of the components from the mixture (e.g. water in a dehydration process) the boiling point of the mixture will increase as shown in Fig. 3.11 for the system n-propanol–

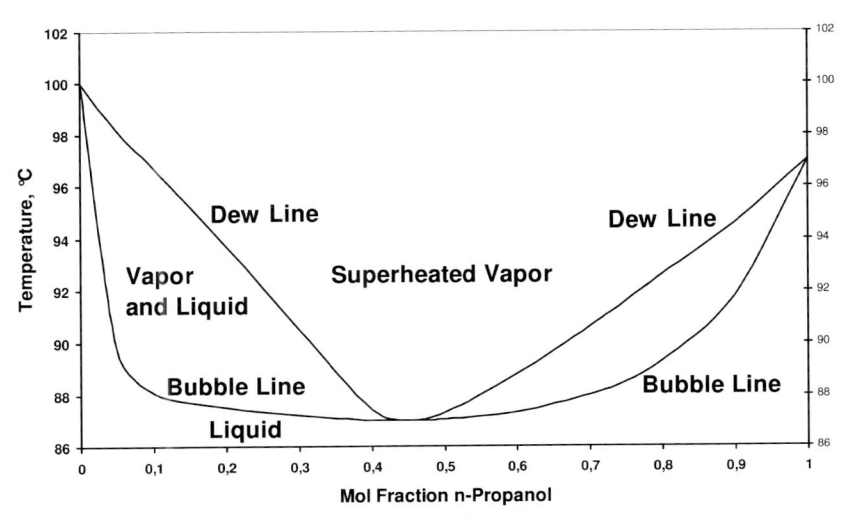

Fig. 3.11 Dew and bubble line of water–n-propanol at constant pressure.

water at atmospheric pressure. When water is removed from the vaporous mixture at azeotropic concentration the boiling temperature increases and at constant pressure (at which a vapor-permeation plant will be operated) the mixture is moved into a region of oversaturation, where only liquid can exist. Consequently, part of the vapor has to condense and liquid and vapor will exist in equilibrium. The heat freed by the partial condensation will increase the temperature of the system to the new equilibrium value. The respective compositions of vapor and liquid are given by the horizontal connection between the dew point and the bubble point at that temperature. Therefore, if a mixture is fed at a composition equivalent to a minimum boiling point azeotrope to a vapor-permeation plant the vapor will increase its temperature when passing over the membrane, and leave the system eventually at the (higher) boiling temperature of the pure organic component. This effect is negligible in the dehydration of azeotropic ethanol where the difference between the boiling point of the azeotrope and that of the pure alcohol is only a few tenths of a degree. In the example of the system water–n-propanol of Fig. 3.11 this temperature increase can be as high as 10 °C, and the effect is even higher for other systems. The rise in temperature between the inlet to the membranes and the outlet can easily be used to monitor the performance of such vapor-permeation plant.

As usually identical membranes are employed for both liquid and vaporous feed mixtures the partial condensation of vapor on the membrane in vapor permeation will have no detrimental effect on the performance of either the membrane or the process. It can even be proven that the highest performance can be obtained with the same membrane, when a mixture of liquid and vapor is directly used as a feed [33]. Condensation of the vaporous portion supplies the heat necessary for the evaporation of the permeate and thus temperature polarization is avoided. As the volume of the vapor phase will exceed that of the liquid a strong mixing effect will occur at the membrane surface, reducing concentration polarization too.

The choice whether to apply the liquid feed mixture directly to the membrane in a pervaporation process or pre-evaporate it and feed a vapor to the membrane, depends mainly on specific site conditions.

Vapor permeation is preferred when:
- The feed is already available as the saturated vapor phase,
  e.g. from a distillation column at the specified temperature and pressure (e.g. 95 to 105 °C).
- Dissolved or undissolved solids are present in the original feed
  (e.g. the feed is a mother liquor), and an additional
  purification step by evaporation has to be performed anyway.
- oncentration change has been specified that otherwise would
  request too many small stages and reheating.
- The additional heat consumption of the plant is not an issue.

Vapor permeation offers the advantage of:
- Simple plant arrangement, the total membrane area is installed in one stage, no need for intermediate heat exchangers, the interconnecting piping, and controls.
- No heat polarization occurs as the evaporation enthalpy has already been supplied to the feed.
- The total membrane area operates at a higher mean temperature, and less membrane area is required.
- Polluted feed streams, containing impurities, can be processed in one plant.

3.3
Membranes

The development of membranes for pervaporation and vapor permeation was highly influenced by the development of desalination and gas-separation membranes and the theoretical knowledge of membrane structure and transport through membranes gained thereby. In the first tests Binning and coworkers employed polymeric films of cellulose esters or polyolefines that were stretched or otherwise post treated. In the late 1970s and early 1980s membranes were investigated originally for reverse osmosis applications. It was, however, fairly quickly understood that, despite the similarities, both processes require different membranes. In both processes the transport through the membrane is a diffusional one, the dense layer responsible for the separation has to be as thin as possible. In desalination, high pressures are applied, up to 100 bar at ambient temperatures, whereas in pervaporation pressure differences across the membrane are in the range of a few bar only. On the other hand, pervaporation membranes have to be stable against aggressive organic components at temperatures of 100 °C and above. In reverse osmosis both sides of the membrane are in contact with a liquid phase and the degree of swelling between the two sides does not differ too much. In pervaporation the feed side of the membrane is highly swollen in contact with the hot liquid (or saturated vapor), whereas the permeate side is "dry" and virtually nonswollen. A high gradient of swelling thus exists over the separating layer of the membrane, demanding additional resistance and stability. It is thus not surprising that specific membranes, made from different materials had to be developed, although the general structures of pervaporation membranes and those for reverse osmosis are very similar.

Two different types of pervaporation membranes based on polymeric materials were developed at about the same time in the beginning of the 1980s:
- Hydrophilic membranes, with a preferential permeation for water, utilized mainly for the removal of water from organic solvents and solvent mixtures, with an emphasis on azeotropic mixtures.
Membranes for the removal of small alcohol molecules like methanol and/or ethanol are of hydrophilic nature as well.

- Organophilic membranes with a preferential permeation for nonpolar compounds utilized for the removal of volatile organic components from aqueous and gas streams.

In both applications a composite membrane structure (Fig. 3.12) is preferred, allowing for very thin defect-free separation layers, but with sufficient chemical, mechanical, and thermal stability. Due to the composite structure flat-sheet configurations are preferred, too. The substructure of both types of flat-sheet pervaporation membranes is very similar: A porous support membrane with an asymmetric pore structure is laid onto a carrier layer of a woven or nonwoven textile fabric and an ultrafiltration membrane is formed. On the free side of this asymmetric porous substructure the pores have diameters of the order of 20 to 50 nanometers that widen up to the fabric side to the micrometer range. Polyester, polyethylene, polypropylene, polyphenylene sulfide, polytetrafluor ethylene, and similar fibers are used for the textile carrier layer. Structural polymers with high resistance against chemical attack and good thermal and mechanical properties, like polyacrylonitrile, polyetherimide, polysulfone, polyethersulfone, and polyvinylidenefluoride, form the porous support. All these structural polymers already have a certain intrinsic separation characteristic; they generally exhibit higher permeability for polar substances like water than for nonpolar organics.

On this substructure a thin dense layer (in the range of 0.5 to 10 μm thick) is coated that has a very high separation capability. Different coating techniques are in use, most commonly a solution of the respective polymer in an appropriate solvent is spread onto the porous substructure. The solvent is evaporated, followed by further treatment to effect crosslinking of the polymer. Photosensitive, solvent-free prepolymers may be used for coatings that are later crosslinked by irradiation, e.g. with UV-light or electrons.

The dense defect-free separating layer of hydrophilic membranes is made from different polymers that have a high affinity towards water. These polymers contain ions, oxygen functions like hydroxyl-, ester-, ether-, or carboxylic moieties, or nitrogen as imino- or imido-groups. They must be crosslinked in order to render them insoluble after the coating process. Preferred hydrophilic poly-

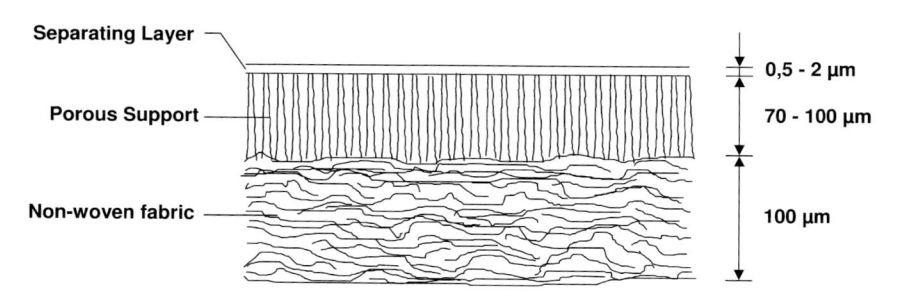

Fig. 3.12 Cross section of a composite membrane.

mers are polyvinylalcohol (PVA) [12], polyimides, natural polymers like chitosan blended with other polymers [34], or cellulose acetate (CA), or alginates, which are crosslinked by various chemical reactions. Other techniques are the deposition of thin layers from a vapor by means of a cold plasma where at least one of the gaseous components contains the above mentioned groups [35]. Ion-exchange polymers have been used as well, either with sulfonic or carboxylic acid groups, the latter mainly in the salt form with an alkali ion as the counterion [36]. Polyelectrolytes formed by blending and internal neutralization of an anion and a cation-exchange polymer are reported in literature, too [37].

Organophilic membranes have the same structure as hydrophilic ones. The dense separating layer is formed by crosslinked silicones, mostly polydimethyl siloxane (PDMS) or polymethyl octyl siloxane (POMS). The methods to apply the dense layer on the porous substructure are similar to those used for hydrophilic membranes.

If instead of flat sheets tubular membranes are manufactured, it is very difficult to coat a dense, defect-free, but very thin layer on such a structure. A lot of effort has been devoted to the development of capillary or hollow fiber membranes, however, so far only organophilic ones are available with a composite structure of a porous support and a dense layer of a siloxane, either on the inside or the outside of the capillary. For organophilic membranes the thickness of the dense separating layer of a siloxane does not need to be controlled within a narrow range, layers of 20 to 100 µm will still provide good fluxes. Hydrophilic hollow fiber membranes are produced as integral structures with dense symmetric walls, without any porous substructure. As these walls are comparably thick, of the order of several ten micrometer, specific fluxes of the membranes are very low, compared to flat-sheet membranes and they do not provide an economical advantage.

In recent years new efforts have been made in academia and industry to develop new membranes for organic–organic separation. So far the only industrial processes in this area are the separation of the light alcohols methanol and ethanol from their mixtures with hydrocarbons, ethers, and esters. The membranes in use are, however, still of the hydrophilic type, in which the more polar small alcohols replace the water [16, 38].

Although a lot of effort has been spent in the development of membranes for the separation of mixtures of nonpolar organic components no large-scale application has yet been reached. Of specific interest is the separation of olefins from paraffins, e.g. propene from propane, aromatics like benzene or toluene from aliphatic hydrocarbons or the separation of the xylene isomers. A number of different membranes are reported in the patent literature [27]. The first pilot plants are being operated and results reported for the separation of sulfur-containing aromatics from gasoline [28], or for the separation of benzene from a mixture of saturated hydrocarbons [29].

In most pervaporation and vapor-permeation processes polymeric membranes are employed. Thermal, mechanical, and chemical stability of the porous substructure as well as of the textile fabric are the main limiting factors for the operation range of this type of membrane, more than the stability of the separating layer. Demand for higher operation temperatures and chemical resistance have stimulated the development of inorganic substructures, preferentially porous ceramics. These can be coated by crosslinked polymeric separating layers similar to those on polymeric substructures; however, the chemical stability of the organic component will limit stability and application.

In more recent developments inorganic separation layers have been developed, either by coating the porous substructure with a layer of zeolites [39], or by reducing the size of the pores to molecular dimensions by deposition of amorphous silica [40].

Zeolites are aluminum silicates with a broad range of the aluminum to silicon ratio. They form crystalline structures with well-defined pores in the range of several Ångstrom. At high aluminum to silicon ratio the crystal and especially the inner lumen of the pore is hydrophilic with a preferential sorption of water inside the pores. At low aluminum content the zeolites are organophilic with preferential sorption of organics inside of the pores.

Again a composite structure has been chosen comprising a porous support, made from ceramics or stainless steel. Usually the pores of the support are too large for a direct coating, thus one or several intermediate layers, mostly of α- or γ-alumina are applied reducing the pore diameter to a range of 10 to 50 nanometers. Zeolite crystals are deposited in a random orientation with a large number of defects between the crystals. Therefore several layers are coated on the porous substructure until all defects between the crystals in one layer are covered by crystals of another layer. Separation through these membranes is effected by adsorption of the more-permeable component (water) inside and transport through very small pores, the size of which is of the order of the size of a water molecule (three to four Å). Zeolites as the effective moiety in the separating layer offer the advantage of uniform pore size, as the diameter of the pore inside a zeolite crystal is fixed by the nature of the zeolite. In particular, NaA-type zeolites with a high ratio of Al/Si are extremely hydrophilic and the pores of the crystals are accessible for water molecules only. High selectivity and high fluxes are reported for hydrophilic zeolite membranes, and they are expected to be stable at temperatures of at least 150 °C. The more hydrophilic the zeolite, however, the higher is the sensitivity against acidic conditions. In particular, NaA-type zeolites are immediately destroyed when they come into contact with acids. More acid-stable zeolites with a lower Al/Si ratio are less hydrophilic, thus the selectivity and the flux of the respective membrane is substantially lower when used in dehydration applications. The first pilot modules are available and the first industrial plants comprising these membranes are in operation.

Coatings of amorphous silica can be applied by a sol-gel technique [41] or through interfacial precipitation. Again, intermediate layers of α- and γ-aluminum oxide are applied to the coarse porous support in order to reduce the pore size to values of a few nanometers before the amorphous silica layer is applied. As the surface of the amorphous silica contains hydroxyl groups, the separating layer is highly hydrophilic, too. Amorphous silica is reported to be stable against acid conditions. It is, however, difficult to obtain a defect-free layer of uniform pore size by a single coating, therefore several coatings have to be applied to this type of membrane as well. Selectivity and flux of silica membrane are comparable to that of a zeolite membrane. However, the long-term stability of the amorphous silica layer is still questionable. It seems that a reformation of the amorphous layer occurs that leads to a flux reduction and eventually loss of selectivity. Replacement of part of the hydroxyl groups by methyl groups does improve the stability. The mechanism of the deterioration of the amorphous silica is not yet fully understood. Other amorphous oxides like those of titanium or zirconium or of rare earth metals may lead to more stable membranes.

The separation mechanism of inorganic membranes is even more complex than that of polymeric separating layers. Compared with polymeric pervaporation membranes these hydrophilic inorganic ones are not dense, but porous. Molecular-sieving effects, caused by shape and size of molecules, and shape and size of the pores determine the separation. The surface of the membrane and the inside of the pore walls are highly hydrophilic, so preferential sorption of water on the membrane and inside the pores and surface diffusion in the adsorbed layer play an additional very important role.

Organophilic inorganic membranes can be prepared as well. Zeolites with a very low Al/Si ratio are no longer hydrophilic but adsorb organic molecules inside the pores. Respective membranes have been tested in the laboratory. A potential application is expected for the removal of methanol and ethanol from larger organic molecules like ethers and esters. When the hydroxyl groups of amorphous silica are replaced to a large extent by methyl or ethyl or even larger groups the material loses its hydrophilicity, too. The groups can be introduced with the starting material or by grafting the respective functional moieties to the free hydroxyl groups at the surface of a hydrophilic membrane. Potential applications are similar to those expected for organophilic zeolite membranes.

Inorganic membranes are so far mostly manufactured as tubes, with the separating layer on the inside or outside surface of the tube. The supports are resistant against temperatures up to 250 °C, and against all neutral organic solvents, especially aprotic ones, like DMF, NMP, DMSO. The inorganic layer does not swell; they will be therefore less sensitive to fast concentration and temperature changes than the polymeric membranes. As long as the presence of any acid can be avoided the zeolite membranes promise to be an alternative to the best polymeric membranes. The long-term stability of amorphous oxide membranes (silica or other oxide) has to be improved and the loss in flux of amorphous silica membranes needs to be understood.

Up to today inorganic membranes are far more expensive than polymeric ones. This is due to the higher cost of the substructure, a sintered ceramic or stainless steel tube, and to the multilayer coating procedure, usually requiring a high-temperature heat treatment between two coating steps. Module assembly with connections between ceramic tubes and the stainless steel of the other module components is complicated and expensive, too. At least partially these higher costs can be outbalanced by the higher flux of inorganic membranes, compared to polymeric ones, especially when operation at higher temperatures becomes possible. It is therefore assumed that with further improvement inorganic pervaporation membranes will find their respective areas of applications.

Zeolites are widely used in adsorption processes due to their high sorption selectivity. When crystals of such a zeolite are embedded in a film made from a polymer that shows a sufficiently high permeability, but not necessarily a high selectivity towards the components of a mixture to be separated, a so-called "mixed-matrix" membrane is formed. The separation characteristic of such a membrane is governed by the preferential sorption of the zeolite [21]. Zeolites with a broad variety of sorption selectivity are available that form potential candidates for the development of mixed matrix membranes especially for the separation of organics. One of the main challenges in the development of such membranes is the choice of the matrix polymer. It has to fully wet the crystals in order to reduce the passage between them, but must not block the entrance to the pores. Permeabilities of the polymer and the zeolite have to match in a narrow range.

3.3.1
Characterization of Membranes

The performance of a membrane is in general characterized by its flux and its selectivity. For practical reasons fluxes for pervaporation membranes are just given in either kg/m² h or in mol/m² h, either as total flux of all components or separated into the partial fluxes of the different components. For comparison of different membranes very often the so-called "pure-component flux" at constant temperature (J_0 in Eqs. (22) and (23)) is calculated by dividing the actual flux of a component by its concentration in the feed.

$$\dot{m}_i''^0 = \frac{\dot{m}_{i,c}''}{x_{i,\text{Feed}}} \tag{25}$$

Although this value can be calculated for all components it is most widely used to characterize hydrophilic membranes by their "pure water flux".

$$\dot{m}_{H_2O}''^0 = \frac{\dot{m}_{H_2O,c}''}{x_{H_2O,\text{Feed}}} \tag{26}$$

The respective "pure water flux" is then depending on the temperature according to Eq. (15) or Eq. (16).

The selectivity of a pervaporation membrane is defined in different ways. Most commonly found in literature is the so-called α-value. This is calculated as the ratio of the more-permeable component (e.g. water) to the less-permeable component (e.g. organic) in the permeate divided by the respective ratio in the feed.

$$a = \frac{\left(\frac{x_{water}}{x_{organic}}\right)_{Permeate}}{\left(\frac{x_{water}}{x_{organic}}\right)_{Feed}} \quad (27)$$

Although the a-value looks fairly simple it is not very informative. For most membranes, especially for dehydration membranes the composition of the permeate is constant over a very broad range of feed compositions (e.g. Fig. 3.13). As a consequence the a-value is not a constant but varies considerably with the feed composition, and cannot really be used for the comparison of two membranes.

A second way to characterize membrane selectivity is the so-called β-value or enrichment factor. This is simply the concentration of the more-permeable component (e.g. water) in the permeate divided by that in the feed.

$$\beta = \frac{x_{water, Permeate}}{x_{water, Feed}} \quad (28)$$

Combination of the a and β values leads to

$$a = \beta \cdot \frac{x_{organic, Feed}}{x_{organic, Permeate}} = \frac{(1 - x_{organic})_{Feed}}{(1 - x_{water})_{Permeate}} \quad (29)$$

when a two-component mixture is assumed. Again, this numerical value is informative for only one feed concentration and not very useful if different membranes have to be compared.

Therefore it has become quite common to drop numerical values of the selectivity. Instead, the composition of the permeate is plotted in a diagram over the

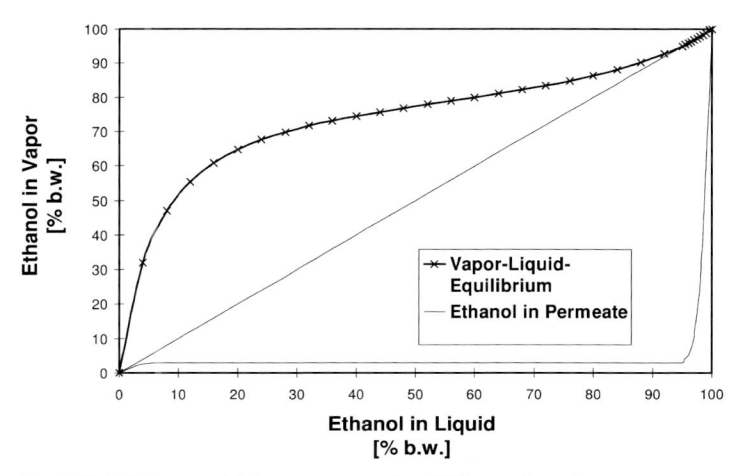

Fig. 3.13 EtOH–water; VLE vs. pervaporation (PVA-membrane).

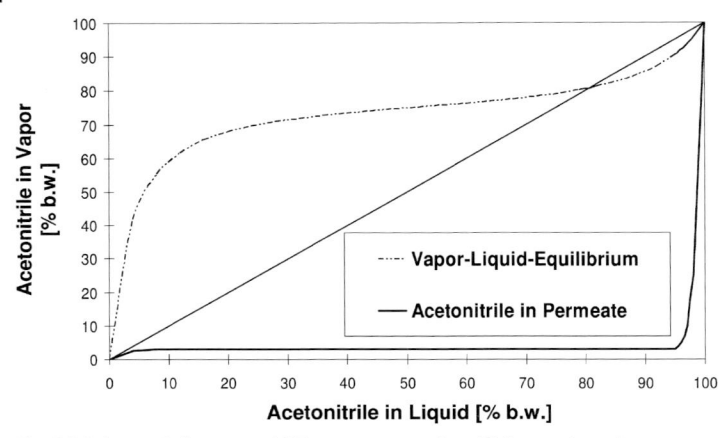

Fig. 3.14 Acetonitrile–water; VLE vs. pervaporation (PVA-membrane).

concentration of the more-permeable component (e.g. water) in the feed (Fig. 3.13 for ethanol-water, Fig. 3.14 for acetonitrile-water) similar to the well-known McCabe-Thiele diagram in distillation. Such diagrams can be used for the comparison of different membranes and provide much better information on the separation performance of a membrane.

Fluxes of organophilic membrane are defined in the same way as for hydrophilic membranes. However, when applied in the removal of low concentrations of volatile organics from water, the flux of the latter can be regarded as constant at constant temperature, as the concentration of water in the feed remains nearly unchanged. Fluxes of the volatile organic components are highly dependent on the concentration in the feed, and thus the same restriction on the information of the α- and β-value exists as for hydrophilic membranes. Furthermore, the fluxes of the organic components depend on the nature of the same, and they may differ by orders of magnitude for different components. For low concentrations of the organic substance in the feed a linear relation between the flux of the component and its concentration in the feed describes the process with sufficient accuracy. As long as the mixture can be regarded as comprising two components only, selectivity values can be specified as outlined above. In multicomponent systems, especially when strong coupling exists between the fluxes of the respective components, relations become much more complex and simple selectivities can no longer be specified.

3.4
Modules

The design of modules for pervaporation and vapor-permeation processes had been based on the experience gained in those for water treatment by membranes, like ultrafiltration and reverse osmosis. However, significant modifications had to be made due to the specific requirements of pervaporation and vapor-permeation processes. Whereas in the water treatment the portion of the feed volume passing into the permeate is small, in pervaporation and vapor permeation the volume of the permeate is much larger than that of the feed.

Pressure losses at the feed side have to be reduced to a minimum in vapor permeation. Otherwise the process would no longer operate at constant pressure, but the feed vapor could reach a region where superheated conditions would exist. Consequently, pressure losses in vapor permeation modules have to be as low as several millibar only. In pervaporation the feed-side pressure losses are not that critical, but in multistage arrangements they will eventually limit the number of applicable stages.

The partial vapor pressure at the permeate side has to be reduced in both processes to fairly low values, especially when low final concentrations of the critical component have to be reached in the retentate. Therefore any pressure losses, even in the range of a few millibar, have to be avoided at the permeate side.

As any feed mixture will contain organic components at high concentration, mostly at elevated temperatures, chemical and mechanical stability of all module components, like spacer, gaskets, potting material and glues are critical. So far, mainly four different types of modules are in use on an industrial scale.

3.4.1
Plate Modules

Plate modules are mainly used for dehydration applications; with permeate channels as open as applicable. A rectangular support plate is provided on both sides with gaskets, which partially cover slots in the plate, acting as distribution channels. A membrane is placed on each gasket, its feed side facing the plate. The permeate side of each membrane is supported by a perforated plate, a grid or spacer is placed between two perforated plates. A membrane, one side of the support plate, and a gasket form a feed chamber, two perforated plates and the space between them a permeate chamber. Each feed chamber is thus adjacent to a permeate chamber, each permeate chamber has a feed chamber at each side. Alternating feed and permeate chambers are arranged in a module. Figure 3.15 depicts an explosion drawing of such a module.

The module package is held together by means of flanges and bolts. The thickness and weight of bolts and flanges limits the maximum internal pressure for such modules to 6 to 10 bar. In order to keep weight and handability of the modules within a reasonable range, the maximum size of these modules does not exceed 30 to 50 m^2 of membrane area or less than 130 support plates. Figure 3.16 shows two physical modules of different membrane area.

Stainless steel is used as a construction material for support plates for the membranes and for spacers. Chemically stable elastomers, like EPDM or perfluorinated polymers are used as gasket material, more widely used is expanded graphite, due to its excellent chemical and thermal resistance. Preferentially the

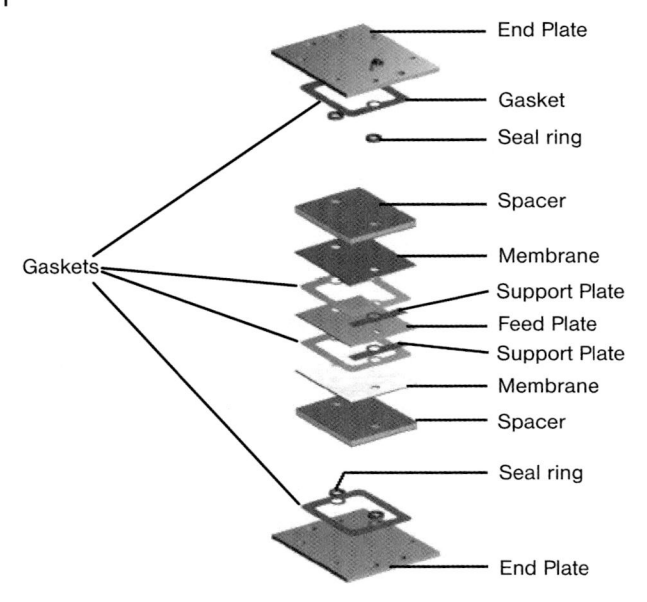

Fig. 3.15 Plate module, explosion drawing.

Fig. 3.16 Plate modules, different sizes (Sulzer Chemtech GmbH).

permeate channels are open over the circumference of the modules that are assembled inside special vacuum vessel. Intermediate heat exchangers and the permeate condenser are sometimes installed inside the vacuum vessel, mostly these items are installed outside for easier access and maintenance.

Usually all membranes in a module are arranged for parallel flow of the feed. The feed channel, between membrane and supporting plate, has a height between 0.5 to 1 mm, linear flow velocities of the liquid feed are of the order of a few centimeters per minute. Serial flow would be desirable in order to allow for higher linear flow velocities and higher Reynolds numbers, but then feed-side pressure losses will become too high. When used with a vaporous feed the feed channels need to be widened and linear velocities over the membrane should be of the order of 1 m per minute.

Alternative designs are very similar to plate heat exchangers, in which the supported membrane replaces the heat-exchanger plates. These modules may be open or closed to the outside on the permeate side, with internal ducts for feed and retentate and, when closed, for permeate removal. It has been proposed to integrate plate heat exchangers as preheaters and permeate condenser into such modules.

3.4.2
Spiral-wound Modules

Spiral-wound modules with stainless steel central tubes, but otherwise similar to those known from the conventional membrane processes ultrafiltration or reverse osmosis, are in use, mainly for organophilic membranes. Due to the larger molecular weight of the substances removed through organophilic membranes the volume of the vaporous permeate is much smaller than in dehydration applications; even at the same permeate-side pressure. As the selectivity of organophilic membranes is low, the total permeate-side pressure can be usually high. Thus pressure losses in the permeate channels are less critical than in water removal. As organophilic applications operate at lower temperatures and low concentrations of organic solvents in the feed, polymers materials can be used as spacer or glue. One or several of the spiral-wound modules are housed inside a pressure tube and assembled in conventional skids, very similar as in water treatment.

Similar considerations are valid for organic–organic separation. Spiral-wound modules have thus been used in pilot plants for the removal of methanol and ethanol from dry organic mixtures or for the removal of aromatic from aliphatic components. The stability of the material for the feed-side spacer and the glue are problems still to be solved.

There has been a development on spiral-wound modules for dehydration applications, too. So far this did not lead to applications in industrial plants. Chemical-stability problems of the components and too high pressure losses in the permeate-side spacer could not be solved satisfactorily, and the costs of the modules and for the installation in a plant were not really lower than those for plate modules.

3.4.3
"Cushion" Module

A special module design that is a hybrid between a plate and a spiral-wound module has been developed by the research institute GKSS in Germany. Here, two membrane sheets are welded together (by heat or ultrasonic welding) to a sandwich structure with a permeate spacer between the two membranes. A multitude of these sandwiches, each with a central hole, is arranged on a central perforated tube that removes the permeate. Each membrane sandwich is sealed from the feed to the permeate side at the central perforated permeate tube by means of a gasket. Around the central hole a perforated ring is inserted into the permeate spacer in order to have an unhindered flow of the permeate into the permeate tube. Feed spacers keep the membrane sandwiches apart from each other. Feed flow over all sandwiches in a module can be in parallel, by means of additional separation plates any number of the sandwiches can be arranged in groups, with the flow parallel in each group, but in serial for the groups. The central tube with the membrane sandwiches around is housed inside a feed vessel, usually of stainless steel. Originally these modules were developed for water treatment, but are now widely used with organophilic membranes in the recovery of organic vapors especially gasoline vapors from air or nitrogen streams.

3.4.4
Tubular Modules

Modules with membranes in the form of tubes were the first to be used in membrane applications. The obvious arrangement is a tube bundle fixed and sealed at both ends, similar to a tubular heat exchanger. With the active separating layer on the inside surface of the tube, and the feed flowing through the inner lumen an even distribution and high velocities of the feed can be reached and thus polarization effects minimized. Depending on the inner diameter of the tube the ratio of feed volume to membrane surface is rather high, and the feed stream cannot be heated inside the module. At high linear velocities this may require partial recirculation of the feed or very small modules in series with the respective large number of intermediate heat exchangers. The burst pressure of the tubes has to be sufficiently high; otherwise additional porous support structures around the outside of the tubes are necessary. Ceramic tubes or tubes made from polymers with small diameters (capillaries or hollow fibers) are sufficiently stable to be employed without an additional support. The removal of the permeate from the outside surface of the tubes is not obstructed, the permeate vapor can be condensed inside the module shell that has to be kept under vacuum. Composite polymeric membranes can be formed into tubes or hoses of relatively large diameter (12 to 22 mm) by winding a strip of a flat-sheet membrane spirally around a mandrel and welding (by heat or ultrasound) or gluing the edges together [42]. These hoses need to be supported in a perforated tube, and sealing each tube individually on both sides is complex and expensive. On the other hand, a module with such tubes could be advantageous and economical for large-scale vapor permeation applications.

A separating layer on the outside of a tubular structure is useful on ceramic or hollow fiber membranes only, other structures would collapse. The pressure loss of the permeate in the bore of a hollow fiber is too large by far, it prevents the use of such structures is pervaporation and vapor permeation. Ceramic tubes need then to be fixed on one side only to a tube sheet, where the inner lumen of the tubes is connected to the permeate volume. Baffle plates are required over the outside of the tube bundle in order to achieve good flow distributions and high Reynolds numbers in the feed. The flow regime is not that well defined and maldistribution and dead ends may occur. Heating of the feed in the module through additional heat-exchanger tubes or through the shell of the module is feasible. The flow through the membrane and the inner diameter of the tubes limit the length of the module, as otherwise the pressure drop at the open permeate side will become too high.

In a more recent development each tubular membrane is again fixed and sealed to a tube sheet, with the inner lumen on one side open to the permeate compartment. Additionally, each membrane's tube is housed inside a heat-exchanger tube. The feed flows through the annulus gap formed by the outside surface of the membrane and the inner surface of the heat-exchanger tube. The feed volume per surface area of membrane can now be controlled by adjusting the size of the annular gap. High linear velocities and thus very high Reynolds numbers can be achieved by this arrangement, without too high volume to surface ratios. From the outside of the heat-exchanger tube the feed can be directly heated and the heat lost by the evaporation of the permeate reintroduced. By specific means the direction of the feed flow can be reversed at the end of the annular gap, and the membranes can be arranged for serial or parallel feed flow, or any combination thereof.

3.4.5
Other Modules

Hollow fibers or capillary modules have not yet found an industrial application in pervaporation or vapor-permeation processes. A few data have been reported where organic capillary structures with an outside diameter of 0.5 to 1 mm have been coated with silicon and used in organophilic separation. With the flow on the shell side permeate pressure losses inside the bore of the fiber control the process. For specific organophilic applications, these pressure losses may be tolerable. For hydrophilic processes, however, the useful length of a module would be of the order of 20 to 30 cm only, even at an inner diameter of the capillary of 1 mm. Such a module, including housing and connection in any industrial application, is more costly than a plate module. So far no potting material is available that combines the necessary chemical and mechanical stability at the operation temperature and pressure of a dehydration plant.

Microfibers with an inner diameter of 20 µm and a wall thickness of 10 to 20 µm have been proposed, too. The specific flux through such a homogeneous membrane would be low, but outbalanced by high packing density and low membrane costs. The fiber length would be of the order of 20 cm, arranged in a modified module, in which the fibers would be potted in the axial direction into the wall of a tube. So far no reports are known on any application of such a module.

3.5
Applications

3.5.1
Organophilic Membranes

Organophilic membranes are mostly applied for the removal of volatile organic components ("VOCs") from gas streams like waste air or nitrogen. The main applications are the treatment of streams originating from the evaporation of solvents in coating processes in film and tape production, the purge of polymers, by which unreacted monomers are removed, or from breathing of storage tanks for solvents and especially from loading and unloading of gasoline tanks in tank farms. Mostly, the feed stream received at atmospheric pressure is compressed in order to increase the feed-side partial vapor pressure of the component to be removed to or above saturation level. Partial condensation of the critical component before entering onto the membrane is a wanted side effect. The permeate is enriched in the critical component, but not necessarily to saturation. It is compressed by means of a vacuum pump and led to the inlet of the feed compressor. The condensate obtained between compressor and membrane is then the only outlet for the separated and recovered organic. In specific cases compression at the feed side is sufficiently high to avoid the application of a vacuum at the permeate side.

The economy of the process is usually determined by the value of the recovered substances. Emission regulations in all industrial countries demand very low final concentrations if the gas stream is released to the atmosphere, therefore the retentate from the gas purification by the membrane is either recycled to the upstream process or further treated by an additional polishing step.

Another more recent application of organophilic membranes is found in the control of the dew point of natural gas. The permeability of higher hydrocarbons and aromatic compounds in siloxane is much higher, up to two orders of magnitude, than that of the light ones. When passing natural or another petroleum gas over such a membrane the low boiling, heavier hydrocarbons will pass much faster through the membrane than, e.g., methane, and the dew point of the gas can be lowered.

Although considerable effort in research and development has been devoted to the removal of VOCs from aqueous streams this technique has not yet been introduced into the industry. Potential mixtures like waste-water streams that could be treated are more complex, the economical value of the recovered substances is low. Even when a pure substance like phenol can be efficiently removed and recovered from water competing processes like biological treatment or adsorption are cheaper and better introduced. Applications may be found in the future in biotechnological processes where high-value products can be separated from a fermentation broth and can be concentrated and purified in the same step.

3.5.2
Hydrophilic Membranes

3.5.2.1 Pervaporation

The largest industrial installations of pervaporation and vapor-permeation processes are equipped with hydrophilic membranes and used for the removal of water from organic solvents and solvent mixtures. Although the first pervaporation plants were installed for the dehydration of bioethanol, already in 1985 the first plant for dehydration of ethyl acetate started its operation in the chemical industry. The first plants were still isolated, or end-of-pipe installations, working from a storage feed tank to a product storage tank (see Fig. 3.8). With relative small capacities of a few tons of solvent per day they could easily be bypassed if any problems occur. With increasing experience and confidence in the new technology, quite soon solvent dehydration by means of pervaporation and vapor permeation became an essential step in a production process. Today the technology is regarded as a reliable state-of-the-art process and, in numerous applications, is even an integrated part of a production process.

Organic solvents are used for a variety of purposes in the chemical industry, e.g. for synthesis of pharmaceuticals, to precipitate materials from aqueous solutions, for cleaning purposes, and for drying of final products. Spent solvents nearly always contain some water. Dehydration is an essential step in their recovery but difficult since most of the more common solvents form azeotropes with water. Final water removal by distillation is then impossible or complicated. Conventional entrainer distillation is not a real option for pharmaceutical or fine-chemical production. The addition and afterwards removal of the entrainer is difficult and the residual concentration will have to be monitored continuously. Furthermore, entrainer distillation systems require a certain minimum capacity to be economical. Quite often this capacity is above the amount of solvent that will have to be treated at a single location. The only solution is then in many cases to ship out the spent solvent and buy fresh one, with all the related problems of logistics and storage.

Pervaporation eliminates the need of an entrainer. It is regarded as a physical process, thus its validation in a process is not too difficult. Due to its modular nature a pervaporation plant is economical even at small capacities, which can be increased by the addition of more membrane area. A well-designed and operated pervaporation plant will recover 90 to 97% of the solvent contained in a feed mixture, thus reducing storage and shipping of hazardous goods. On site solvent recovery using pervaporation and vapor permeation is becoming standard practice in the pharmaceutical and chemical industries.

The most important solvents to be treated are the light alcohols, ethanol, the propanols, and butanols. Other solvents are esters like ethyl and butyl acetate, ketones like acetone, butanone (MEK) or methyl isobutyl ketone, ethers like tetrahydrofuran (THF) or methyl tertiary butyl ether, or acetonitrile, or mixtures of these solvents. The selectivity of the polymeric membranes for all components in such mixtures is high and fairly similar; multicomponent mixtures can thus

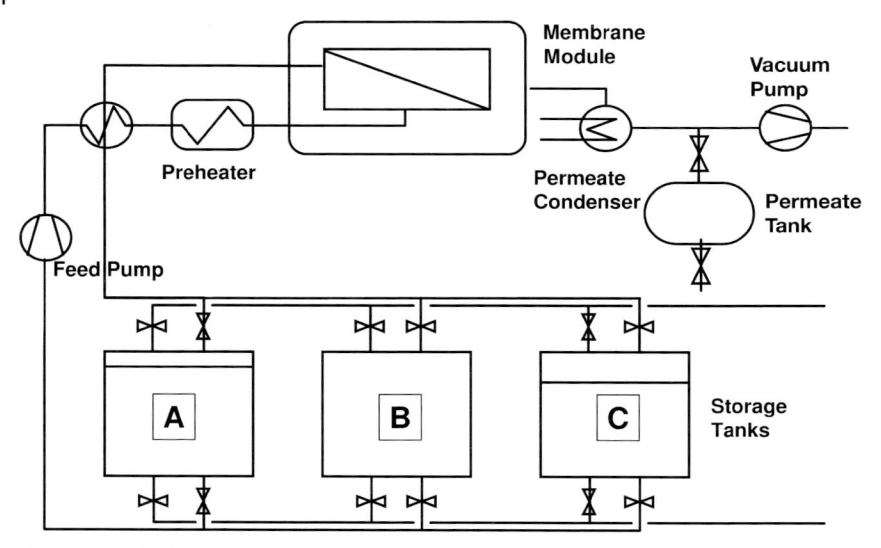

Fig. 3.17 Batch plant.

often be treated as binary ones, comprising water and one organic compound. The final water concentrations to be reached vary between 1% to below 500 ppm for the alcohols to below 100 ppm for THF; in extreme cases concentrations below 50 ppm are reached. Methanol is rarely treated by pervaporation as it does not form an azeotrope with water and can easily be purified by distillation. Selectivity and flux of the polymeric membranes are generally only in favor for the dehydration of methanol in the range of between 5 to 1% water.

The installed plants are either of the type shown in Fig. 3.8 or batch plants as shown in Fig. 3.17. In the first case the plant and the number of stages is optimized for a specific separation and capacity, and consumes the minimum of energy. If a different stream has to be treated, the plant will be operated outside optimal conditions, and compromises with respect to capacity or final product quality will have to be accepted. A batch plant comprises usually only one stage and one preheater by which the feed is brought to the operation temperature. The feed stream is circulated back to the storage tank and passed over the membrane several times until the whole content of the tank has finally reached the specified concentration. Due to the lower efficiency caused by the unavoidable redilution of the product and the fact that not all the sensible heat of the circulating stream can be recovered, such plant consumes more energy and requires more membrane area than the straightforward plant of Fig. 3.8. However, it offers more flexibility with respect to the final product quality by additional passes of the feed. Capacity can be adapted by the same means, and streams of different nature and composition can be treated with the same plant. Equation (23) is a useful tool in estimating plant capacity and product quality, when the pure water flux of the installed membrane is known for different feed streams.

3.5.2.2 Vapor Permeation

The criteria to choose between pervaporation or vapor permeation have been discussed in Section 3.2.6. In Fig. 3.10 the principal features of a standalone vapor-permeation plant are shown. The liquid feed stream from a storage tank is completely evaporated; the composition of the vapor entering the membrane modules equals that of the feed entering the evaporator. If the feed is an azeotrope the composition of liquid feed, vapor, and evaporator content are identical. If the feed is not an azeotrope the composition of the evaporator content will vary from the other two streams. Depending on the concentration of the impurities and their solubility in the liquid phase in the evaporator a bleed stream will have to be removed from the evaporator. In specific applications, e.g. treating a mother liquor, this bleed stream may be as high as 10% of the total feed. It was found that even from fairly pure solvents, e.g. those used in the electronics industry, low-volatility impurities accumulate in the evaporator, necessitating draining and cleaning the evaporator periodically.

Coupling of a distillation column with a vapor permeation unit, the latter treating the vapor from the top of the column, is shown in Fig. 3.18. The distillation column may have to be operated under pressure in order to allow the membrane system to be run at a temperature of around 100 °C. The permeate can be recycled to the inlet of the distillation column, which will result in nearly 100% recovery of the organic component. The only additional energy input in this scheme for the final dehydration of the predistilled product is that for the condensation of the permeate, as the dehydrated product would have to be con-

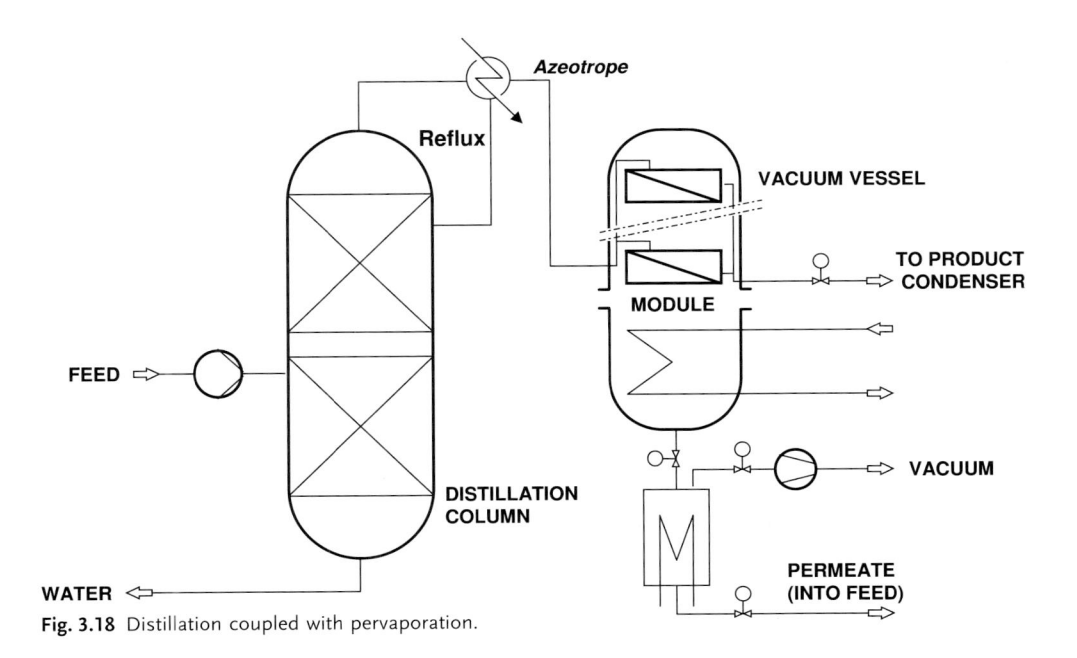

Fig. 3.18 Distillation coupled with pervaporation.

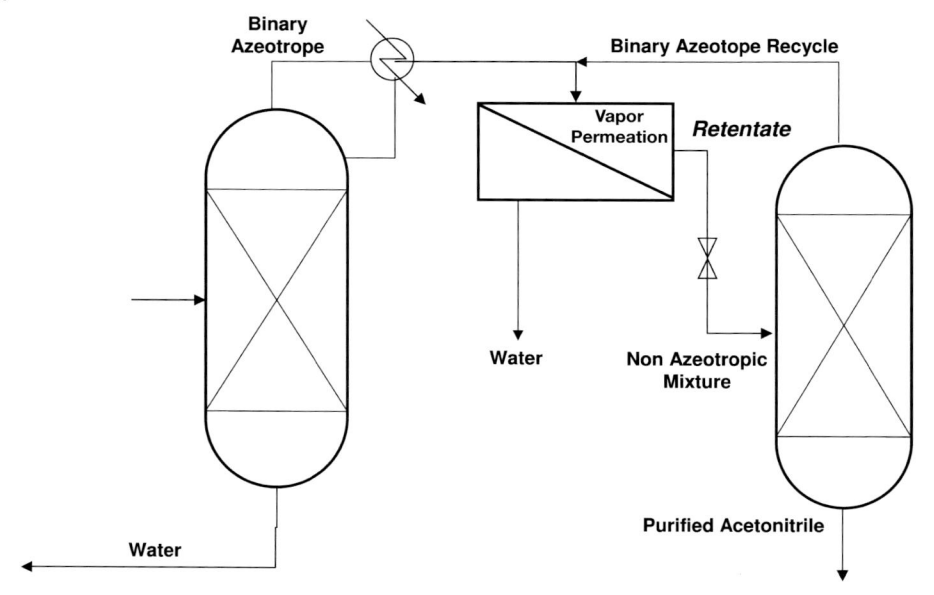

Fig. 3.19 Vapor permeation between two columns.

densed anyway. Balancing between the concentration of the vapor (azeotropic or nonazeotropic composition) entering the membrane system and thus between the size and energy consumption of the column, on the one hand and the size of the membrane system on the other hand will lead to an optimal hybrid system.

In specific cases when the vapor–liquid equilibrium favors distillation at the side of the organic component of an azeotrope and a high purity of the organic component is specified the membrane system may be used just to split the azeotrope. For the separation of the system acetonitrile–water a hybrid system as shown in Fig. 3.19 may be economically advantageous. Here the membrane system is used to cross the azeotropic point; the partially dehydrated vapor enters the second column in which final dehydration is effected. Again it is necessary to determine the economical optimum between the size of both columns, the energy consumption of the first one and the volume of the recycle stream from the second column at one side, and the size of the membrane system and its outlet concentration on the other side.

Figure 3.20 depicts a similar arrangement, however, here two organic components, an alcohol and an ester have to be separated and purified from their ternary mixture with water. The ternary vapor mixture from the first column is passed over the membrane of the vapor permeation unit, and nearly all water is removed. The permeate is recycled to the inlet of the first column and all the water is removed from its bottom. The two organic components can now be separated in the second column, with any residual water leaving together with the alcohol.

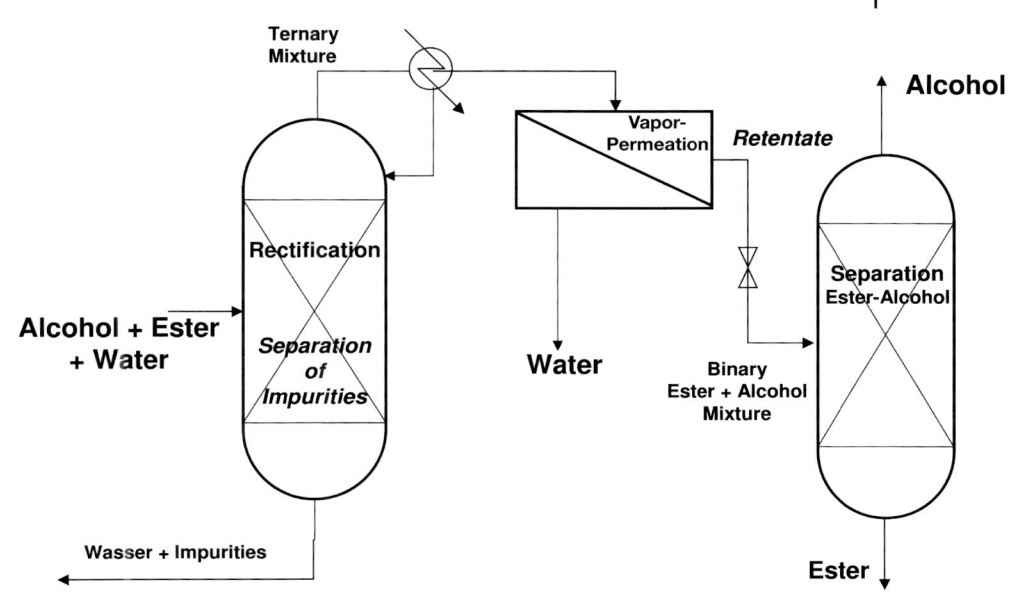

Fig. 3.20 Separation of a ternary mixture.

When the original feed composition is on the organic side of the azeotrope an arrangement as in Fig. 3.21 may have its advantages. The column separates the feed into the high-boiling organic at the bottom and a low-boiling mixture close to the azeotrope at the top. This vapor from the top is passed through a vapor-permeation plant that removes water, preferentially to a residual concentration close to that of the original feed. This retentate is recycled to the inlet of the column. All the water from the feed has to permeate through the membrane, but the most economical concentration range can be chosen. A certain drawback of this arrangement is the fact that any impurities of the original feed will remain in the purified organic stream.

Existing azeotropic distillation plants for the dewatering of organic solvents can simply be revamped by the addition of a vapor-permeation or pervaporation unit. Such plants comprise usually a first distillation column by which the feed is distilled close to the binary aqueous–organic azeotrope. In a second column the entrainer is added, which forms a ternary azeotrope with water and the organic. At the bottom of the second column the dry organic component is obtained, whereas the ternary azeotrope from the top of this column is condensed and split into two phases, an aqueous one, which is fed to a third column for further purification and entrainer recovery, and an organic one containing most of the entrainer, which is returned to the second column. The upper part of this second column limits the capacity of the plant as all water has to pass through it as the ternary azeotrope. When the stream from the first column is increased, the additional water can be removed by a pervaporation–vapor-permeation system, without overloading the second column. Calculations and tests have proven

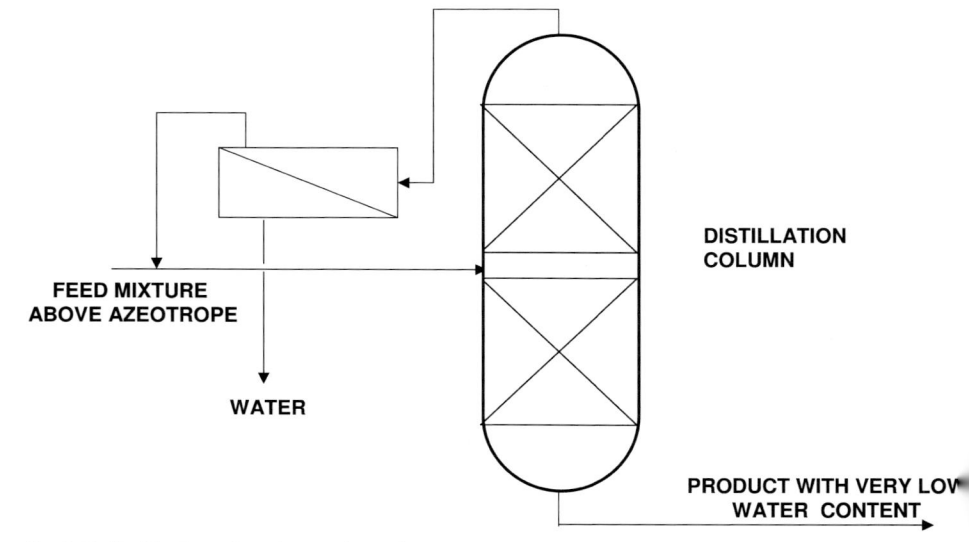

Fig. 3.21 Purification of a mixture above the azeotrope.

that the capacity of an existing azeotropic distillation can be increased by up to 40% by reducing the reflux ratio of the first column and removal of the excess water by pervaporation between the first and the second column.

Any of the commercially available chemical engineering programs (e.g. Aspen®, Chemcad®, and PROII®) can be used for calculation and simulation of hybrid system. An additional modulus has to be introduced into the program that in rather simple terms describes the membrane system. In a first approach it is necessary only to check the relation between feed flow and concentration on the membrane area, and the product quality and amount. Even a modified version of Eq. (21), relating feed flow and concentration, product flow and concentration, and membrane area would be sufficient for a first design.

3.5.3
Removal of Water from Reaction Mixtures

In many chemical condensation reactions like esterification, acetalization, ketalization, or etherification water is produced as an unwanted byproduct. All these reactions of the form

$$A + B \Leftrightarrow C + H_2O \tag{30}$$

lead to an equilibrium that limits the maximum conversion of the initial components. Removal of the water from the mixture will shift the reaction equilibrium to the side of the wanted product. If one of the educts is used at a surplus over the stoichiometry nearly full conversion of the other, usually the more valuable educt, can be achieved. This will result in a much higher yield of the wanted product. Furthermore, the wanted product has no longer to be separated and purified from a four-component mixture (the two educts, the wanted product, and water), but from a two-component mixture. Water has been removed through the pervaporation membrane, one of the educts is nearly totally converted, and thus the product has to be separated from the surplus educt only. The facilitated downstream purification may be even at least as economically important as the higher yield and conversion ratio.

The simplest arrangement of a pervaporation system coupled to a reactor is schematically shown in Fig. 3.22. A batch reactor is filled with the reaction mixture, one of the educts at precalculated stoichiometric surplus. The mixture is passed continuously over the pervaporation membrane, until the water introduced with the raw materials and freed in the reaction has been removed to the wanted extent. As indicated in Fig. 3.22 the more volatile portion, e.g. an aqueous azeotrope, can be alternatively evaporated from the reactor and passed through a vapor permeation unit. By arranging several reactors and membrane systems in a cascade and passing a bleed stream downstream of the first membrane system to a second reactor, a continuous operation is possible. Depending on the type and nature of the reaction, reactors, membrane systems, and distillation columns can be combined in different arrangements for an optimum yield and downstream purification. One of the first industrial plants, combining pervaporation and an esterification reaction, operating continuously with a cascade of reactors and pervaporation units has been described in the literature [14].

For a simulation and optimization of the coupled process the kinetics of the reaction and the performance of the membrane have to be known. An esterification reaction as in Eq. (30) can be described as a second-order reaction.

$$\frac{d[C]}{dt} = k_1 \cdot [A] \cdot [B] - k_2 \cdot [C] \cdot [H_2O] \tag{31}$$

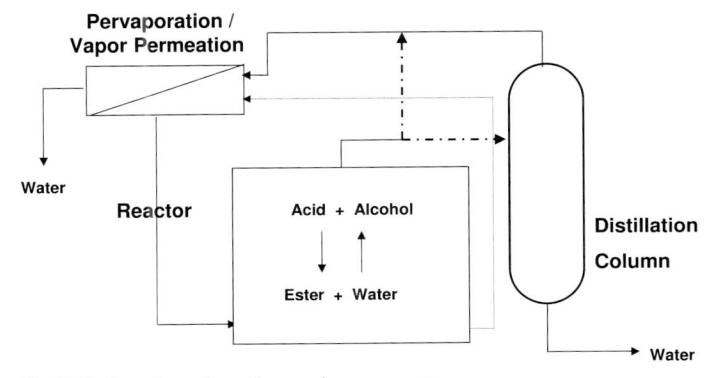

Fig. 3.22 Coupling of reaction and pervaporation.

The symbols in brackets relate to the concentration of the respective substances, k_1 and k_2 are the reaction rate constants for the forward (esterification) and backward (hydrolysis) reaction, including their dependence on temperature and catalyst activity. For a given reaction the equilibrium constant is then given by

$$K = k_2/k_1 \tag{32}$$

For the formation of water an equation similar to Eq. (31) is valid

$$\frac{d[H_2O]}{dt} = k_1 \cdot [A] \cdot [B] - k_2 \cdot [C] \cdot [H_2O] \tag{33}$$

When water is removed from the mixture the kinetic of water removal can be written as

$$-\frac{d[H_2O]}{dt} = [H_2O] \cdot Q \cdot \frac{A}{m} \tag{34}$$

with A the membrane area, m the mass of the reaction mixture and Q the permeability of the membrane. For simplicity it is assumed that Q depends on temperature only, not on any concentration. Introducing the water removal into Eq. (33) one gets

$$\frac{d[H_2O]}{dt} = k_1 \cdot [A] \cdot [B] - k_2 \cdot [C] \cdot [H_2O] - [H_2O] \cdot Q \cdot \frac{A}{m} \tag{35}$$

or, combining all constants in Eq. (34)

$$\frac{d[H_2O]}{dt} = k_1 \cdot [A] \cdot [B] - k_2 \cdot [C] \cdot [H_2O] - [H_2O] \cdot D \tag{36}$$

Calculation of the increase of the product C is more tedious. The interdependent differential equations (31) and (35) have to be solved numerically, which with today's computer is not too difficult.

In Fig. 3.23 the conversion ratio of the wanted product (ester) and the water present in the reaction mixture are plotted over the reaction time for a given membrane area and two ratios of the educts. Without removal of water from the mixture by means of a membrane the wanted product C and water are produced at the same rate, and both concentrations in the reaction mixture increase until equilibrium is reached. When water is continuously removed through the membrane at a certain time the water content passes through a maximum, when the water is removed as fast as it is formed. The time to reach this point depends on the membrane area installed. The water content then goes down and eventually reaches a value close to zero, when the water is removed much faster than formed.

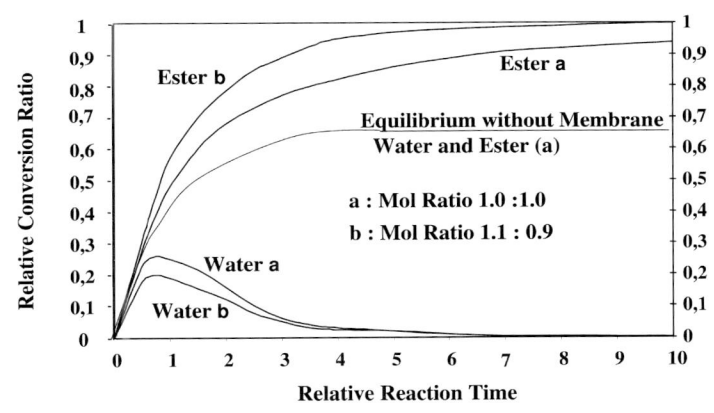

Fig. 3.23 Coupling of pervaporation and reaction, kinetics.

At the stoichiometric ratio of the educts (case a) the conversation ratio of component C approaches only asymptotically the 100% value even when water is continuously removed. This is easily understood from Eq. (31). When virtually all the water has been extracted from the mixture, the rate of formation of component C depends on the rate of the forward reaction only. In the second-order reaction the concentrations of component A and B have become small, their product is even smaller, and the overall reaction rate could only be improved by an increase in the reaction rate constant k_1, e.g. by increasing the reaction temperature or catalyst activity.

Starting at a nonstoichiometric ratio of the educts (case b), the concentration of the surplus educt is higher at the end of the reaction and remains nearly constant. The second-order reaction of Eq. (31) then becomes a first-order reaction, and full conversion can be reached in finite time.

With the knowledge of the kinetic parameters for a given reaction, which are relative easily accessible by a test or even found in the relevant literature, and the membrane performance the optimum ratio Q/m of membrane area to mass of the reaction mixture can be determined for that reaction, with the initial ratio of the educts as an adaptable parameter.

3.5.4
Organic–Organic Separation

Specific modification of hydrophilic membranes can be used to remove the light alcohols methanol and ethanol from their mixtures with other organics. The selectivity of these membranes is not as high as in dehydration processes, but sufficient for effective and economical large-scale industrial applications. One such plant for the removal of methanol from an organic azeotrope has been described [16].

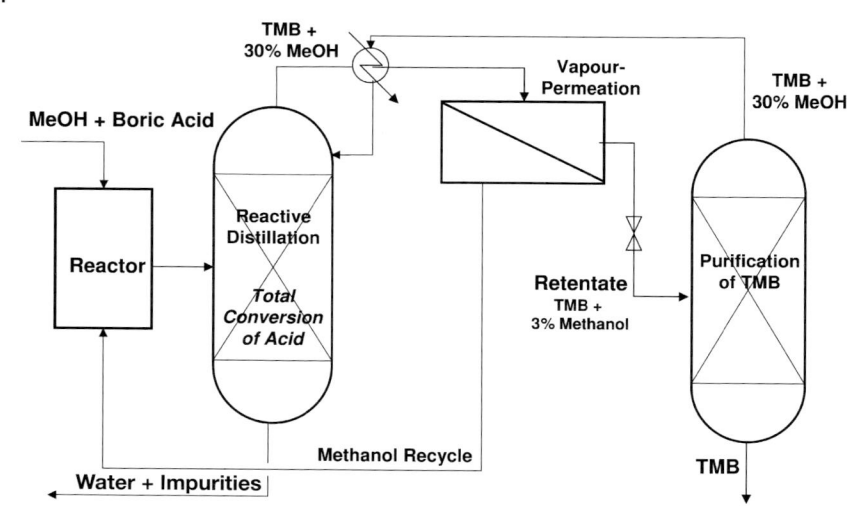

Fig. 3.24 Production and purification of TMB.

In the production of trimethyl borate (TMB) methanol and boric acid are fed to a reactor followed by a reactive distillation column (Fig. 3.24). Methanol is used in a surplus in order to convert all the boric acid and avoid the pollution of the bottom product of the reactive distillation column with residual acid. From the top of this first column an azeotropic mixture of 30% methanol and 70% TMB is obtained under higher than atmospheric pressure. This azeotrope cannot be separated by water wash, as the TMB will immediately hydrolyze when in contact with water. The azeotropic vapor is thus led to a vapor permeation unit, equipped with membranes that permeate methanol, but retain TMB. The concentrated TMB, which contains approximately 3% of methanol, is introduced into a second distillation column, which separates the feed into pure TMB at the bottom and a nearly azeotropic mixture at the top. This mixture is returned into the reflux of the first column, the permeate of the membrane system, mainly methanol, is recycled to the reactor.

In their publication the operator states that the investment for the membrane system is lower than for a competing absorption system, combined with considerable savings in energy, personal and maintenance costs, resulting, as they say, in a negative pay-back period.

In other plants methanol is separated from other methylesters of simple organic acids like methyl acetate, which generally form an azeotrope with methanol, or from acetone.

A specific application is found in transesterification reactions, where a methylester is reacted with another alcohol, e.g. one containing an amino group. It is desired to convert all of the amino alcohol; therefore a surplus of the methyl ester is applied. Again, an azeotropic mixture of methanol and the respective methyl ester is obtained as a byproduct. For recirculation of the ester

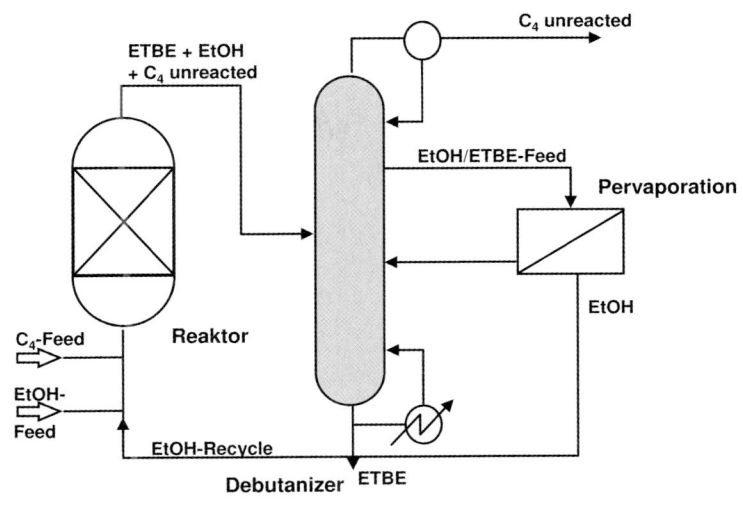

Fig. 3.25 Treating a side stream of a distillation column.

the methanol content has to be reduced significantly, which can be effected by pervaporation.

In the production of methyl tertiary butyl ether (MTBE) and ethyl tertiary butyl ether (ETBE) a C4 cut is reacted with a surplus of the respective alcohol to the ether. Only the isobutene is selectively converted to the ether. From the reaction mixture the unreacted C4 and the surplus of alcohol has to be separated. Unfortunately, both ethers form azeotropes with the respective alcohol, so their separation is effected by several distillation columns, operated at different pressures. When a side stream of the first debutanizer column (Fig. 3.25) is extracted and passed over a pervaporation membrane, the alcohol can be removed through the membrane and returned to the reactor. No significant residues of alcohol will then be present in either the bottom or the feed product. A more detailed engineering study [43] has shown that only a relatively small membrane area is required for a large-scale production plant, combined with significant savings in operation costs. The same scheme, treating a side stream of a distillation column by a pervaporation and removing one component from a three-component system, may be effective and economical in a multitude of applications.

The first effort in the development of pervaporation membranes was aimed at their potential to separate organic mixtures, especially those of hydrocarbons [2]. Following the results of the latest research [27], and the operation of pilot plants [29], this goal may be reached in the next couple of years.

3.6
Conclusion

During the past 15 years removal of water from organic liquids and liquid mixtures by means of pervaporation/vapor permeation has developed into a mature and state-of-the-art technology. More than 150 industrial plants have been installed around the world, with capacities between 20 kg/h to several tons per hour. Nearly all of these plants are equipped with polymeric membranes, which are also used in the plants for removal of methanol from its azeotropic mixtures. Membranes and modules have proven their stability and reliability, and the economy of the processes. With increasing energy cost one of the main advantages of the pervaporation processes, their superior energy efficiency, will lead to further utilization. The introduction of ceramic membranes will complement the area of application of the process to higher temperature where existing polymeric membranes cannot be used. However, ceramic membranes still have to prove their long-term stability under the industrial operation conditions of high temperatures, the presence of acid or alkali, and of high water content of the feed in dehydration. One more question not yet resolved is the potential susceptibility of the (porous) ceramic membranes to fouling.

One major obstacle to a wider use of pervaporation and vapor permeation has been in the past the prejudice and lack of information of many people in the field of chemical engineering. With a new generation of engineers having been educated in membrane technology this situation is hopefully going to change. Pervaporation processes can be operated as standalone installations, their optimal use, however, is found in hybrid systems, and mostly combined with distillation. This requires the rethinking and eventually the new design of the overall process.

References

1 KOBER PA (1917), Pervaporation, Perstillation, and Percrystallisation, *J. Am. Chem. Soc.*, **39**, 9444.
2 BINNING RC, LEE RJ, JENNING JF, MARTIN EC (1961), Separation of liquid mixtures by permeation, *Ind. Eng. Chem.*, **53**, 45.
3 BINNING RC, JAMES FE (1958), How to separate by membrane permeation, *Petroleum Refiner*, **37**, no. 5, 214.
4 BINNING RC, JAMES FE (1958), Permeation. A new commercial separation tool. *The Refiner Engineer*, **30**, no. 6, C14.
5 BINNING R.C (1961), Separation of mixtures, US Patent 2981680.
6 LEE RJ (1961), Permeation process using irradiated polyethylene membranes, US Patent 2984623.
7 APTEL P, CUNY J, JOSEFOWICZ J, NÉEL J (1972), Liquid transport through membranes prepared by grafting of polar monomers onto polytetrafluoroethylene films, Parts I, *J. Appl. Polym. Sci.*, **16** 1061, Part II, *J. Appl. Polym. Sci.*, **18** (1974), 351, Part III, *J. Appl. Polym. Sci.*, **18** (1974), 365.
8 FRIES R, NÉEL J (1965), Transfer sélectif á travers des membranes actives, *J. Chim. Phys.*, **62**, 494.

9 Aptel P, Cuny J, Morel G, Josefowicz J, Néel J (1973), Pervaporation á travers des films de polytérafluoroéthylene modifiés par greffage radiochimique de N-vinylpyrrolidone. *Europ. Polym. J.*, **9**, 877.
10 Brüschke HEA, Schneider WH, Tusel GF (1982), Pervaporation membrane for the separation of water and oxygen-containing simple organic solvents. European Workshop on Pervaporation. Nancy, France, Sept. 21–22.
11 Ballweg AH, Brüschke HEA, Schneider WH, Tusel GF, Böddeker KW, Wenzlaff A (1982), Pervaporation membranes. An economical method to replace conventional dehydration and rectification columns in ethanol distilleries. Fifth Intern. Sympos. on Alcohol Fuel Technology, Auckland, New Zealand, May 13–18.
12 Brüschke HEA (1982), Verwendung einer mehrschichtigen Membran zur Trennung von Flüssigkeitsgemischen nach dem Pervaporationsverfahren; Europ Patent EP 096339.
13 Sander U, Soukop P (1988), Design and operation of a pervaporation plant for ethanol dehydration, *J. Memb. Sci.*, **36**, 463.
14 Rapin JL (1988), The BETHENIVILLE pervaporation unit. The first large-scale productive plant for the dehydration of ethanol. Third Intern. Confer. on Pervaporation Processes in the Chem. Industry, Nancy, France, Sept. 19–22.
15 Brüschke HEA (1995), Optimierung einer Kopplung Pervaporation und Reaktion zur Esterherstellung. Preprints 5. Aachener Membran Kolloquium, Aachen.
16 van der Ent L (1999), Succesverhaal rond Damppermeatie, npt Processtechnologie, September–October 1999, p. 25–28 (in Dutch).
17 Eustache H, Histi G (1981), Separation of aqueous organic mixtures by pervaporation and analysis by mass spectrometry or a coupled gas chromatograph-mass spectrometer. *J. Membr. Sci.* **8**, 105.
18 Böddeker KW, Bengtson G (1987), Phenolanreicherung durch Pervaporation, *Erdöl und Kohle* **40**, 439.
19 Bengtson G, Böddeker KW, (1988), Pervaporation of low volatiles from water, Proc. 3rd Int. Conference on Pervaporation Processes, Nancy (R. Bakish, ed.) Englewood, 439.
20 Brüschke HEA, Schneider WH, Tusel GF (1994), Verfahren zur Reduktion des Alkoholgehaltes alkolischer Getränke, Europ. Patent 0332738.
21 te Hennepe HJ, Mulder MHV, Smolders CA, Bargemann D, Schröder G AT (1990), Pervaporation Process and Membrane, US Patent 4025562.
22 Gundernatsch W, Kimmerle K, Stroh N, Chmiel H (1988), Recovery and concentration of high vapor pressure bioproducts by means of controlled membrane separation, *J. Membr. Sci.* **36**, 331.
23 Baker R.W, Blume I, Helm V, Khan A, Maguire J, Yoshioka N (1991), Membrane Research in Energy and Solvent Recovery from Industrial Effluent Streams, Energy Conservation, DOE/ID/12379-TI (DE 84016819).
24 Ohlrogge K, Wind J, Scholles C, Brinkmann T (2005), Membranverfahren zur Abtrennung organischer Dämpfe in der chemischen und petrochemischen Industrie. *Chemie Ingenieur Technik* **77** (5): 527.
25 Black L (1989), Selective permeation of aromatic hydrocarbons through polyethylene glycol impregnated regenerated cellulose or cellulose acetate membrane, US Patent 4802987.
26 Schucker R.C (1992), Multiblock polymer comprising a first amide acid prepolymer, chain extended with a compatible second prepolymer, the membrane made thereof and its use for the separations, US Patent 5130017.
27 Ren J, Staudt-Bickel C, Lichtenthaler R.N (2001), Separation of aromatics/aliphatics with crosslinked 6FDA-based copolyimides. *Sep. Purif. Technology* **22/23**, 37.
28 Brüschke H.E.A, Wynn N, Balko J (2003), Desulfurization of Gasoline. Preprints 9. Aachener Membran Kolloquium, Aachen.
29 Schwake M, Brickwede F (2005) Produktionsintegrierter Umweltschutz mit

innovativer Membrantechnik. *Chemie Ingenieur Technik* **77** (5), 600.

30 Hauser J, Heintz A, Reinhard G.A, Schmittecker B, Wesslein M, Lichtenthaler RN (1987), Sorption, Diffusion, and Pervaporation of water–alcohol mixtures in PVA-membranes. Proc. of the 2nd Int. Conference on Pervaporation Processes, San Antonio (R. Bakish ed.) Englewood, 15.

31 Hauser J, Reinhard GA, Stumm F, Heintz A (1989), Experimental study of solubilities of water containing organic mixtures in Polyvinylalcohol using Gas Chromatographic and Infrared spectroscopic analysis. *Fluid Phase Equil.*, **49**, 195.

32 Hauser J, Reinhardt GA, Stumm F, Heintz A (1988), Sorption Non-Ideal Behavior of Liquid Mixtures in PVA and its Influence on the Pervaporation Process. Third International Conference on Pervaporation Processes in the Chemical Industry. R Bakish (ed.). Bakish Material Corporation, Englewood, NJ, USA, 134.

33 Brüschke HEA, Schneider WH (1993), Membrane Process for Separating Fluid Mixtures, US Patent 5512179.

34 Ageev YP, Kotova SL, Zesin AB, Skorikova EE (1995), PV – Membranes based on polyelectrolyte complexes of Chitosan and Poly-(acrylic acid), Proc. of 7th Internat. Conferences on Pervaporation Processes, Reno, Nevada (R. Bakish, ed.) Englewood, NJ 52.

35 Steinhauser H.A, Brüschke H.E.A, Ellinghorst G (1996), Verfahren zur Abtrennung von C1–C3 Alkoholen aus Gemischen dieser Alkohole mit andren organischen Flüssigkeiten, Europ. Patent EP 0593011.

36 deV. Naylor T, Zelaya F, Bratton GJ (1989), The BP-Kalsep pervaporation system, Proc. of 4th internat. Conferences on Pervaporation Processes, Ft. Lauderdale, Florida, (R. Bakish, ed.) Englewood, NJ 428.

37 Schwarz HH, Apostel R, Richau K, Paul D (1992), Separation of water-alcohol mixtures through High Flux Polyelectrolyte Complex Membranes, Proc. of 6th internat. Conferences on Pervaporation Processes Ottawa, Canada (R. Bakish, ed.) Englewood, NJ 223.

38 Maus E.M, Brüschke HEA (1998), Entfernung von Methanol und Ethanol aus nicht-wässrigen Systemen. DECHEMA Jahrestagung Wiesbaden.

39 Kita H, Horii K, Tanaka K, Okamoto K-I (1995), Pervaporation of Water-Organic liquid mixtures using a Zeolite NaA membrane, Proc. of 7th Internat. Conferences on Pervaporation Processes, Reno, Nevada (R. Bakish, ed.) Englewood, NJ 364.

40 Burggraaf AJ, Keizer K, Uhlhorn RJR, De Lang RSA (1996), Manufacturing Ceramic Membrane, Europ. Patent EP 0586745.

41 Burggraaf AJ, Cot L (1996) Fundamentals of Inorganic Membranes and Technology. Elsevier, Amsterdam.

42 Brüschke HEA, Wynn N, Marggraff F (2003) Tubularmodul DE 10323440.

43 Hömmerich U (1996) Integration der Pervaporation in den MTBE – Herstellungsprozess, IVT Information der RWTH Aachen, 26, Nr. 2.

4
Organic Solvent Nanofiltration

A. G. Livingston, L. G. Peeva and P. Silva

Summary

In recent years, solvent-stable nanofiltration membranes with molecular weight cutoffs (MWCOs) ranging from 200–1000 g mol^{-1} have emerged. This new generation of organic solvent nanofiltration (OSN) membranes have been employed in a wide range of applications at full process and lab scale. Perhaps the best-known and most successful application of OSN to date has been in the MAX-DEWAX process developed by W. R. Grace and ExxonMobil and employed for crude-oil dewaxing at the 72 000 barrels per day scale. Concomitantly, a growing number of publications concerning the transport of organic solvents in OSN membranes have appeared. While some studies support the use of pore-flow models, others suggest using a solution-diffusion approach. The effects of concentration polarization and osmotic pressure during OSN have not, as yet, received a great deal of attention in academic studies. In this chapter, we summarize work in which continuous crossflow nanofiltration cells have been used to study the performance of STARMEM^{TM} 122 (W. R. Grace and Co) and MPF 50 (Koch Membrane Systems) OSN membranes over periods of weeks. We summarize the results of experimental and modeling work on work on two problems: (i) binary mixtures of solvents, and prediction of mixture fluxes based on fluxes of pure solvents, and; (ii) the influence of mass transfer phenomena (concentration polarization) and osmotic pressure on the solvent flux and solute rejection for systems with concentrated (up to 20 wt%) high molecular weight solutes.

4.1
Current Applications and Potential

In recent years, solvent-stable nanofiltration membranes with molecular weight cutoffs (MWCOs) ranging from 200–1000 g mol^{-1} have emerged [1–3]. Applications have been proposed for a variety of industries including refining – e.g. hy-

Membrane Technology in the Chemical Industry.
Edited by Suzana Pereira Nunes and Klaus-Viktor Peinemann
Copyright © 2006 WILEY-VCH Verlag GmbH & Co. KGaA, Weinheim
ISBN: 3-527-31316-8

drocarbons separation [4], food industry – e.g. deacidification of vegetable oil [5, 6], fine-chemical and pharmaceutical synthesis – e.g. organometallic catalyst separation [7–16] and solvent exchange [17], etc. During the 1990s the first large-scale application of OSN was realized for solvent recycle in lube oil production, the MAX-DEWAX process [18, 19]. Solvent lube oil dewaxing processes are practiced worldwide in refineries. In these lube dewaxing processes, waxy lube raffinate feedstocks are processed in block operations to produce lube base stocks having distinct viscosity characteristics. The raffinates are first dissolved in a light hydrocarbon solvent mixture, for example a blend of methyl ethyl ketone (MEK) and toluene. This mixture is cooled to precipitate the wax component, and is filtered to remove the wax. The solvent mixture is then recovered from the lube oil using (energy-intensive) distillation and the solvents are recycled to the process. For most feeds, overall production and dewaxed oil yield is limited by solvent circulation and/or refrigeration capacity. A membrane process (Fig. 4.1) can be used to substantially debottleneck the refrigeration and recovery sections of a solvent lube plant. The process recovers the solvent at or near dewaxing temperatures, which allows the recovered solvent to recycle to the dewaxing process without the need for additional cooling. The process can selectively recover up to 50% of the cold solvent in the dewaxed oil filtrate. It results in significant equipment and energy savings by reducing the amount of solvent subjected to the heating and cooling in the solvent-recovery section.

A commercial polyimide membrane packaged as spiral-wound modules by Grace Davison Membranes was installed in 1998 as the separation membrane

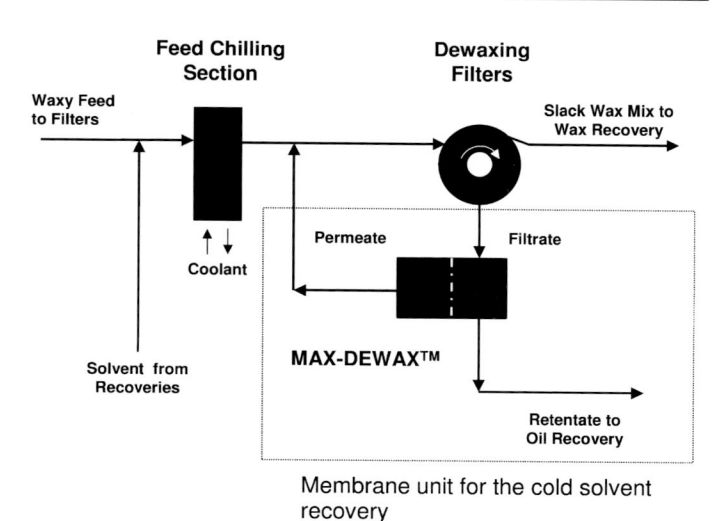

Fig. 4.1 Schematic of the MAX-DEWAX™ process.

in a process trademarked MAX-DEWAX at ExxonMobil's Beaumont refinery, Texas. This unit, designed for an ultimate feed rate of ~11 500 m^3 d^{-1} (~72 000 barrels per day) is the largest membrane separator of liquid-phase organics in the world. It is currently processing ~5800 m^3 d^{-1} (~36 000 barrels per day) of feed, consistent with existing permeate demands. The capacity of the unit can be expanded by adding more membrane area as demand increases. The capital cost of the membrane unit was $5.5 million. The MAX-DEWAX process, combined with selected ancillary equipment upgrades, increased base oil production by over 25 vol.% and improved dewaxed oil yields by 3–5 vol.%. Total energy consumption at the higher throughput was essentially the same as before the expansion. Consequently, energy use per unit volume of product was reduced by nearly 20%. The installed cost of the membrane unit was about one-third of that which would have been required for equivalent process improvements using conventional technology. The net annual membrane uplift from the membrane unit is over $6 million, making it very attractive technology. The capital expenditure was paid back in less than 1 year by the increase in the net profitability of the lube dewaxing plant. This highly successful application at large scale clearly shows the potential for OSN to impact the energy and chemicals sectors.

4.2
Theoretical Background to Transport Processes

Due to the large and still largely unexploited potential of OSN, research in this field has become an area of intensive study in the last decade, and there is a growing body of information available on the processes controlling solvent fluxes and solute rejections [20–34]. Although OSN systems have been studied for several years and much knowledge has been gained, they are not yet well understood. For example, there is currently no universally accepted model describing transport processes in OSN and various approaches are reported in the literature. While some studies support the use of pore-flow models, others suggest using a solution-diffusion approach. In what follows we will briefly review the currently available literature models and their basic equations.

4.2.1
Pore-flow Model

Stable pores are assumed to be present inside the membrane and the driving force for transport is the pressure gradient across the membrane. Assuming a system at constant temperature where there are no external forces except pressure, one can derive, from the Stefan-Maxwell equations [34], the following equations describing the total volumetric flux through a membrane:

Hagen-Poiseuille equation – if the membrane is assumed to be composed of more or less cylindrical pores

$$N_v = -\frac{d_{pore}^2}{32\eta}\frac{\varepsilon}{\tau}\nabla p \tag{1}$$

Carman-Kozeny equation – if the membrane consists of a packed bed of particles

$$N_v = -\frac{d_{particle}^2}{180(1-\varepsilon)^2\eta}\frac{\varepsilon^3}{\tau}\nabla p \tag{2}$$

4.2.2
Solution-Diffusion Model

In the solution-diffusion model, it is assumed that each permeating molecule dissolves in, and diffuses through, the membrane phase in response to a concentration gradient. There is no pressure gradient inside the membrane, and starting again from the Stefan-Maxwell equation the following equation can be derived [32, 34] for the molar flux of each specie though the membrane:

$$N_i = P_{i,m}^{molar}\left(x_{i,F} - \frac{\gamma_{i,P}}{\gamma_{i,F}}x_{i,P}\exp\left(-\frac{\bar{V}_i\Delta p}{RT}\right)\right) \tag{3}$$

The above equation assumes that the swelling of the membrane separating layer is negligible. It is similar to the well-known equation presented by Wijmans and Baker [35], differing only by the ratio of $\frac{\gamma_{i,P}}{\gamma_{i,F}}$, which has been shown to be important when there are significantly different component mole fractions on each side of the membrane and the system is nonideal. For cases where the separating layer is a rubbery material, the assumption of low swelling is unlikely to be true and in fact the membrane will often be highly swollen – in that case the analysis developed by Paul through a series of papers in the 1970s and recently reformulated [36, 37] is more appropriate.

These models have been used in previous work. Robinson et al. [30] reported that their experimental data for the permeation of n-alkanes, i-alkanes and cyclic compounds in a polydimethylsiloxane (PDMS) composite OSN membrane was consistent with the Hagen-Poiseuille pore-flow model, Eq. (1). Whu et al. [22] also suggest this pore-flow model for fluxes through the commercial OSN membrane MPF-60 (Koch Membrane Systems). The membranes employed in both papers comprised rubbery materials attached to a support, and at least in the case of Robinson et al. [30] were highly swollen under operation. Application of the Hagen-Poiseuille model implies that a pressure gradient exists across the thin PDMS layer. The careful argument presented by Paul and Ebra-Lima [36] based on mechanics, suggests that such a pressure gradient is not possible in a swollen rubber phase, and so the exact physical picture in the pore-flow interpretation of the data of [22, 30] is not clear.

Bhanushali et al. [24] suggested that solvent viscosity and surface tension are dominant factors controlling solvent transport through NF membranes, and a solution-diffusion approach was proposed to predict pure solvent permeation. Stafie et al. [31] employed the solution-diffusion model to describe sunflower oil/hexane and polyisobutylene/hexane permeations through a composite PDMS membrane with poly(acrylonitrile) (PAN). Some of the work reported by Bhanushali et al. [24] and all of the data presented by Stafie et al. [31] employed membranes with separating layers based on rubbery materials, for which the swollen rubber models of Paul are probably more appropriate than the simple solution-diffusion model actually employed based on [35].

White [25] investigated the transport properties of a series of asymmetric polyimide OSN membranes with normal and branched alkanes, and aromatic compounds. His experimental results were consistent with the solution-diffusion model presented in [35]. Since polyimides are reported to swell by less than 15%, and usually considerably less, in common solvents this simple solution-diffusion model is appropriate. However, the solution-diffusion model assumes a discontinuity in pressure profile at the downstream side of the separating layer. When the separating layer is not a rubbery polymer coated onto a support material, but is a dense top layer formed by phase inversion, as in the polyimide membranes reported by White, it is not clear where this discontinuity is located, or whether it will actually exist. The fact that the model is based on an abstract representation of the membrane that may not correspond well to the physical reality should be borne in mind when using either modelling approach.

Finally, Machado et al. [21] developed a resistances-in-series model and proposed that solvent transport through the MPF membrane consists of three main steps: (1) transfer of the solvent into the top active layer, which is characterized by surface resistance; (2) viscous flow through NF pores and (3) viscous flow through support layer pores, all expressed by viscous resistances, i.e.

$$N_v = \frac{\Delta p}{R_s^0 + R_\mu^1 + R_\mu^2} \qquad (4)$$

Where R_s^0, R_μ^1 and R_μ^2 are the surface resistance and viscous resistances through NF active layer and support layers, respectively. The surface resistance is proportional to the surface-tension difference between the solvent and the OSN top layer, and viscous resistances are proportional to solvent viscosity.

4.2.3
Models Combining Membrane Transport with the Film Theory of Mass Transfer

Almost all reported OSN data have been obtained at the lab scale and with dilute solutions (<1 wt% solute in solvent), whereas in actual applications, solutes will be more concentrated (>5 wt%). Under these conditions, concentration polarization and osmotic pressure may contribute to the solvent flux, as they do in well-studied aqueous systems. There are several studies on concentration polarization in aqueous systems, mainly concerning ultrafiltration [38–42]. We assume a concentration gradient on the feed side, but not on the permeate side, reasonable when the permeate-side solute concentration is low. The film theory of mass transfer, for component i, gives:

$$N_v c_i - D_i \frac{dc_i}{dz} - N_v c_{i,P} = 0 \tag{5}$$

with boundary conditions:
$z=0 \quad c_i = c_{i,\,FM}$
$z=\delta \quad c_i = c_{i,\,F}$

For the total volumetric flux one can obtain (see Fig. 4.2 and [32] for details)

$$\frac{N_v}{k_i} = \ln\left(\frac{c_{i,FM} - c_{i,P}}{c_{i,F} - c_{i,P}}\right) \tag{6}$$

Concentration polarization for liquid film mass transfer can be coupled with the model for membrane transport (for example the solution-diffusion model Eq. (3)) [32, 38, 43, 44], to describe membrane transport in a mass transfer limited system.

Combined solution-diffusion–film-theory models have been presented already in several publications on aqueous systems, however, either 100% rejection of the solute is assumed [38], or detailed experimental flux and rejection results are required in order to find parameters by nonlinear parameter estimation [43, 44]. Consequently, it is difficult to apply these models for predictive purposes. In OSN, it is also important to account for the effect of different activities of the species on both sides of the membrane. We have proposed a set of equations [32], Eqs. (7) to (13), taking these factors into account. We assume a binary system, although the equations could be generalized for a system of n components. In this analysis component 1 is the solute and component 2 is the solvent. The only parameters to be estimated, other than physical properties, are

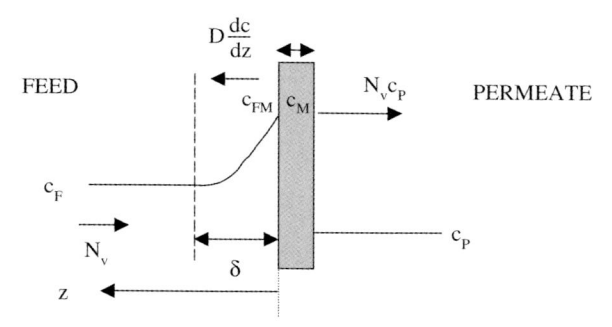

Fig. 4.2 Schematic representation of the concentration polarization phenomena in membrane processes (e.g. UF, NF) under steady-state conditions. (Reprinted from [32] with permission from Elsevier.)

the mass transfer coefficients, which may be measured, and the permeabilities, $P_{i,m}^{molar}$, which may be calculated from flux data.

$$\frac{N_v}{k_1} = \ln\left(\frac{c_{1,FM} - c_{1,P}}{c_{1,F} - c_{1,P}}\right) \quad (7)$$

$$\frac{N_v}{k_2} = \ln\left(\frac{c_{2,P} - c_{2,FM}}{c_{2,P} - c_{2,F}}\right) \quad (8)$$

$$N_v = N_1 \bar{V}_1 + N_2 \bar{V}_2 \quad (9)$$

$$N_1 = P_{1,m}^{molar}\left(x_{1,FM} - \frac{\gamma_{1,P}}{\gamma_{1,FM}} \frac{N_1}{N_1 + N_2} \exp\left(-\frac{\bar{V}_1 \Delta p}{RT}\right)\right) \quad (10)$$

$$N_2 = P_{2,m}^{molar}\left(x_{2,FM} - \frac{\gamma_{2,P}}{\gamma_{2,FM}} \frac{N_2}{N_2 + N_1} \exp\left(-\frac{\bar{V}_2 \Delta p}{RT}\right)\right) \quad (11)$$

$$\gamma_1 = f(x_1), \gamma_2 = f(x_2) \quad \text{or} \quad \gamma_1 = \gamma_2 = 1 \quad (12)$$

$$rej_{calc} = 1 - \frac{N_1}{N_v c_{1,F}} \quad (13)$$

Equations (7) to (13) constitute a system of 7 nonlinear algebraic equations, which allow the prediction of permeate flux and solute rejection, when the membrane permeability for a given component and the mass transfer characteristics of the equipment are known. Equations (7) and (8) describe the diffusion in the liquid film adjacent to the membrane, while Eqs. (9) to (12) describe membrane transport and Eq. (13) defines the rejection.

More detailed discussion on the above models, as applied to experimental evidence, will be provided in the following sections and can be found in more detail in Refs. [32] and [34], from which this chapter is largely drawn, with permission from Elsevier. The above models can be used to gain insight into transport processes, and also for design calculations. While most of the above mentioned references have focused on the former use, we have been concerned more with the latter, i.e. (i) the design problem of how to best predict fluxes from over a wide range of solvent mixtures from a limited data set of the pure solvent fluxes; and (ii) how best to integrate concentration polarization and nonideal solution behavior into design models for OSN.

To investigate (i) we have measured the fluxes of pure solvents, and used this data to obtain the parameters of Eqs. (1) and (3). These parameters were then used to predict the data for mixtures of the two solvents across the whole concentration range. Since for both solvent systems studied there were only relatively small changes in viscosity across the concentration ranges, and there were major differences in solvent flux, we concluded that solvent viscosity was not the only variable determining solvent flux.

To investigate (ii), we obtained data for toluene in a crossflow filtration cell, and for two different large organic molecules at concentrations >1 wt%. The experimental data can be described using a combined solution-diffusion–film-theory model, since the data suggest that at these high concentrations, concentration polarization becomes significant. Careful attention was paid to the role of the activity coefficients of the solute and the solvent, which seem to markedly affect the permeate flux.

4.3
Transport of Solvent Mixtures

4.3.1
Experimental

Data was obtained for three organic solvents, ethyl acetate, toluene and methanol – these solvents are commonly used in the pharmaceutical and chemical industries. STARMEM^{TM} 122, an asymmetric OSN membrane with an active layer of polyimide, in a "dry" form but with a lube oil soaked into the membrane as a preserving agent, with a nominal MWCO of 220 g·mol^{-1} (manufacturer's data) was supplied by Membrane Extraction Technology Ltd (UK).

4.3.1.1 Filtration Equipment and Experimental Measurements
The crossflow nanofiltration rig [34] used to obtain this data consisted of four crossflow nanofiltration cells connected in series, a solution reservoir, a backpressure regulator, and a piston pump. The solution entered the crossflow cell tangentially from the cell wall and exited the cell from the top center, providing turbulent hydrodynamic conditions to minimize the effect of concentration polarization during filtration. The applied pressure was controlled at 30 bar using the backpressure regulator and the temperature at 30±0.5 °C using a water bath and heat exchanger.

4.3.2
Results for Binary Solvent Fluxes

Both solution-diffusion and pore-flow models have been used to analyze the experimental solvent flux data.

For the solution-diffusion model it can be assumed that concentration polarization does not exist and the activity coefficients in the permeate and in the retentate are equal for each specie. Mass fraction units (easier to determine) are used instead of the commonly used mole fraction. Eqs. (7)–(13) for a binary mixture in mass term units can be simplified [34] to

$$n_1 = P_{1,m}^{mass}\left(w_{1,F} - \frac{n_1}{n_1+n_2}\exp\left(-\frac{\bar{V}_1\Delta p}{RT}\right)\right) \quad (14)$$

Table 4.1 Model parameters.

Solvent	Permeabilities		
	Solution diffusion [kg m^{-2} s^{-1}]	Solution diffusion [mol m^{-2} h^{-1}]	Pore-flow [×10^{15} m]
Methanol	0.481	54056	3.91
Toluene	0.054	2107	1.32
Ethyl acetate	0.226	9251	1.62

(Reprinted from [34] with permission from Elsevier.)

$$n_2 = P_{2,m}^{mass}\left(w_{2,F} - \frac{n_2}{n_1 + n_2}\exp\left(-\frac{\bar{V}_2 \Delta p}{RT}\right)\right) \tag{15}$$

where n_i is the mass flux and w_i the mass fraction of species i. It should also be noted that the solvent mixtures are not assumed to be ideal solutions; rather it is assumed that the activity coefficients are equal in the permeate and in the retentate for each solvent.

It can be seen that the permeability for each solvent has to be determined if Eqs. (14) and (15) are used to predict solvent fluxes. It was assumed that the permeability for each solvent is independent of solvent compositions in the solution. The permeabilities for each solvent were determined by using pure solvent flux data, and are shown in Tab. 4.1.

The toluene permeability is similar to that reported previously by White [25], i.e. 2803 mol m^{-2} h^{-1} for toluene in a polyimide (Lenzig P84) nanofiltration membrane, corresponding to a mass permeability of 0.072 kg m^{-2} s^{-1}.

Equations (14) and (15) were then used to calculate mass flux for each solvent mixture. The predicted mass flux for each solvent was converted to give a total volumetric flux by using

$$N_v = \frac{n_1}{\rho_1} + \frac{n_2}{\rho_2} \tag{16}$$

The comparison of predicted and experimental solvent flux data is shown in Figs. 4.3 and 4.4. It is clear that the model provides a reasonable fit to the experimental data for both systems.

Two Hagen-Poiseuille models (a one-parameter model and a two-parameter model) were also used to describe the experimental solvent flux data shown in Figs. 4.3 and 4.4. The permeability term $\left(\frac{\varepsilon d_{pore}^2}{32l\tau}\right)$ in Eq. (1) is determined by the physical properties of the membrane. When membrane geometry is assumed constant, i.e. there is no membrane compaction or membrane swelling, this term should be independent of the solvent mixtures investigated, and one parameter value should describe fluxes of all solvents. In fact, the derived values

Methanol-Toluene

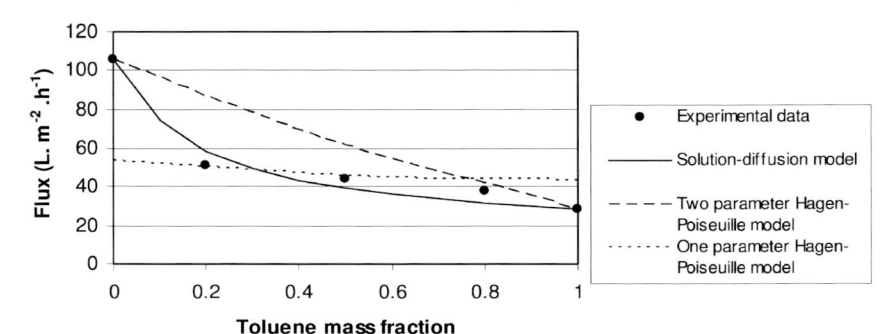

Fig. 4.3 Comparison of experimental methanol/toluene flux data in STARMEM™ 122 (at 30 bar applied pressure and at 30 °C) and the calculated values by solution-diffusion and Hagen-Poiseuille models. (Reprinted from [34] with permission from Elsevier.)

Ethyl Acetate-Toluene

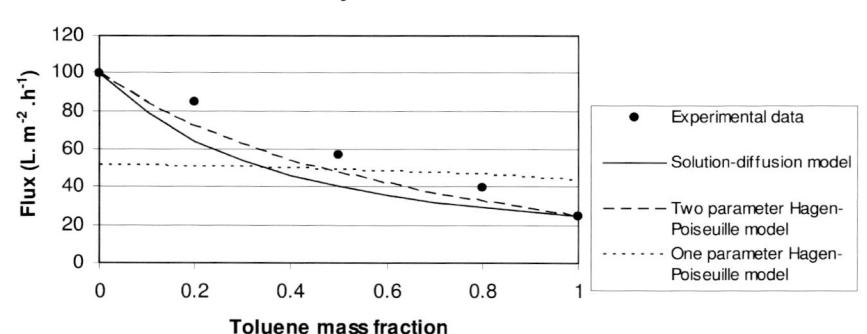

Fig. 4.4 Comparison of experimental toluene/ethyl acetate flux data in STARMEM™ 122 (at 30 bar applied pressure and at 30 °C) and the calculated values by solution-diffusion and Hagen-Poiseuille models. (Reprinted from [34] with permission from Elsevier.)

for this term are very different (Tab. 4.1) for pure methanol (3.91×10^{-15} m), toluene (1.32×10^{-15} m) and ethyl acetate (1.62×10^{-15} m). Nevertheless, for the one-parameter Hagen-Poiseuille model we take an arithmetic average of these specific pemeabilities (2.28×10^{-15} m) and have the viscosity as the only composition-dependent parameter. For the two-parameter Hagen-Poiseuille model, we incorporate the idea that the physical properties of the membrane change with the solvent due to different solvent-polymer interactions, i.e. different degrees of swelling. For this two-parameter Hagen-Poiseuille model, an approximate approach was used to describe the term $\left(\frac{\varepsilon d_{pore}^2}{32l\tau}\right)_{mix}$ for the solvent mixtures. Considering a concentration average of the pure solvent values and assuming no viscous selectivity and a linear pressure profile inside the membrane the following relation using these two pure solvent parameters is obtained [34] to describe the total flux of a binary mixture

$$N_v = \frac{\Delta p}{\eta}\left(\bar{V}_1 c_{1(m)} \left(\frac{\varepsilon d_{pore}^2}{32l\tau}\right)_1 + \bar{V}_2 c_{2(m)} \left(\frac{\varepsilon d_{pore}^2}{32l\tau}\right)_2\right) \quad (17)$$

Together with the viscosities for solvent mixtures, Eq. (17) was used to predict solvent flux data. The predicted values using both one-parameter and two-parameter Hagen-Poiseuille models are also shown in Figs. 4.3 and 4.4. It is clear that the one parameter model fits the data poorly, while the two parameter model provides much better predictions. The reason for the one-parameter Hagen-Poiseuille model predictive failure is that the permeability change due to swelling of the polymer matrix is ignored by the use of a constant average permeability. Therefore viscosity is clearly not the determining factor in the transport.

The most appropriate comparison is between the two-parameter Hagen-Poiseuille model and the solution-diffusion model, since the solution diffusion is also effectively a two-parameter model (i.e. it uses two permeabilites). It can be seen from Fig. 4.3 that the solution-diffusion model better predicts solvent fluxes of methanol/toluene mixtures than the two-parameter Hagen-Poiseuille models, while the two-parameter Hagen-Poiseuille model gives a slightly better description for solvent fluxes of toluene/ethyl acetate mixtures in Fig. 4.4. These results indicate that the solution-diffusion model gives moderately better results for predicting mixtures of solvent fluxes from pure solvent permeability data in STARMEM[TM] 122 than the two-parameter Hagen-Poiseuille model.

4.4
Concentration Polarization and Osmotic Pressure

4.4.1
Experimental

A quaternary ammonium salt, tetraoctylammonium bromide (TOABr) and docosane were used as solutes and toluene was used as the solvent. The membrane used in this study was again the solvent-resistant polyimide membrane, STARMEM[TM] 122. A dual-cell crossflow filtration rig, similar to the one described in Section 4.3.1 was used in all the experiments with an effective membrane area of 78 cm^2. The stage cut was between 0.01 and 0.3% over the whole concentration and pressure range. Ideal mixing is assumed throughout the system.

Two sets of experiments were performed. One set of experiments with toluene solutions of TOABr was performed to study the influence of the feed flow rate (crossflow velocity) on the permeate flux at a constant pressure of 30 bar.

The other set studied the influence of solute concentration and applied pressure on the permeate flux and solute rejection. A wide range of experiments was performed using toluene solutions of docosane (MW = 310 Da) and TOABr (MW = 546 Da) using a range of pressures: 0–50 bar, and concentrations: 0–20 wt% (0–0.35 M, 0–0.04 mole fraction) for TOABr in toluene and 0–20 wt% (0–0.67 M, 0–0.09 mole fraction) for docosane in toluene. The construction of the crossflow rig made it difficult for exactly the same flow rate to be maintained through the cells at different pressures, however, it was always kept in the range of 40–80 L h^{-1} in one of the cells, 120–150 L h^{-1} in the other.

4.4.2
Results for Concentration Polarization and Osmotic Pressure

The combined solution-diffusion film theory model (Eqs. (7)–(13)) is used to describe the experimental results. The equations were solved numerically.

4.4.2.1
Parameter Estimation

The mass transfer coefficients in the crossflow cell were estimated from independent measurements of dissolution of a plate of benzoic acid into water at two different crossflow rates: 50 L h^{-1} and 120 L h^{-1}, at 30 °C. Mass transfer coefficients for docosane and TOABr were estimated based on the experimentally measured benzoic acid mass transfer coefficients values and the Chilton-Colburn mass transfer coefficient correlation. Details of the procedure applied are described elsewhere [32].

The molar volumes of toluene and docosane were taken from the literature [25]. The molar volume of TOABr was estimated based on Fedors method [45].

The mass transfer coefficients calculated for docosane and TOABr, using the Chilton-Colburn correlation, are presented in Tab. 4.2.

Although all of the mass transfer coefficient data available in the literature are for aqueous systems and most are for ultrafiltration [46–49], the values are in the same order of magnitude as those obtained in these experiments ~10^{-5} m s^{-1}, indicating that the technique used gives reasonable estimates.

The activity coefficients for docosane and toluene were calculated applying the modified UNIFAC method [50]. From these results it was possible to develop a simple algebraic function describing the activity coefficient as a function of mole fraction of docosane and toluene, respectively:

Toluene: $\gamma_T = 0.99 + 0.30 x_T - 0.29\, x_T^2$ (18)

Docosane: $\gamma_D = 3.57 - 2.63 x_D / (0.01 + x_D)$ (19)

Table 4.2 Summary of the mass transfer coefficient values used in the model.

Compound	Concentration [mol L^{-1}]	Mass transfer coefficient at 120–150 L h^{-1} flow rate $\times 10^5$ [m s^{-1}]		Mass transfer coefficient at 40–80 L h^{-1} flow rate $\times 10^5$ [m s^{-1}]	
		From Chilton–Colburn	Best fit of experimental data	From Chilton-Colburn	Best fit of experimental data
Docosane	0.33	5.3	5.3	1.9	1.9
	0.67	4.8	4.8	1.7	1.7
TOABr	0.21	2.2	1.1	0.8	0.8
	0.33	1.7	1.7	0.6	0.9

(Reprinted from [32] with permission from Elsevier.)

where x_T and x_D are the mole fractions of toluene and docosane, respectively.

This function was applied to both the permeate and retentate sides in the model.

The activity coefficients for toluene in the TOABr-toluene system at different mole fractions of TOABr were calculated using a model [32] that combines a modified Debye-Huckel term, accounting for the long-range (LR) electrostatic forces, with the original UNIFAC [51] group contribution method for the short-range (SR) physical interactions.

Again, a function describing the data was developed:

Toluene: $\gamma_T = -4.16 x_T^2 + 7.29 x_T - 2.13$ (20)

This activity coefficient function was applied to toluene on the retentate/feed side in the model. The activity coefficient of toluene on the permeate side was assumed to be unity, because the solute mole fraction is sufficiently low. For simplicity, all the TOABr activity coefficients were assumed to be unity since the solute mole fraction on the permeate side is close to zero and so this term does not contribute significantly to the results.

The membrane permeability for toluene was determined from independent measurements of the pure toluene flux at different applied pressures. Docosane and TOABr membrane permeabilities were determined from the nanofiltration data assuming a concentration driving force and a solute flux experimentally determined at a low applied pressure of 4 bar, to avoid the influence of the exponential term in the solution-diffusion model and, the effect of concentration polarization. The model parameter values are summarized in Tab. 4.3.

Table 4.3 Summary of the model parameter values.

Compound	Docosane	TOABr	Toluene-docosane	Toluene-TOABr
Diffusion coefficient [m^2 s^{-1}]	1.23×10^{-9}	0.88×10^{-9}	1.23×10^{-9}	0.88×10^{-9}
Molar volume [m^3 mol^{-1}]	398×10^{-6}	766×10^{-6}	106×10^{-6}	106×10^{-6}
Membrane permeability [mol m^{-2} s^{-1}]	0.0007	3.10^{-5}	1.1	1.1
Activity coefficient [–]	Eq. (19)	1	Eq. (18)	Eq. (20)

(Reprinted from [32] with permission from Elsevier.)

4.4.2.2 Nanofiltration of Docosane-Toluene Solutions

The first experiments were conducted with the docosane-toluene system. This is considered an 'easy' binary system with which to verify the model due to the fact that nanofiltration data are available in the literature for comparison [25], and the change in viscosity with concentration is negligible [32].

Two concentrations of docosane (0.33 M and 0.67 M) in toluene were tested at various pressures and flow rates. The results for the permeate flux and docosane rejection are presented in Fig. 4.5 (A, A', B, B'). As can be seen from the figures, both docosane rejection and permeate flux decrease with decreasing pressure at both concentrations. The fluxes and rejections are lower at the higher docosane concentrations. This type of result is not surprising and has been observed previously with other systems [43, 44]. Experimentally, the flow rate through the crossflow cell does not have a significant effect on the flux or the rejection performance.

The design model was then applied to the docosane system. The results were calculated firstly assuming that the activity coefficients of the solvent and solute were equal to unity, and subsequently by applying the activity-coefficient functions derived from the UNIFAC data (Eqs. (18) and (19)). The comparisons of the model results with the experimental values for the permeate flux and docosane rejection are shown in Fig. 4.5 (A, A', B, B').

For the flux data (Fig. 4.5 A, A'), the calculated values correspond better with the experimental data at higher pressures. When activity coefficients are taken as unity, the model predicts almost no flux at pressures lower than 8 bar for

Fig. 4.5 A, A'. Experimental and calculated values for permeate flux of 0.33 M (A) and 0.67 M (A') docosane solution. B, B'. Experimental and calculated values for rejection of 0.33 M (B) and 0.67 M (B') docosane solution. Results shown in both cases for activity coefficients: $\gamma_T = \gamma_D = 1$, and $\gamma_T = 0.99 + 0.30 x_T - 0.29 x_T^2$, $\gamma_D = 3.57 - 2.63 x_D/(0.01 + x_D)$ (Reprinted from [32] with permission from Elsevier.)

▼ Experimental results at flow rate 40–80 L h-1
● Experimental results at flow rate 120–150 L h-1
— Calculated flux with activity coefficient functions, Eqs. (18), (19)
---- Calculated flux with all activity coefficients = 1

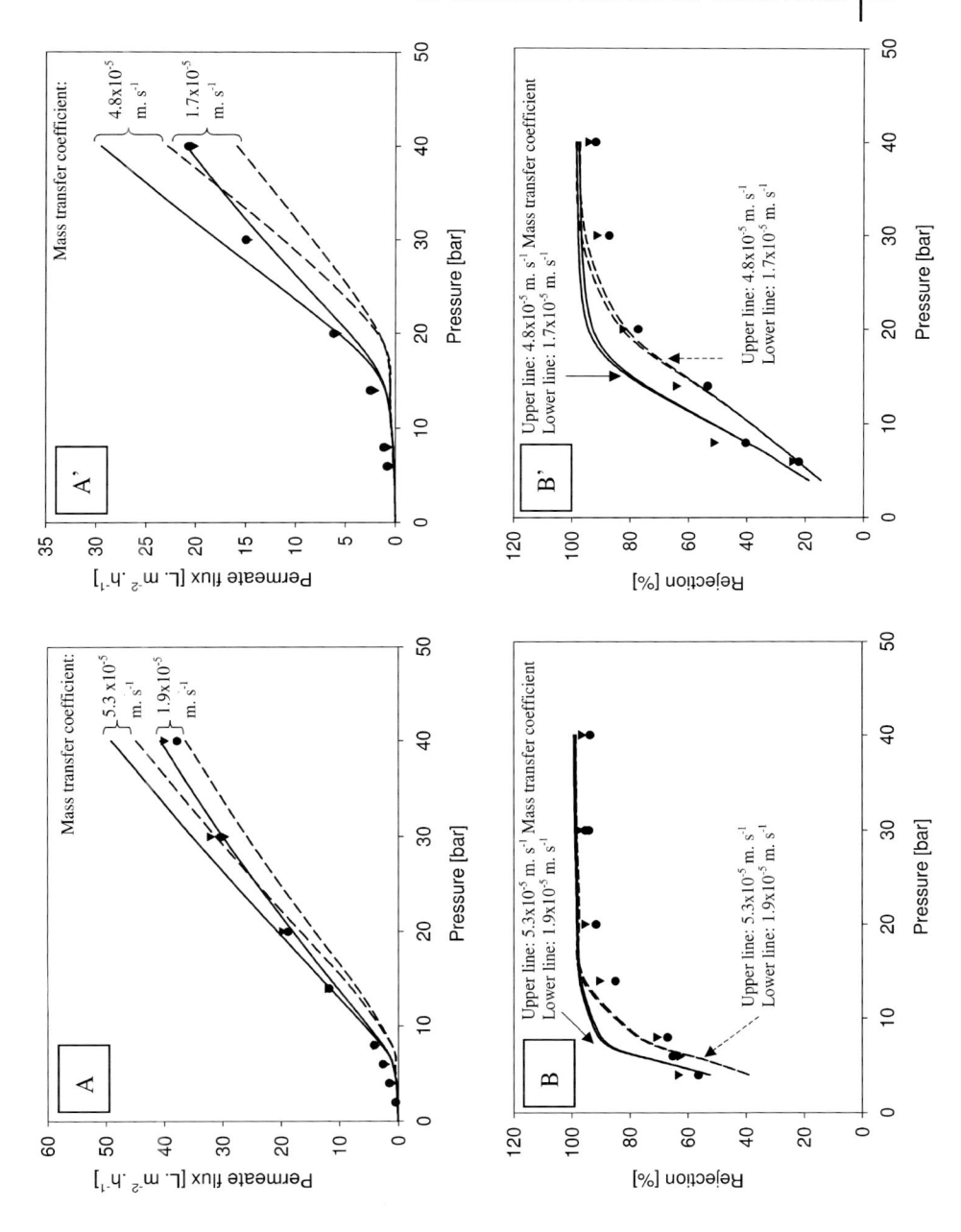

4.4 Concentration Polarization and Osmotic Pressure | 217

0.33 M concentration, and ~18 bar for 0.67 M concentration, whereas, experimentally, flux is seen at all pressures. Since the predicted rejection corresponds reasonably well with the experimental values over this pressure range, the existence of flux experimentally suggests that the effective osmotic pressure is lower than predicted and that the system deviates from ideality. Introduction of the activity coefficient ratios improves the fit of the model to the permeate flux data. At pressures higher than 20 bar the model predicts some influence of the flow rate on the permeate flux, however, none is seen experimentally. This could be the influence of membrane compaction at higher pressures, which contributes to the membrane performance as follows: if we consider the mass transport from the resistances-in-series point of view, the overall resistance for nanofiltration consists of 3 components: the liquid boundary layer resistance, the top layer resistance and the porous support resistance. However, if due to membrane compaction, the membrane resistance increases as a result of pore size reduction in the porous support and/or decrease in the free volume in the top active layer, then the influence of the boundary layer resistance will be minimized compared with these increased resistances. This effect is difficult to quantify for use in the model. An alternative explanation is that the mass transfer coefficient values were estimated considering the diffusion coefficient in dilute conditions: at the high concentrations that we are working with, the variation of the mass transfer coefficients at different flow rates could be less significant. The diffusion coefficient is a more complicated function depending on concentration, pressure, viscosity and activity of the components of the system [52]. Therefore it is not surprising that a discrepancy is observed between experimental results and the results calculated from the model on the basis of a single value of the diffusion coefficient, and this is an interesting area for further study.

For the rejection data (Fig. 4.5 B), no influence of the mass transfer coefficient (i.e. flow rate) is predicted for 0.33 M docosane. For 0.67 M docosane (Fig. 4.5 B'), a slight variation is predicted due to a more significant concentration polarization effect at higher concentrations. As for the flux data, the shape of the predicted curve improves when the activity coefficient ratios are not constrained to unity, but the model values are higher than the experimental values, especially at high pressures. This discrepancy could be due to the simplified approach used to estimate the membrane permeability for docosane. It should also be noted that the membrane permeability is assumed to be constant, independent of pressure and concentration of the components. However, a detailed analysis of the factors contributing to the membrane permeability term suggests that this assumption is not always true. The three contributing terms are the component diffusion in the membrane, the partition coefficient and the membrane thickness. The diffusion coefficient in the membrane is unlikely to change with pressure and concentration. However, the partition coefficient is the ratio between the activities of the component in the feed and the membrane, which is not necessarily a constant independent of concentration. The membrane thickness may also vary due to membrane compaction, or mem-

4.4.2.3 Nanofiltration of TOABr-Toluene Solutions

Following the work with the docosane-toluene system, a more "difficult" binary mixture was chosen, TOABr-toluene, in which system there are significant changes in viscosity with concentration [32].

The results of the influence of the flow rate on the permeate flux are presented in Fig. 4.6. These experiments were performed in order to understand whether concentration polarization is important in this process, and also its range of influence. As can be seen from the figure, the effect of concentration polarization is significant at all except very low concentrations ~0.005 M. This behavior is markedly different from that observed in the docosane-toluene system, where the flow rate has a negligible effect on the permeate flux. The difference could be attributed to two factors.

Firstly, the diffusion coefficient of docosane in toluene is higher than that of TOABr (1.23×10^{-9} m^2 s^{-1} compared with 0.88×10^{-9} m^2 s^{-1}) when calculated at infinite dilution, but in concentrated solutions, considering the viscosity change, this difference could be even higher. Secondly, the rejection of TOABr remains in the range 98–99% (unlike docosane), which increases the build up of solute in the boundary layer.

After performing several experiments varying pressure, flow rate and solute concentrations an attempt was made to fully describe the process using the

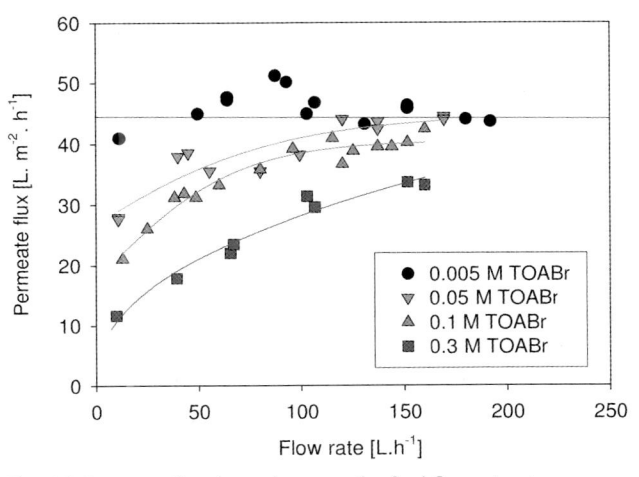

Fig. 4.6 Permeate flux dependence on the feed flow rate at different TOABr concentrations. The crossflow unit was operated at 30 °C and 30 bar pressure. (Reprinted from [32] with permission from Elsevier.)

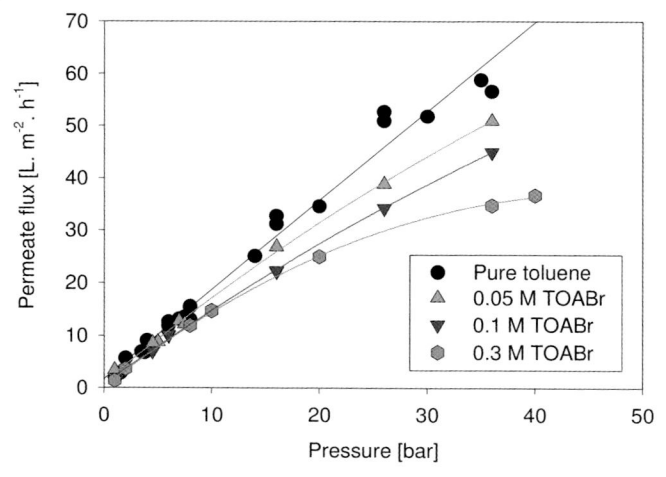

Fig. 4.7 Permeate flux for various concentrations of TOABr in toluene, as a function of pressure: pure toluene, 0.05 M, 0.1 M and 0.33 M at crossflow rate 120–150 L h^{-1}. (Reprinted from [32] with permission from Elsevier.)

transport model. Experimental data is shown in Fig. 4.7. Although the solute rejection was very high (~99%) over the whole pressure range, the shapes of the permeate flux versus applied pressure were completely different from those for the salt-water solutions of the same concentrations where the permeate flux, as a function of pressure, was a straight line and the x-axis intercept at each concentration corresponded well to the osmotic pressure calculated from the Van't Hoff equation [32].

Similar types of curves have been reported in the literature with macromolecular solutions [40] where the activity of the system components differs from unity. The observed divergence of the dependence of flux on pressure from linearity at higher concentrations also suggests the existence of concentration polarization.

Initially, we investigated the influence of the mass transfer coefficient, k, on the permeate flux, assuming all the activity coefficients were unity. However, as shown in Fig. 4.8A, for 0.21 M TOABr in toluene, (dashed lines), the data could not be described in this way, no matter what the mass transfer coefficient values were. Even when the mass transfer coefficient value → ∞ (line 4 on Fig. 4.8A), the model predicts an osmotic pressure of around 6 bar, at a concentration of 0.21 M, which is not observed experimentally. Activity differences could be responsible for this difference. Further evaluation of the influence of model parameters confirmed that the activity coefficient of toluene in the boundary layer does indeed have an important effect in this system. Even a very small change in the activity coefficient has a significant effect on the permeate flux.

Since the activity coefficient has been shown to have an important role in this system, the activity coefficient function (Eq. (20)), was included in the model.

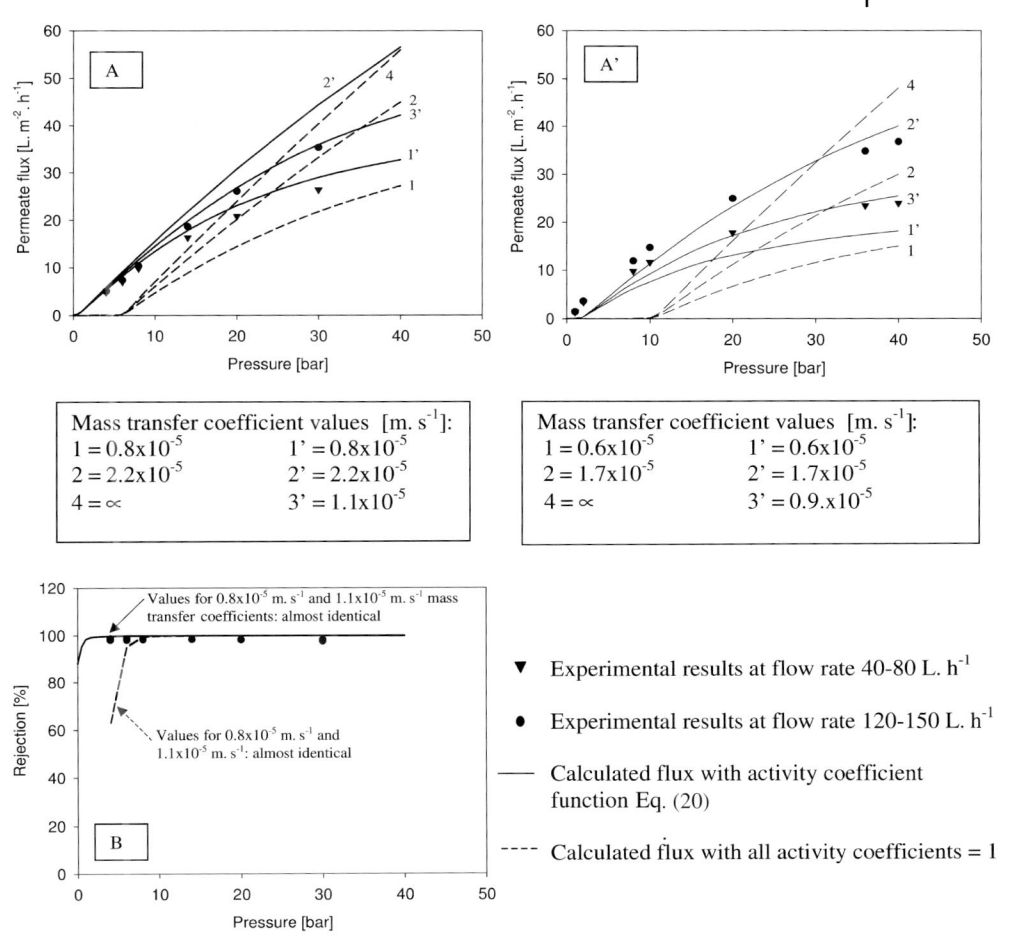

Fig. 4.8 A, A'. Experimental and calculated values for permeate flux of 0.21 M (A) and 0.33 M (A') TOABr solution. B. Experimental and calculated values for rejection of 0.21 M TOABr solution. Results shown in both cases for activity coefficients: $\gamma_T = \gamma_{TOABr} = 1$, and $\gamma_T = -4.16 x_T^2 + 7.29 x_T -2.13$, $\gamma_{TOABr} = 1$. (Reprinted from [32] with permission from Elsevier.)

Mass transfer coefficient values [m. s^{-1}] (A):
1 = 0.8x10^{-5}
1' = 0.8x10^{-5}
2 = 2.2x10^{-5}
2' = 2.2x10^{-5}
4 = ∞
3' = 1.1x10^{-5}

Mass transfer coefficient values [m. s^{-1}] (A'):
1 = 0.6x10^{-5}
1' = 0.6x10^{-5}
2 = 1.7x10^{-5}
2' = 1.7x10^{-5}
4 = ∞
3' = 0.9.x10^{-5}

▼ Experimental results at flow rate 40-80 L. h^{-1}

● Experimental results at flow rate 120-150 L. h^{-1}

—— Calculated flux with activity coefficient function Eq. (20)

---- Calculated flux with all activity coefficients = 1

Figure 4.8A shows a comparison of the experimental and model data. The figure demonstrates that it is only possible to describe the low pressure flux behavior of the system with the inclusion of activity coefficients, indicating that the system is not ideal. Other parameters in the model, such as the permeability of the solvent and the solute, were varied to check whether they could be responsible for this effect. However, it was found to be impossible to describe the data by alteration of the two permeabilities or the mass transfer coefficient. Equally, it is only possible to describe the high-pressure, concentration polarization effect with the inclusion of mass transfer effects. The overall system requires both activity coefficients and mass transfer coefficients in order to obtain a satisfactory description of the experimental data. The mass transfer coefficient values estimated from Chilton-Colburn correlation (see Tab. 4.2) describe the experimental data well at the lower flow rate (40–80 L h^{-1}), corresponding to a mass transfer coefficient of 0.77×10^{-5} m s^{-1}. The flux values at the higher flow rate range 120–180 L h^{-1}, are overestimated, using the mass transfer coefficient of 2.2×10^{-5} m s^{-1} predicted by Chilton-Colburn correlation and the experimental data are better described by a mass transfer coefficient of 1.1×10^{-5} m s^{-1}, as shown in Fig. 4.8 A. This difference can be attributed to the fact that the Chilton-Colburn correlation, as for many other mass transfer correlations is developed for nonporous smooth duct flow and its application to membrane operations may be limited [53]. It does not account for the change of physical properties such as viscosity and diffusivity across the boundary layer. Also, as mentioned earlier, the flow pattern in the cell is likely to be a mixed-tangential one, which makes hydrodynamics difficult to describe. Therefore, the values estimated from the Chilton-Colburn correlation should be considered an approximation.

The model predicts very high rejection (Fig. 4.8 B) for both the ideal and nonideal cases above about 10 bar, as observed experimentally. The mass transfer coefficient seems to have a negligible effect on the rejection, as observed for docosane. There is a discrepancy between nonideal and ideal model data for pressures under 10 bar. If activity coefficients are included, the model predicts ~100% rejection for nearly all pressures, only deviating slightly from 100% at very low pressures (~2 bar). If activity coefficients are not included, the rejection begins to deviate from 100% at around 8 bar and decreases to ~60% as the pressure decreases to 4 bar, where the total flux becomes nearly zero. This behavior for the ideal solution case is due to the fact that the model predicts that the solvent flux drops considerably at pressures lower than 6 bar, while the solute flux does not change so dramatically, thus forcing the rejection to drop.

The divergence of the system from ideality increases at higher concentrations of TOABr, as illustrated for the flux data in Fig. 4.8 A′, for 0.33 M TOABr in toluene. Note that, as for the 0.21 M case, the model without activity coefficients also predicts an osmotic pressure, this time about 10 bar, even at infinite mass transfer coefficient (line 4 on Fig. 4.8 A′). As before, this phenomenon is attributed to activity differences. Again the mass transfer correlation slightly under predicts the permeate flux values, but this time at the lower flow rate (40–80 L h^{-1}), with a mass transfer coefficient value of 0.6×10^{-5} m s^{-1} versus 0.9×10^{-5} m s^{-1} as estimated by comparing the model to the experimental data. More surprising is the fact that the mass transfer coefficient values describing the experimental data at 0.33 M TOABr are slightly higher than those describing 0.21 M TOABr solutions. This could be due to nonideality of the system, the unpredictable changes of the diffusion coefficient with concentration or the build up of a gel layer at the membrane surface. The latter is investigated further below.

The extent of concentration polarization at 0.33 M is demonstrated by Fig. 4.9 (A, B), which shows the ratio of the predicted concentration at the membrane/liq-

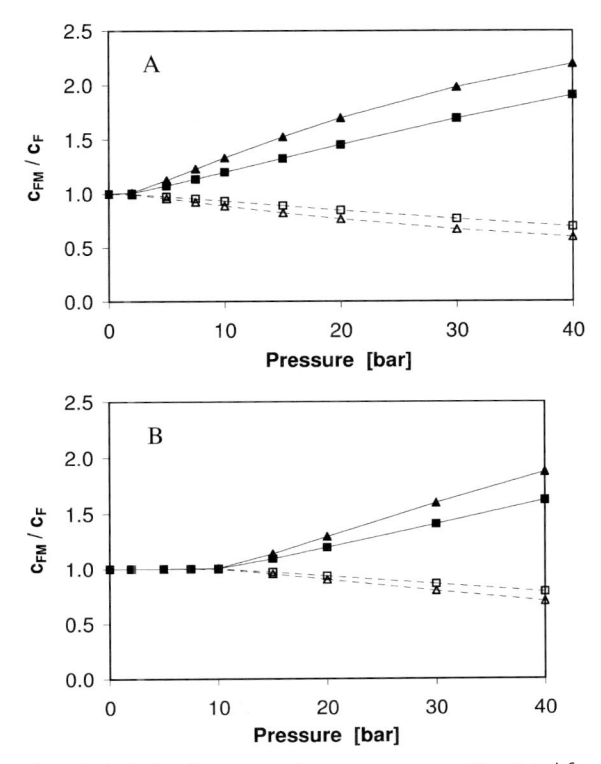

Fig. 4.9 A. Ratio of concentration at membrane surface to bulk concentration for 0.33 M TOABr solution, $\gamma_T = -4.16 x_T^2 + 7.29 x_T - 2.13$, $\gamma_{TOABr} = 1$. B. Ratio of concentration at membrane surface to bulk concentration for 0.33 M TOABr solution, $\gamma_T = \gamma_{TOABr} = 1$. (Reprinted from [32] with permission from Elsevier.)
■ 1.7×10^{-5} m s^{-1} TOABr
▲ 0.9×10^{-5} m s^{-1} TOABr
□ 1.7×10^{-5} m s^{-1} toluene
△ 0.9×10^{-5} m s^{-1} toluene

uid interface to the bulk liquid concentration. At 40 bar pressure, the TOABr concentration at the membrane/liquid interface is over twice the bulk concentration (Fig. 4.9A, for the nonideal case), illustrating that the mass transfer limitation in the system is severe. This represents a concentration of over 0.72 M at the membrane surface, causing concern that a gel-layer might be formed. The solubility of TOABr in toluene at 30 °C was measured to be 0.76 M. Hence, the TOABr should not precipitate out of solution at the membrane surface, but clearly is approaching the range where this might occur. At low pressures, this effect is much less significant due to the lower solvent flux. When nonideality is not accounted for (Fig. 4.9B), the mass transfer limitation is less severe (the solvent flux is lower due to the higher osmotic pressure effect): the concentration at the membrane surface is about 1.9 times the bulk concentration at 40 bar. For both the ideal and the nonideal case, the concentration polarization effect appears over the whole pressure range, up to the point where the permeate flux becomes ~0.

If the concentrations at the membrane surface really are as high as 0.72 M, the viscosity at the membrane surface would also be high, thus inhibiting mass transfer even further. This questions whether it is valid to use a constant mass transfer coefficient in the system. An extension of this study could be to include variation of the diffusion coefficient (and thus the mass transfer coefficient) with position in the boundary layer.

An interesting comparison is the variation in the solvent flux for the two different systems under exactly the same conditions: 40 bar, 0.33 M, and cell flow rate 120–150 L h^{-1}. The toluene flux in the docosane-toluene system is 20.7 L m^{-2} h^{-1} and in the TOABr-toluene system is 36.7 L m^{-2} h^{-1}. This is in spite of the higher viscosity and lower mass transfer in the TOABr-toluene system. Thus it can be seen that the nonideality of the TOABr-toluene system actually assists the filtration process by reducing the osmotic pressure difference across the membrane and thus allowing a higher flux.

4.5
Conclusions

From this brief summary of the data for crossflow OSN transport processes, we can conclude that:
- The data for binary systems of mixed solvents can be described well using a two-parameter model, when the two parameters are obtained from the fluxes of each of the pure solvents. The solution-diffusion model provides a slightly better fit, but there is not much difference between this and the pore-flow model.
- This approach should be able to be extended to the more challenging ternary and quaternary mixtures of solvents, when 3- and 4-parameter models would be used with parameters based on the fluxes of each of the 3 or 4 pure solvents, respectively. In fact it should of course be possible to generalize either approach to n solvents and use the data for n pure solvents to calculate mixture fluxes.
- Osmotic pressure and concentration polarization can both be important in OSN, especially in systems such as TOABr-toluene where viscosity increases significantly with concentration.
- Unsurprisingly, properly accounting for the driving forces in OSN involves calculating the activity coefficients of the organic species present in the fluid streams upstream and downstream of the membrane. This is analogous to the thermodynamic calculations one must undertake for other separation processes, such as distillation, and techniques and methods (such as UNIFAC) can be employed.
- Obtaining the fluxes of pure solvents is straightforward experimentally, and one can easily imagine that a designer will in fu-

ture have a database of the membrane permeabilities of different solvents for specific commercial membranes. These should make a-priori calculations of the fluxes of mixtures of solvents possible.
- Obtaining the membrane permeabilities of larger, solute-like molecules (one could consider these as molecules that would either be solids in pure form, or be oils with no measurable membrane flux) is more complex. Up to now we have had to infer these from data for solutes present in solvents. This is unsatisfactory, since it means that we must always make experimental measurements for specific systems.

While there are many challenges faced in developing good design models, methods and databases for OSN, the technology itself has already been used at large process scale in lube oil refining and will, we believe, grow into more and further applications in the coming years.

Nomenclature

c	Molar concentration (mol m^{-3})
k	Mass transfer coefficient (m s^{-1})
l	Membrane thickness (m)
n	Mass flux (kg m^{-2} s^{-1})
N	Molar flux (mol m^{-2} s^{-1})
N_v	Total volumetric flux (m s^{-1})
p	Pressure (Pa)
$P_{i,m}^{molar}$	Molar permeability (mol m^{-2} s^{-1})
$P_{i,m}^{mass}$	Mass permeability (kg m^{-2} s^{-1})
R	Ideal gas constant (Pa m^3 mol^{-1} K^{-1})
T	Temperature (K)
$\bar{V}$	Partial molar volume (m^3 mol^{-1})
w	Mass fraction (–)
x	Molar fraction (–)
d_{pore}	Pore diameter (m)
$d_{particle}$	Particle diameter (m)

Greek letters

δ	Boundary layer thickness (m)
ε	Porosity (–)
γ	Molar activity coefficient (–)
η	Viscosity (Pa s)
ρ	Density (kg m^{-3})
τ	Tortuosity factor (–)
$\nabla_{T,p}$	Gradient at constant temperature and pressure (m^{-1})

Subscripts

$i, 1, 2$	Species
$m, (m)$	Membrane
mix	Mixture
F	Feed side
FM	Feed side, at the membrane/liquid interface
P	Permeate side

References

1 L. S. White, I. Wang, B. S Minhas, Polyimide membrane for separation of solvents from lube oil, US Patent 5264166 (1993).
2 C. Linder, M. Nemas, M. Perry, R. Katraro, Silicone-derived solvent stable membranes, US Patent 5205934 (1993).
3 L. S. White, Polyimide membranes for hyperfiltration recovery of aromatic solvents, US Patent 6180008 (2001).
4 S. S. Kulkarni, E. W. Funk. N. N. Li, Hydrocarbon separations with polymeric membranes, *AICHE Symp. Series* No 250, **82** (1986) 78–84.
5 L. P. Raman, M. Cheryan, N. Rajagopalan, Deacidification of soybean oil by membrane technology, *JAOSC* **73** (1996) 219–224.
6 H. J. Zwijenburg, A. M. Krosse, K. Ebert, K. V. Peinnemann, F. P. Cuperus, Acetone-stable nanofiltration membranes in deacidifying vegetable oil, *JAOSC* **76** (1999) 83–87.
7 L. W. Gosser, W. H. Knoth, G. W. Parshall, Reverse osmosis in homogeneous catalysis, *J. Molec. Catal.* **2** (1977) 253–263.
8 S. S. Luthra, X. Yang, L. M. Freitas dos Santos, L. S. White, A. G. Livingston, Phase-transfer catalyst separation and re-use by solvent resistant nanofiltration membranes, *Chem. Commun.* (2000) 1468–1469.
9 D. Nair, J. T. Scarpello, L. S. White, L. M. Freitas dos Santos, I. F. J. Vankelecom, A. G. Livingston, Semi-continuous nanofiltration-coupled Heck reactions as a new approach to improve productivity of homogeneous catalysts, *Tetrahedr. Lett.* **42** (2001) 8219–8222.
10 K. De Smet, S. Averts, E. Cuelemans, I. F. J. Vankelecom, P. A. Jacobs, Nanofiltration-coupled catalysis to combine the advantages of homogeneous and heterogeneous catalysis, *Chem. Commun.* (2001) 597–598.
11 S. S. Luthra, X. Yang, L. M. Freitas dos Santos, L. S. White, A. G. Livingston, Homogeneous phase transfer catalyst recovery and re-use using solvent resistant membranes, *J. Membr. Sci.* **201** (2002) 65–75.
12 D. Nair, S. S. Luthra, J. T. Scarpello, L. S. White, L. M. Freitas dos Santos, A. G. Livingston, Homogeneous catalyst separation and re-use through nanofiltration of organic solvents. *Desalination* **147**, 1–3 (2002) 301–306.
13 H. P. Dijkstra, G. P. M. Van Klink, G. Van Koten, The use of ultra- and nanofiltration techniques in homogeneous catalyst recycling, *Acc. Chem. Res.* **35** (2002) 798–810.
14 A. Datta, K. Ebert, H. Plenio, Nanofiltration for homogeneous catalysis separation: Soluble polymer-supported palladium catalysts for Heck, Sonogashira, and Suzuki coupling of aryl halides, *Organometallics* **22** (2003) 4685–4691.
15 J. Krockel, U. Kragl, Nanofiltration for the separation of nonvolatile products from solutions containing ionic liquids, *Chem. Eng. Technol.* **26** (2003) 1166–1168.
16 A. G. Livingston, L. Peeva, S. Han, S. S. Luthra, L. S. White, L. M. Freitas dos Santos, Membrane separation in

green chemical processing – solvent nanofiltration in liquid-phase organic synthesis reactions, Advanced Membrane Technology, Volume. 984, Ann. New York Acad. Sci., New York, 2003.
17 J. Sheth, Y. Qin, K.K. Sirkar, B.C. Baltzis, Nanofiltration-based diafiltration process for solvent exchange in pharmaceutical manufacturing, *J. Membr. Sci.* **63** (1999) 93–102.
18 L.S. White, A.R. Nitsch, Solvent recovery from lube oil filtrates with a polyimide membrane, *J. Membr. Sci.* **179** (2000) 267–274.
19 N.A. Bhore, R.M. Gould, S.M. Jacob, P.O. Staffeld, D. McNally, P.H. Smiley, C.R. Wildemuth, New membrane process debottlenecks solvent dewaxing unit, *Oil Gas J.* **97** (46) (1999) 67–74.
20 D.R. Machado, D. Hasson, R. Semiat, Effect of solvent properties on permeate flow through nanofiltration membranes. Part I. Investigation of parameters affecting solvent flux, *J. Membr. Sci.* **163** (1999) 93–102.
21 D.R. Machado, D. Hasson, R. Semiat, Effect of solvent properties on permeate flow through nanofiltration membranes, Part II. Transport model, *J. Membr. Sci.* **166** (2000) 63–69.
22 J.A. Whu, B.C. Baltzis and K.K. Sirkar, Nanofiltration studies of larger organic microsolutes in methanol solutions, *J. Membr. Sci.* **170** (2000) 159–172.
23 X.J. Yang, A.G. Livingston, L. Freitas dos Santos, Experimental observations of nanofiltration with organic solvents, *J. Membr. Sci.* **190** (2001) 45–55.
24 D. Bhanushali, S. Kloos, C. Kurth, D. Battacharyya, Performance of solvent-resistant membranes for non-aqueous systems: solvent permeation results and modeling, *J. Membr. Sci.* **189** (2001) 1–21.
25 L.S. White, Transport properties of a polyimide solvent resistant nanofiltration membrane, *J. Membr. Sci.* **205** (2002) 191–202.
26 E. Gibbins, M.D. Antonio, D. Nair, L.S. White, L.M. Freitas dos Santos, I.F.J. Vankelecom, A.G. Livingston, Observation of solvent flux and solute rejection across solvent resistant nanofiltration membranes, *Desalination* **147** (2002) 307–313.
27 B. Van der Bruggen, J Geens, C. Vandecasteele, Influence of organic solvents on the performance of polymeric nanofiltration membranes, *Sep. Sci. Technol.* **37** (2002) 783.
28 B. Van der Bruggen, J. Geens, C. Vandecasteele, Fluxes and rejection for nanofiltration with solvent stable polymeric membranes in water, ethanol and n-hexane, *Chem. Eng. Sci.* **57** (2002) 2511–2518.
29 D. Bhanushali, S. Kloos, D. Battarcharyya, Solute transport in solvent-resistant nanofiltration membranes for non-aqueous systems: experimental results and the role of solute-solvent coupling, *J. Membr. Sci.* **208** (2002) 343–359.
30 J.P. Robinson, E.S. Tarleton, C.R. Millington, A. Nijmeijer, Solvent flux through dense polymeric nanofiltration membranes, *J. Membr. Sci.* **230** (2004) 29–37.
31 N. Stafie, D.F. Stamatialis, M. Wessling, Insight into the transport of hexane-solute systems through tailor-made composite membranes, *J. Membr. Sci.* **228** (2004) 103–116.
32 L.G. Peeva, E. Gibbins, S.S. Luthra, L.S. White, R.P. Stateva, A.G. Livingston, Effect of concentration polarization and osmotic pressure on flux in organic solvent nanofiltration, *J. Membr. Sci.* **236** (2004) 121–136.
33 I.F.J. Vankelecom, K. DeSmet, L.E.M. Gevers, A.G. Livingston, D. Nair, S. Aerts, S. Kuypers, P.A. Jacobs, Physico-chemical interpretation of the SRNF transport mechanism for solvents through dense silicone membranes, *J. Membr. Sci.* **231** (2004) 99–108.
34 P. Silva, S. Han and A.G. Livingston, Solvent transport in organic solvent nanofiltration membranes, *J. Membr. Sci.* **262** (2005) 49–59.
35 J.G. Wijmans, R.W. Baker, The solution-diffusion model: a review, *J. Membr. Sci.* **107** (1995) 1–21.
36 D.R. Paul, O.M. Ebra-Lima, Pressure induced diffusion of organic liquids through highly swollen polymer mem-

37 D. R. Paul, Reformulation of the solution-diffusion theory of reverse osmosis, *J. Membr. Sci.* **241** (2004) 371–386.
38 J. G. Wijmans, S. Nakao, C. A. Smolders, Flux limitation in ultrafiltration: osmotic pressure model and gel layer model, *J. Membr. Sci.* **20** (1984) 115–124.
39 J. G. Wijmans, S. Nakao, J. W. A. Van der Berg, F. R. Troelstra, C. A. Smolders, Hydrodynamic resistance of concentration polarization boundary layers in ultrafiltration, *J. Membr. Sci.* **27** (1985) 117–135.
40 S. Nakao, J. G. Wijmans, C. A. Smolders, Resistance to the permeate flux in unstirred ultra-filtration of dissolved macromolecular solutions, *J. Membr. Sci.* **26** (1986) 165–178.
41 G. B. Van der Berg, C. A. Smolders, The boundary-layer resistance model for unstirred ultrafiltration. A new approach. *J. Membr. Sci.* **40** (1989) 149–172.
42 W. N. Gill, Effect of viscosity on concentration polarization in ultrafiltration, *AIChE J.* **34**, no. 9 (1998) 1563–1567.
43 Z. V. P. Murthy, S. K. Gupta, Estimation of mass transfer coefficient using a combined nonlinear membrane transport and film theory, *Desalination* **109** (1997) 39–49.
44 Z. V. P. Murthy, S. K. Gupta, Sodium cyanide separation and parameter estimation for reverse osmosis thin-film composite polyamide membrane, *J. Membr. Sci.* **154** (1999) 89–103.
45 R. H. Perry, D. W. Green, Perry's Chemical Engineers' Handbook, 6th edn, McGraw-Hill, 1984, 3–266 & 3–273.
46 P. P. Prádamos, J. I. Arribas, A. Herrádez, Mass-transfer coefficient and retention of PEGs in low pressure crossflow Ultrafiltration through asymmetric membranes, *J. Membr. Sci.* **99** (1995) 1–20.
47 M. Um, S. Yoon, C. Lee, K. Chung, J. Kim, Flux enhancement with gas injection in crossflow ultrafiltration of oily waste water, *Wat. Res.* **35** (2001) 4095–4101.
48 H. M. Yeh, Y. K. Chen, I. H. Tseng, The effect of aspect ratio on solvent extraction in crossflow parallel plate membrane modules, *Sep. Purif. Technol.* **28** (2002) 181–190.
49 S. Platt, M. Mauramo, S. Butylina, M. Nyström, Retention of PEGs in crossflow ultrafiltration through membranes, *Desalination.* **148** (2002) 417–422.
50 B. L. Larsen, P. Rasmussen, A. A. Fredenslund, A modified UNIFAC group contribution model for prediction of phase equilibria and heat of mixing. *Ind. Eng. Chem. Res.* **26** (1987) 2274–2286.
51 A. Fredenslund, J. Gmehling, P. Rasmussen, Vapor–liquid equilibria using UNIFAC, Elsevier, Amsterdam, 1977.
52 B. E. Poling, J. M. Prausnitz, J. P. O'Connel, The properties of gases and liquids, McGraw-Hill, Fifth edition, 2001, 11.34–11.41.
53 V. Gekas, B. Hallström, Mass transfer in the membrane concentration polarization layer under turbulent crossflow. I Critical literature review and adaptation of existing Sherwood correlations to membrane operations, *J. Membr. Sci.* **30** (1987) 153–170.

5
Industrial Membrane Reactors

M. F. Kemmere and J. T. F. Keurentjes

5.1
Introduction

Over the past decades, membrane processes have found broad application for a wide range of separations. The first large-scale applications of membrane technology can be found in brackish-water desalination using reverse osmosis and hemodialysis. Based on the different driving forces applied, the range of separations can be divided into various filtration processes (microfiltration, ultrafiltration, nanofiltration and reverse osmosis), gas and vapor separation, pervaporation and electromembrane processes (including electrodialysis, membrane electrolysis and bipolar membrane processes). Additionally, based on preferential wetting properties, porous membranes have been used as a support for liquid membranes and for various contactor applications (including membrane-based solvent extraction and gas absorption). These processes usually focus on a desired separation of a gas or liquid mixture.

When it comes to combination with a reaction or conversion, membranes have mainly found application in a sequential mode, i.e. reaction followed by separation. In this chapter, we will focus on the integration of conversion and separation in so-called membrane reactors. As the separation function of the membrane can be used in various modes of operation, this leads to a broad variety of process options. In the past few years, several review papers have emerged, usually covering parts of this huge research field [1–5]. The general advantages of membrane reactors as compared to sequential reaction-separation systems are:
- increased reaction rates
- reduced byproduct formation
- lower energy requirement
- possibility of heat integration.

These advantages potentially lead to compact process equipment that can be operated with a high degree of flexibility. Because of the reduced byproduct forma-

Membrane Technology in the Chemical Industry.
Edited by Suzana Pereira Nunes and Klaus-Viktor Peinemann
Copyright © 2006 WILEY-VCH Verlag GmbH & Co. KGaA, Weinheim
ISBN: 3-527-31316-8

tion and the more efficient use of energy, the development of membrane reactors clearly fits into the scope of developing sustainable processes for the future.

First applications of membrane reactors can be found in the field of bioprocess engineering using whole cells in fermentations or enzymatic bioconversions [6, 7]. Most of these processes use polymeric membranes, as temperatures seldomly exceed 60 °C. The development of inorganic membrane materials (zeolites, ceramics and metals) has broadened the application potential of membrane reactors towards the (petro)chemical industry [8]. Many of these materials can be applied at elevated temperatures (up to 1000 °C), allowing their application in catalytic processes.

The basic functions of the membrane in membrane reactors can be divided into (Fig. 5.1):
- selective and nonselective addition of reactants
- selective and nonselective removal of reaction products
- retention of the catalyst

As the membrane acts as a separating medium between two flow compartments, these basic functions can be applied to liquid/liquid, gas/liquid and gas/gas systems, respectively. The physical shape of the membrane strongly depends on the membrane material used. For polymeric systems, these can be flat sheets in a plate-and-frame configuration, spiral-wound modules, and tubular mem-

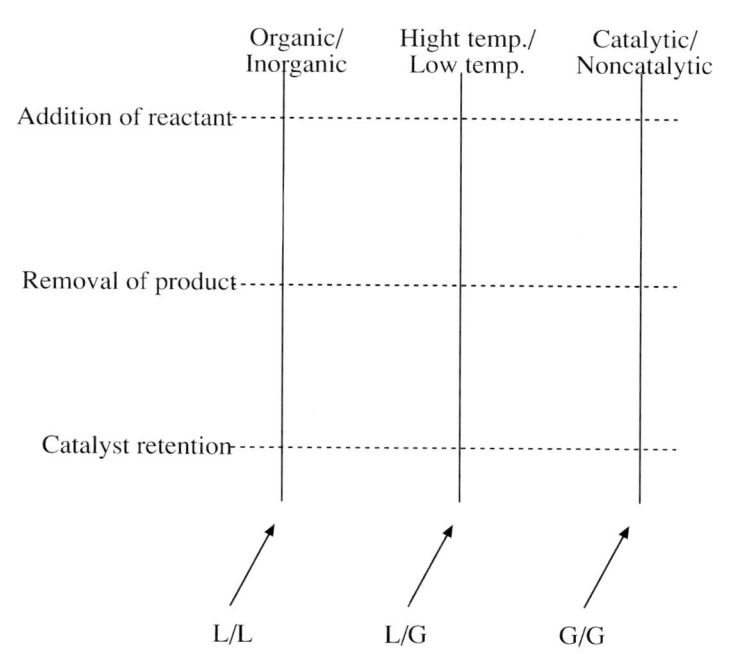

Fig. 5.1 Schematic overview of the basic functions membranes can have in membrane reactors.

branes and hollow fibers in a shell-and-tube configuration, respectively [9]. The first two systems will not easily allow for independent flow of both compartments, as the permeate chamber can not be flushed. On the other hand, spiral-wound modules allow for a large surface area per volume (typically 1000 m^2/m^3). Even more area per volume can be obtained in hollow-fiber units, for which a typical value is around 10 000 m^2/m^3. Most polymeric membranes can not be applied at temperatures above 100–150 °C, which implies that for these conditions inorganic membranes have to be used. These can be produced in several shapes: flat plates, tubes, multihole elements and hollow fibers. Although the hollow-fiber systems are still in an early stage of development [10–13] they represent a promising group of materials, especially due to the high surface area per volume that can be obtained.

With respect to catalytic membrane reactors, processes can be divided in homogeneously and heterogeneously catalyzed reactions (see Fig. 5.2). In homogeneously catalyzed processes, the membrane modules can be used in loop reactors. For heterogeneously catalyzed reactions several configurations are possible:
- the membrane units and fixed-bed catalysts can be applied in series
- the fixed-bed catalysts can be integrated in the membrane module
- the membrane itself can have the desired catalytic activity.

In the following sections, we will discuss the developments in the field of membrane reactors. Emphasis will be on the application potential of the various options on a large scale. The setup will be along the lines of the three basic functions the membrane can have in these systems. Finally, some developments that have led to applications on an industrial scale, or that are relatively close to this, will be described in more detail.

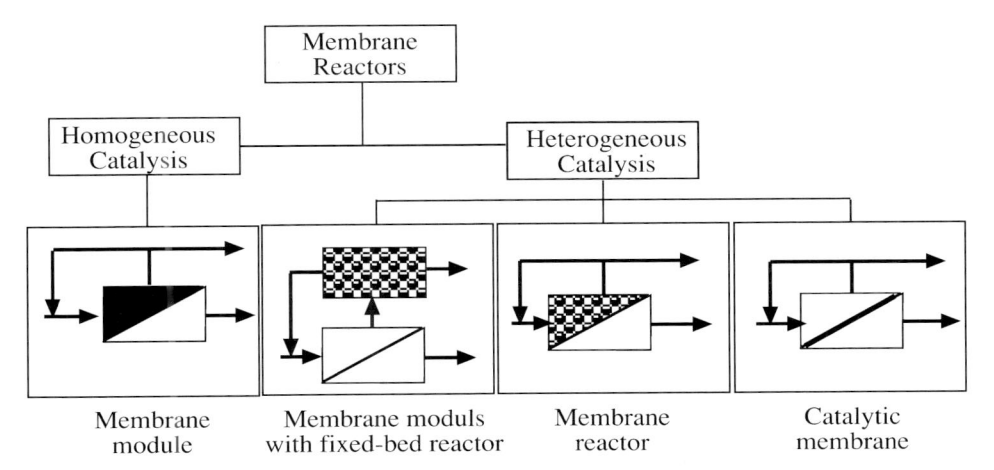

Fig. 5.2 Classification of catalytic membrane reactors [14].

5.2
Membrane Functions in Reactors

The most generic distinction in the wide variety of membrane reactors can be made according to the possible functional roles of the membrane in the reactor, being controlled addition of reactants, separation of products from the reaction mixture and retention of the catalyst. Additionally, membrane processes can be divided based on the physical state of the retentate and permeate, respectively:
- liquid/liquid systems such as microfiltration, ultrafiltration, nanofiltration, reverse osmosis, liquid/liquid contactors and dialysis
- liquid/gas systems like pervaporation or gas/liquid contactors
- gas/gas systems such as gas permeation.

Based on a major division by membrane function in the reactor, a number of examples of membrane reactors are given below, illustrating the importance of the use of membranes for combining reaction and separation. Obviously, the list of membrane-based processes described here will not be exhaustive, although the following paragraphs will give an overview of the applications of membranes for chemical reactions.

5.2.1
Controlled Introduction of Reactants

The major advantage of using membranes for the addition of reactants comprises the independent control of the concentration levels of each reactant in the reaction zone. One reactant can be fed along the length of a reactor, as schematically shown in Fig. 5.3. This is commonly done in a tube-and-shell configuration. An additional advantage is the possibility to apply a permselective membrane for purification of a reactant from a mixed stream before addition into the reaction zone, e.g. utilizing pure O_2 from an air stream. Also, the membrane can be used for the coupling of two reactions by physically separating the two reaction media and introducing the product of one reaction as a reactant for the second reaction.

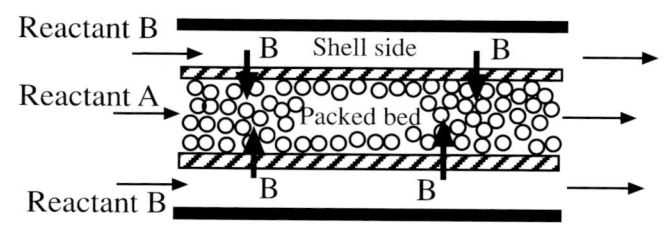

Fig. 5.3 Schematic representation of a membrane reactor for controlled reactant feed [1].

Gas-phase Reactions

For gas-phase reactions, the controlled addition of reactants A and B can effectively be applied to systems with two competing reactions, e.g. partial oxidation of hydrocarbons:

$$A + \alpha B \xrightarrow{k_1} P \tag{1}$$

$$A + \beta B \xrightarrow{k_2} S \tag{2}$$

In this reaction scheme, P is the desired product and S is the undesired byproduct. In the case the reaction rates are proportional to the partial pressure of reactant B ($r_1 = k_1 p_B^{n1}$ and $r_2 = k_2 p_B^{n2}$, respectively) the kinetics are favourable if $n1 < n2$. Also, a lower p_B will slow reaction 2 more than reaction 1, inducing an increased selectivity for the desired product P. For this purpose, mainly porous membranes are used. To control the uniformity of the distribution of B, the membrane should have sufficient resistance to equalize the pressure on the reactant side, i.e. a constant transmembrane pressure drop along the tube [15, 16]. Another problem to tackle in this type of systems is back diffusion of reactant A and product(s) P and S. Here also, an increased pressure drop across the membrane will be advantageous, although it also decreases the permeation rate of B, which potentially leads to problems balancing the feed rate to the reaction rate 1.

In addition to the use of porous membranes, dense membranes can also be applied for controlled addition of reactants. In this respect, focus has mainly been on hydrogen and oxygen supply. For hydrogenation reactions, Pd-based membranes have been used [17, 18], resulting in improved yields. Nevertheless, this can not be attributed to the kinetics considerations given above, but is due to a better availability of H^+ at the membrane surface. Most of the work on dense membranes for controlled addition of reactants, however, has been done for oxygen supply. Despite the high temperatures required (>700 °C), currently the main focus is on the application of solid oxides, with much emphasis on the use of various perovskites [19–23]. Controlled supply of oxygen has mainly been studied for the oxidative coupling of methane. Yields in excess of 50% have been predicted theoretically, however, the experimental results have not exceeded 25–30%, similar to fixed-bed results. Both dense [19–23] and porous membranes [16, 24–26] have been used, both leading to similar results.

In addition to oxidative coupling of methane, several oxidative dehydrogenations have been investigated, including the conversion of ethane to ethene [15, 27, 28], propane to propene [29] and butane to butene [30]. In these systems, controlling the hydrocarbon to oxygen ratio was found to be crucial for selectivity (with respect to byproducts like CO_2, CO, etc.). At low and moderate ratios, the ethene yield in the membrane reactor exceeds that in a cofeed plug flow reactor operated under the same conditions by a factor of 3. In the membrane reactor the reactant feed ratio near the inlet of the reactor is relatively high, resulting in a high selectivity. Continuous addition of oxygen along the tube ensures high conversion as well.

Non-permselective membranes can also be used to provide a location for a reaction zone. One reactant is fed on the tube side of the membrane, and the other reactant is fed on the shell side. The partial-pressure gradients have to be chosen such that the two reactants permeate towards each other inside the membrane, where they can react. Usually, the membrane itself contains a suitable catalyst. In this type of membrane reactor, reactions are performed at a strict stoichiometric ratio. For fast reactions, this results in a reaction plane, whereas for slower reactions a reaction zone will be formed. This is shown schematically in Fig. 5.4. Balancing reaction rate and permeability can result in a reaction zone entirely located inside the membrane. When breakthrough of reactants can be avoided, and the product diffuses out on one side only, this can simplify the further separations required.

An example of this setup is the dehydrogenation of methanol and butane using microporous γ-alumina and Ag-modified γ-alumina membranes as the catalyst [31–34]. The methanol and oxygen are fed on different sides of the membrane, thus minimizing undesired gas-phase reactions. Additionally, the catalytic activity of the membranes appeared to be 10 times higher than the activity of the same catalyst when packed. This is attributed to the effective regen-

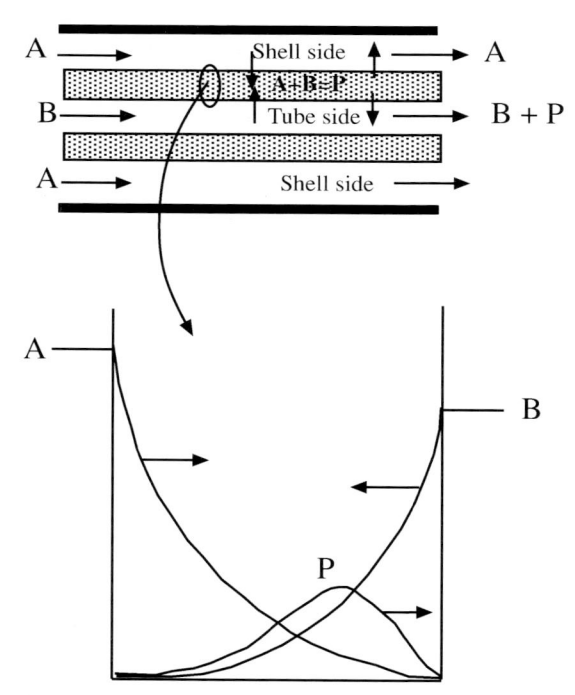

Fig. 5.4 Catalytic non-permselective membrane reactor with separated feed, showing transmembrane concentration profiles of the different species involved and the direction of the permeation [1].

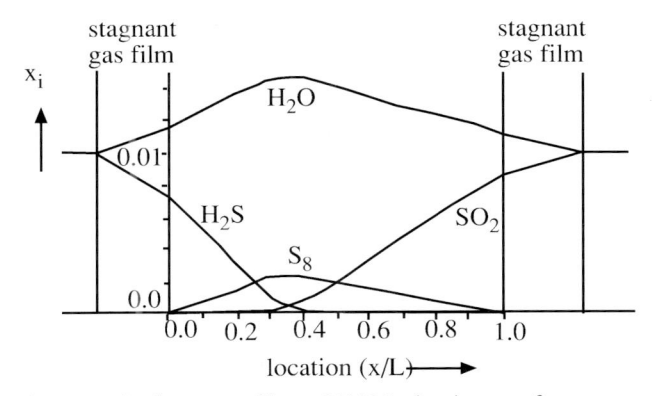

Fig. 5.5 Molar fraction profiles at 200 °C in the absence of a pressure difference over the (non-permselective) membrane used for the Claus reaction [38, 39].

eration of the catalyst. Since all oxygen passes through the catalyst layer, this allows for effective burning of the carbon deposit. A similar system was used for the reduction of nitric oxide with ammonia [34–37]. The ability of the membrane to act as a barrier between reactants has also been shown to be effective for the Claus reaction (Fig. 5.5) [38, 39]. In general, using the membrane as the reaction zone is of particular interest for fast, exothermic heterogeneously catalyzed reactions, since runaway is prevented due to mass transport being the rate-limiting step [40].

Liquid-phase Reactions

The main application of membrane reactors for liquid–liquid systems is based on intimate phase contacting, without the formation of an emulsion. This avoids troublesome phase-splitting afterwards. Microporous membranes have proven to be particularly useful for this type of applications, since the two immiscible liquids can be kept on two sides of the membrane with their interface immobilized at the membrane surface [41–43]. The general advantages of these systems are: no dispersion is formed, thus avoiding coalescence; no density difference required between the two phases; large and known interfacial area (typically 10 000 m^2/m^3); no loading and flooding, thus allowing for widely different phase flow ratios. The application of membrane contactors for reactive systems has been explored for three types of systems: fermentor-extractor, enzymatic reactions (enzymatic hydrolysis/esterification and enzymatic resolution of isomers) and phase transfer catalysis (PTC).

In the group of Sirkar, the application of microporous hollow fibers in the fermentative production of ethanol, acetone-butanol-ethanol (ABE), etc. has been explored. In these systems, the role of the membrane is twofold. First, oxygen or nitrogen is supplied and the reaction products CO_2 and H_2 are removed. Secondly, an organic solvent is passed through the fibers, extracting the products (ethanol, ABE) [44–47]. This reduces product inhibition and can therefore lead to considerable increased volumetric fermenter productivity.

When lipases are used for enzymatic conversions, the enzyme is mainly active at a phase boundary, which can effectively be provided by a membrane. Additionally, for conversions requiring two phases (e.g. fat splitting [48–50] and esterifications [51]), the membrane also keeps the two liquid phases (an oil and an aqueous phase, respectively) separated. This is schematically depicted in Fig. 5.6. The equilibrium reactions involved are

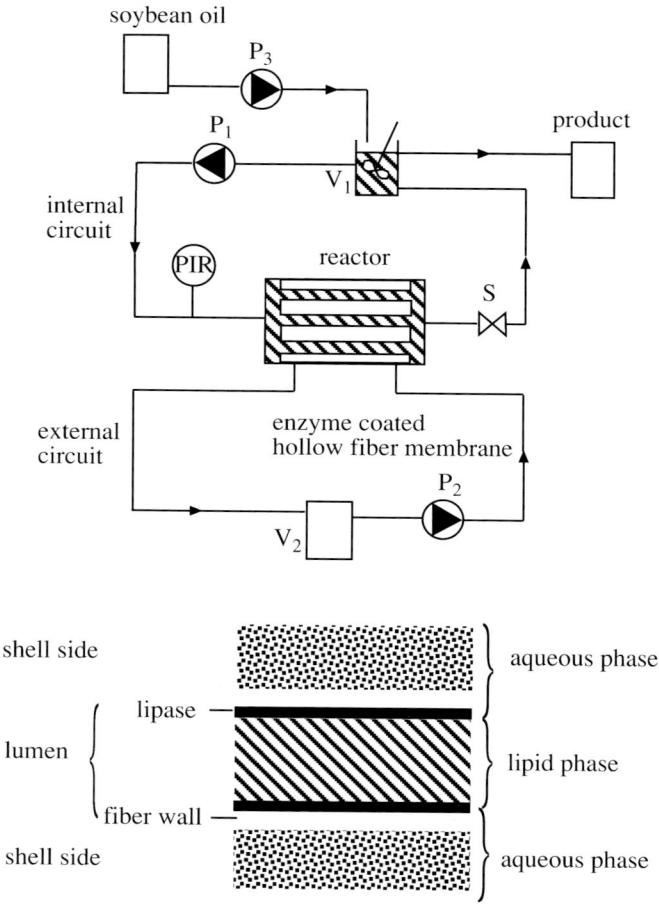

Fig. 5.6 Experimental setup (top) and schematic representation of a cross section through a hollow fibre (bottom) for the enzymatic conversion of triglycerides into fatty acids [49]. For this purpose a hydrophilic membrane is used, coated with the enzyme (lipase) on the lipid side.

Triglycerides + water <−> diglyceride + fatty acid (3)

Diglyceride + water <−> monoglyceride + fatty acid (4)

Monoglyceride + water <−> glycerol + fatty acid (5)

The use of both hydrophilic [49, 51] and hydrophobic [48, 50] membranes have proven to be efficient in binding the enzyme. The main advantage of this system as compared to emulsion systems lies in the ease of the downstream processing, as no enzyme-stabilized emulsion has to be broken.

In addition to the enzymatic fat splitting and esterifications, a multiphase extractive enzyme membrane reactor is being used for the industrial production of a diltiazem chiral intermediate. This process will be described in more detail in Section 5.3.5.

Phase-transfer catalysis has been investigated for the model displacement reaction of bromooctane with iodide. The first is dissolved in an organic solvent (chlorobenzene) and the latter in an aqueous phase. The phase-transfer catalyst used was the tetrabutylammonium ion, dissolved in the organic phase [52]. Using a hydrophobic membrane-contactor device, conventional coalescence problems were avoided. Additionally, as a result of the interfacial area being known, operation of the reactor can be performed with greater flexibility.

Gas–liquid Reactions

Reactions requiring both a gaseous and a liquid reactant are usually performed in trickle-bed reactors in which the gas and liquid are pumped counter- or co-currently through a bed of catalyst particles [53, 54]. Many of these systems encounter mass-transfer limitations as a result of intraparticle mass-transfer resistance, liquid-film resistance, liquid maldistribution and channelling. To overcome these problems, membrane reactors have been used for chemical reactions as well as biological conversions.

Gas-liquid reactions investigated are the hydrogenation of a-methylstyrene to cumene using a porous γ-alumina membrane impregnated with a Pd catalyst [55] and the hydrogenation of nitrobenzene to aniline in a Pt-impregnated porous γ-alumina membrane [56]. From both studies is was concluded that the membrane reactors can be performed without any operational problems and that reaction rates increased significantly (up to factor of 20) as a result of easy access of the gas to the catalytically active sites.

Biotreatment of large air streams in membrane contactors has been evaluated widely (Tab. 5.1 [57]). The removal of organic compounds (e.g. propene, dichloromethane, etc.) and inorganic substances (SO_2, NO_x etc.) has proven to be highly efficient in membrane contactors. The gas stream to be treated is led on one side of the membrane, whereas an appropriate aqueous solution containing a single or a mixture of bacteria is circulated on the other side. This allows for

Table 5.1 Membrane bioreactors for biological waste gas treatment in historical order [57].

Ref.	Compound removed from air	Conc. [ppm]	Type of membrane	Nutrient supply in liquid	Bio-film	Inoculant
58	Xylenes	30–140	Silicone, tubes	Minerals	Yes	Sludge
58	n-Butanol	40–180	Silicone, tubes	Minerals	Yes	Sludge
58	Dichloromethane	60–220	Silicone, tubes	Minerals	No	Sludge
59	Toluene	20	HM[a], sheet	Minerals	NA[b]	Pseudomonas, GJ40
59	Dichloromethane	47	HM, sheet	Minerals	NA	Strain DM21
60	Toluene	56	HM, sheet	Minerals	Yes	Pseudomonas GJ40
60	Dichloromethane	69	HM, sheet	Minerals	Yes[c]	Strain DM21
61	n-Hexane	32	Silicone, tubes[d]	Minerals	?	?
61	Toluene	32	Silicone, tubes	Minerals	?	?
62	NO	5	H, sheet	Alcohols, minerals	Yes	Methylobacter
63	Mixture	'low'	HM, sheet	Minerals	Yes	Various strains
64	Dichloroethane	150	Silicone	Minerals[e]	Yes	Xanthobacter GJ10
65	Propene	250–300	Spiral-wound HM, sheet	Minerals	Yes	Xanthobacter Py2
66	Trichloroethene	20	Polysulfone fibers	Acetate, minerals[f]	Yes	Sludge
60	Propene	330–2700	HM, fibers	Minerals	Yes	Xanthobacter Py2

a) H, hydrophobic material; M, microporous material.
b) Experiments lasted less than 1 day.
c) Severe sloughing observed after 4 days.
d) Reactor in a combination of a membrane bioreactor and a bubble-column and was designed for simultaneous degradation of both hydrophilic and hydrophobic contaminants from the gas phase.
e) Gas phase did not contain oxygen.
f) Liquid phase was kept anaerobic.

an easy adjustment of conditions (pH and nutrients). The bacteria used can either grow as a film onto the membrane surface or can be homogeneously dispersed in the liquid. This approach has led to large-scale operations, e.g. for the treatment of traffic-tunnel vent streams.

5.2.2
Separation of Products

In general, a reversible reaction such as Eq. (6) is often limited in conversion or yield by the reaction equilibrium. Removal of one or both products by a membrane can increase the conversion as the reversible reaction is shifted to the right.

$$A + B \leftrightarrow C + D \tag{6}$$

Additionally, undesirable side reactions such as the formation of component E in Eq. (7) can be avoided by the separation of product C via a membrane. In consecutive catalytic reactions like Eq. (8), the desired intermediate product B can be obtained by selective removal of B from the reaction zone. Inhibition effects by one of the formed products, as is often the case in fermentation processes, can be reduced by removal of the products from the reaction.

$$B + D \leftrightarrow E \tag{7}$$

$$A \rightarrow B \rightarrow C \tag{8}$$

A comparison between membrane reactors and conventional plug-flow reactors can be made on the basis of the two rates governing the latter performance: the reaction rate and the rate of reactant feed per catalyst volume to the reactor. The ratio of these two is the Damköhler number (Da), which also involves tube dimensions. The membrane reactor brings in a third rate constant: the permeation rate of the fastest-permeating species. For a comparison between the two reactor types, the Damköhler-Peclet product can effectively be used (DaPe; maximum reaction rate per volume/maximum permeation rate per volume) [67]. For proper performance of a reactor, these three rates will have to be properly balanced. At too low a permeation rate, the membrane has little effect and the reactor behaves like a plug-flow reactor, whereas at too high a permeation rate the shell and tube side will equilibrate too quickly [68]. Bernstein and Lund [67] recommend $0.1 < \mathrm{DaPe} < 10$ as covering the optimal range.

Gas-phase Reactions

Most studies on selective product removal in gas-phase reactions have been focused on hydrogen removal. A detailed summary of the early studies (up to 1994) has been given by Saracco and Specchia [2]. Examples are decomposition reactions (HI and H_2S) [69–71] and relatively simple alkane dehydrogenations. For the dehydrogenation of ethane catalytically active tubular membranes have been used [72], whereas cyclohexane dehydrogenation was performed in packed-bed membrane reactors [73, 74]. Given the industrial importance, the conversion of ethylbenzene to styrene has been studied extensively also [75–78]. More recent studies also focus on hydrogen removal, but tackle more complicated reactions, e.g. the dehydrogenation of propane [79–82], isobutane [83, 84], n-butane [85], methane steam reforming and the water-gas-shift reaction [86, 87].

From a critical review by Armor [8] a number of problem areas can be defined for the industrial application of dehydrogenation membrane reactors. These are: defects in metallic membranes at elevated temperatures, phase transitions in metallic membranes, leakage, low surface area per volume, severe

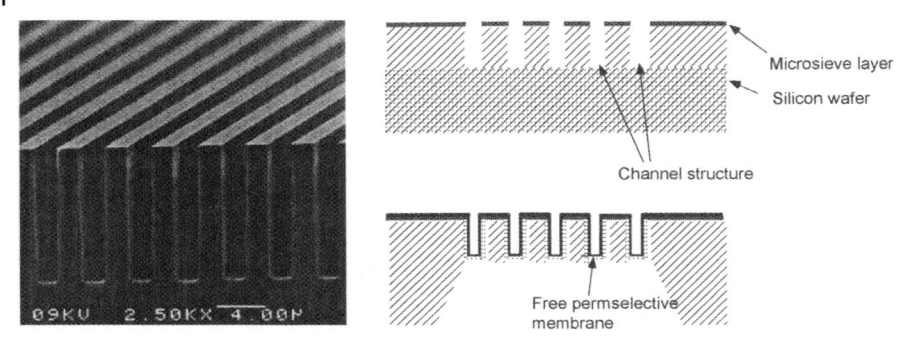

Fig. 5.7 Effective surface area enlargement by vertical etching in a silicon wafer (left). This can be applied to produce high surface area membranes via several deposition and etching steps (right).

mass-transfer limitations, very low feed flow rates, carbon deposition, and the low turnover number of commercially available dehydrogenation catalysts.

A new development in this field is the use of fluidized-bed systems instead of a packed bed. For this purpose, steam reforming of methane has been used as a model reaction [88]. From experimental and theoretical work it can be concluded that fluidized-bed membrane reactors potentially represent a promising system as problems of heat transfer and equilibrium limitations can be addressed simultaneously. As one of the major problems encountered is to provide sufficient membrane area per volume, possible solutions are the use of hollow-fiber systems [13] or membranes based on microsystem technology. In Fig. 5.7 an indication can be obtained for the potential of this approach to enlarge the effective membrane area versus the superficial area of the wafers used [89].

Apart from the hydrogen-removal studies, reactions in which O_2 has to be removed, e.g. NO and CO_2 decomposition, are of environmental interest. The membrane materials used for this purpose are mixed oxides such as zirconias [90, 91] and perovskites [92].

Liquid-phase Reactions

Reverse osmosis (RO), nanofiltration (NF) and ultrafiltration (UF) can effectively be used to remove a single component or a group of components from a liquid mixture. As an example, a RO membrane has been used in the yeast-catalyzed conversion of glucose into ethanol [93]. The membrane retains the yeast cells as well as the unreacted glucose, thus providing an efficient separation between substrate and product. Ultrafiltration has often been used in enzyme and whole-cell bioreactors [6, 94]. Many systems have been described, including protein and carbohydrate hydrolysis (e.g. starch and cellulose [95]), in which the low molecular weight products are removed through the membrane.

5.2 Membrane Functions in Reactors

Esterifications and etherifications are industrially relevant chemical reactions. These reactions are often severely limited in conversion, due to an unfavorable reaction equilibrium. In industry, these reactions are forced to completion by adding a large excess of the alcohol, which induces a highly inefficient use of reactor space. Pervaporation has been investigated to remove water selectively from the reaction mixture [96–98], which also avoids the energy-consuming distillation of the excess alcohol. Most of the work published in the literature focuses on the use of polymeric membranes for this purpose, e.g. based on poly(vinyl alcohol) (PVA). Although the principle has proven to be efficient, no large-scale applications have come out until now. The development of processes using ceramic pervaporation membranes, however, seems to lead to industrial applications in the near future [99, 100]. This will be discussed in more detail in Section 5.3.1.

5.2.3
Catalyst Retention

In the previous sections, a number of catalytic systems have already been described. Here we will summarize the possible roles a membrane can have in a catalytic process. For this purpose, three basic types of catalytic systems can be distinguished (Fig. 5.8 [101]):

a) A membrane can be used to retain a mobile catalyst, thus keeping the catalyst in the reaction fluid. Ultrafiltration and nanofil-

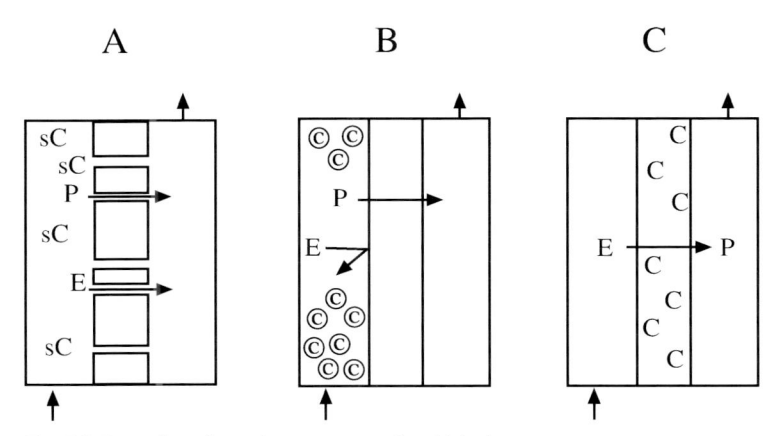

Fig. 5.8 Examples of membrane reactors in which the membrane acts in different ways: (a) The soluble catalyst is retained by a membrane, through which products can pass. (b) Selective removal of product by a selective membrane – the immobilized catalyst is present in a fixed or fluidized bed. (c) Catalytically active membrane, where the membrane material itself is catalytically actice or the catalyst is immobilized within the membrane [101].

tration are often applied to retain mobile catalysts such as enzymes, whole cells and homogeneous catalysts.
b) A catalyst can be immobilized in a porous membrane structure. Examples of such catalysts are enzymes and whole cells for biocatalysis and oxides and metals for non-biological synthesis.
c) The membrane itself can act as the catalyst. This form of catalytic system often applies for inorganic membranes such as palladium and zeolite membranes.

As most of the polymeric membranes available are not stable in organic solvents, the main focus of catalyst retention has been in the field of aqueous-phase bioconversions, either by enzymes or by whole cells. This has led to commercial processes for the production of fine chemicals, e.g. L-methionine or various amino acids (see Section 5.3.4) [102–104]. As most common homogeneous catalysts have molecular weights in the range of 100–1000, this requires the development of solvent-resistant nanofiltration membranes, either polymeric or inorganic in nature. Nevertheless, these membranes can be characterized as "development products". Therefore, the common solution is to enlarge the homogeneous catalyst, allowing for retention by a solvent-resistant ultrafiltration membrane (e.g. aromatic polyamides) [105]. This has been done for the enantioselective addition of diethyl zinc to benzaldehyde using a soluble polymer (a copolymer of 2-hydroxy ethyl methacrylate and octadecyl methacrylate of MW = 96 000) to enlarge the low molecular weight chiral ligand (a,a-diphenyl-L-prolinol) [106, 107].

5.3
Applications

In this section, several applications of membrane reactors on the commercial scale will be highlighted as well as some membrane-based processes that have potential for industrial application. Membrane-assisted esterifications and dehydrogenations will be discussed as well as the OTM process for the production of syngas. Additionally, typical membrane bioreactors such as used in the acylase process developed by Degussa AG, and membrane extraction systems such as the MPGM system and the Sepracor process are described.

5.3.1
Pervaporation-assisted Esterification

In industry, esterifications represent an important class of chemical reactions. As esterifications are equilibrium reactions (9), high yields can be obtained by adding an excess of one reactant or by constant removal of the produced water from the reaction mixture in order to shift the reaction to the product side.

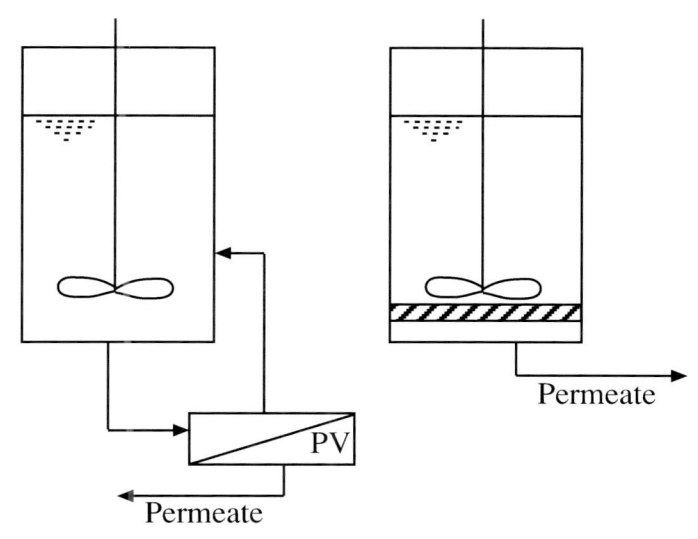

Fig. 5.9 Configuration of a pervaporation reactor with an external pervaporation unit (1) and with an internal pervaporation unit (2), respectively [108].

$$R_1\text{-CO-OH} + R_2\text{-OH} <-> R_1\text{-CO-OR}_2 + H_2O \tag{9}$$

Application of pervaporation processes to selectively separate water from the reacting mixture forms an interesting alternative to conventional distillation, especially in the case of azeotrope formation and low-boiling reactants.

Both polymer and ceramic membranes are applied in pervaporation-based reactors, for which Fig. 5.9 shows the two basic configurations [108]. Table 5.2 gives an overview of the performance of various pervaporation membranes and Tab. 5.3 shows some examples of membrane-assisted esterification reactions. In addition to these low molecular weight esters, pervaporation can also be used for the production of polycondensation esters (resins) [99, 100].

A process performance study has been conducted by David et al. [96] taking the coupling of the esterification reactions of 1-propanol and 2-propanol with propionic acid to pervaporation as a model system. Toluene sulphonic acid was applied as the homogeneous acid catalyst. A poly(vinyl alcohol)-based composite membrane, supplied by Carbone Lorraine-GFT, was used. Figure 5.10 shows the comparison between the esterification reaction with and without pervaporation. Without pervaporation, the conversion factor reaches a limit, which corresponds to the equilibrium of the esterification reaction. Coupling of the esterification to pervaporation allows the reaction to reach almost complete conversion.

The influence of four different operating parameters on the conversion were evaluated [96], which can be divided into three groups:

Table 5.2 Overview of several pervaporation membranes and their performance for the system water/isopropanol [115]. PSI is defined as the product of flux and selectivity.

Ref.	Membrane type or material	Temp. [°C]	Flux kg/(m³ h) 10 wt%	Flux kg/(m³ h) 10 wt%	α (–) 10/5 wt%	PSI kg/(m³ h) 10/5 wt%	Comments
109	CMC-CE-01	65	0.11	0.055	370/520	80/30	PSI drops with with increasing temperature
	CMC-CE-02	55		0.09	800	70	PSI increases with temp.
110	Carboxymethylated poly(vinyl alcohol)	80	0.5	0.20	1800/3700	900/900	
111	Chitosan (crosslinked)	30	0.15	0.09	1100/2000	160/180	PSI is roughly the same at 60°C
113	Sodium alginate	70		1.0	2500	2500	
114	Silica	70		0.3	500	150	
115	Silica	70		2.1	600	1250	after stabilisation PSI=1800

- Factors that influence directly the esterification reaction (the catalyst concentration and initial molar ratio)
- Factors that influence the pervaporation kinetics directly (the ratio of membrane area to reactor volume)
- Factors that influence simultaneously the esterification as well as the pervaporation kinetics (the temperature).

For a rapid conversion of lab-scale results into an economically viable reaction-pervaporation system, an optimum value can be determined for each parameter. Based on experimental results as well as a model describing the kinetics of the system, it has been found that the temperature has the strongest influence on the performance of the system as it affects both the kinetics of esterification and of pervaporation. The rate of reaction increases with temperature according to an Arrhenius law, whereas the pervaporation is accelerated by an increased temperature also. Consequently, the water content fluctuates much faster at a higher temperature. The second important parameter is the initial molar ratio. It has to be noted, however, that a deviation in the initial molar ratio from the stoichiometric value requires a rather expensive separation step to recover the unreacted component afterwards. The third factor is the ratio of membrane area to reaction volume, at least in the case of a batch reactor. For continuous operation, the flow rate should be considered as the determining factor for the contact time of the mixture with the membrane and subsequently the permeation

Table 5.3 Overview of pervaporation-assisted esterifications, adapted from [14].

Ref.	Reaction	Membrane material	Membrane type	Membrane area [m^2]	Temp. [°C]
116	methanol + acetic acid <–> methyl acetate + H$_2$O	Nafion	tube	5.0×10^{-3}	25
117	ethanol + acetic acid <–> ethyl acetate + H$_2$O	polyvinyl alcohol	flat cell	1.2	90
117	ethanol + acetic acid <–> ethyl acetate + H$_2$O	Nafion 117	flat cell	1.2	90
117	ethanol + acetic acid <–> ethyl acetate + H$_2$O	Nafion 324	flat cell	1.2	90
118, 119	ethanol + acetic acid <–> ethyl acetate + H$_2$O	polyvinyl alcohol	tube	1.1	80
120	ethanol + acetic acid <–> ethyl acetate + H$_2$O	polyether imide	flat cell	1.9	75
120	ethanol + oleic acid <–> oleic acid ethyl ester + H$_2$O	polyvinyl imide	flat cell	1.9	60
121	1-propanol + propionic acid <–> propionic acid propyl ester + H$_2$O	polyvinyl alcohol	flat cell	2.0	50
122	1-propanol + propionic acid <–> propionic acid propyl ester + H$_2$O	polyvinyl alcohol	flat cell	2.0	50
122	1-propanol + propionic acid <–> propionic acid propyl ester + H$_2$O	PSSH-polyvinyl alcohol	flat cell	2.0	50
121	2-propanol + propionic acid <–> propionic acid propyl ester + H$_2$O	polyvinyl alcohol	flat cell	2.0	55
121	2-propanol + propionic acid <–> propionic acid propyl ester + H$_2$O	polyvinyl alcohol	flat cell	2.0	65
120	2-propanol + propionic acid <–> propionic acid propyl ester + H$_2$O	polyether imide	flat cell	2.5	85
123	1-butanol + acetic acid <–> butyl acetate + H$_2$O	polyvinyl acetate	channel reactor	–	155
116	1-butanol + acetic acid <–> butyl acetate + H$_2$O	Nafion	tube	5.0	25

flux. The catalyst concentration exhibits the weakest influence on the pervaporation-esterification system. The esterification reaction is a first-order reaction with respect to the catalyst for both the 1-propanol and 2-propanol esters, in which the apparent reaction rate increases linearly with the catalyst concentration.

The system developed by David et al. [96] shows that application of a membrane process in combination with an equilibrium reaction to continuously remove one of the products formed is an interesting approach to obtain complete conversion of the reactants. For optimization of the process a predictive model

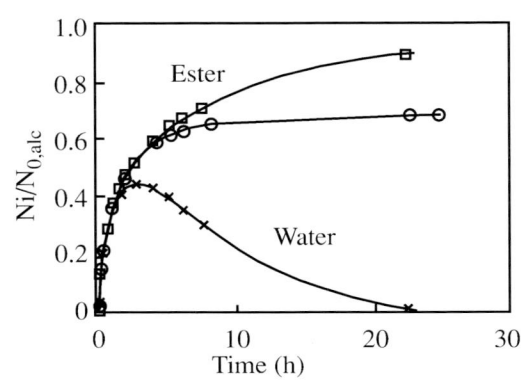

Fig. 5.10 Variation of the molar amounts of ester and water as a function of time for esterification with (□, X) and without (■) pervaporation. N_i: molar amount of component i; $N_{0,alc}$: initial molar amount of alcohol [96].

has proven to be very useful in order to determine the influence of various operating parameters.

The catalytic esterification of ethanol and acetic acid to ethyl acetate and water has been taken as a representative example to emphasize the potential advantages of the application of membrane technology as compared to conventional distillation [14] (see Fig. 5.11). From the McCabe-Thiele diagram for the separation of ethanol-water mixtures it follows that pervaporation can reach high water selectivities at the azeotropic point in contrast to the distillation process. Considering the economic evaluation of the membrane-assisted esterifications as compared to the conventional distillation technique [14], a decrease of 75% in energy input and 50% lower investment and operation costs can be calculated. The characteristics of the membrane and the module design mainly determine the investment costs of membrane processes, whereas the operational costs are influenced by the lifetime of the membranes.

Keurentjes et al. [98] studied the esterification of tartaric acid with ethanol using pervaporation. The equilibrium composition could be shifted significantly towards the final product diethyltartrate by integration of pervaporation with hydrophilic poly(vinyl alcohol)-based composite membranes in the process. Based on the kinetic parameters, an optimum membrane surface area could be calculated that results in a minimal reaction time for the esterification reaction. Where the membrane surface area to volume ratio is too low, the water removal is rather slow, whereas at high surface area to volume ratios significant amounts of ethanol are removed as well.

Although low molecular weight esterifications (and etherifications) can benefit substantially from an integration with pervaporation, to our knowledge no large-scale applications have emerged so far. Probably closer to final application is the development by Akzo Nobel on the application of ceramic pervaporation membranes in polycondensation reactions [99, 100]. In the production of alkyd coating

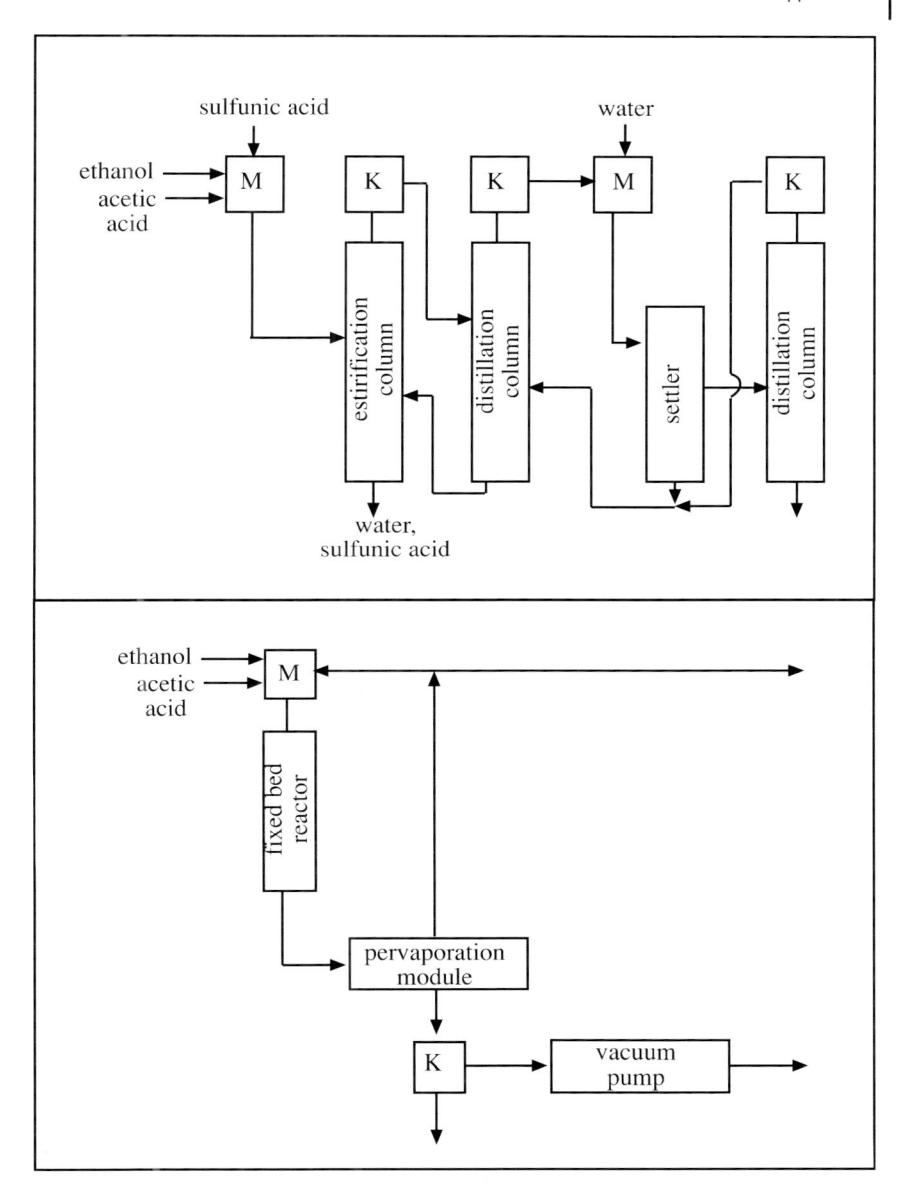

Fig. 5.11 Process scheme of the conventional reaction distillation process (top) versus the membrane-assisted esterification (bottom) for the production of ethyl acetate [14].

resins a mixture of acids, acid anhydrides and alcohols react under the formation of a resin and water. Reaction equation (10) represents a simplified form of this reaction, typically performed at temperatures between 150 and 300 °C. The reaction mixture is relatively viscous and in some cases even heterogeneous.

$$R_1\text{-}(COOH)_x + R_2\text{-}(OH)_y \iff \text{coating resin} + H_2O \tag{10}$$

With pervaporation membranes the water can be removed during the condensation reaction. In this case, a tubular microporous ceramic membrane supplied by ECN [124] was used. The separating layer of this membrane consists of a less than 0.5 mm film of microporous amorphous silica on the outside of a multilayer alumina support. The average pore size of this layer is 0.3–0.4 nm. After addition of the reactants, the reactor is heated to the desired temperature, the recycle of the mixture over the outside of the membrane tubes is started and a vacuum is applied at the permeate side. In some cases a sweep gas can also be used. The pressure inside the reactor is a function of the partial vapor pressures and the reaction mixture is non-boiling. Although it can be anticipated that concentration polarization will play an important role in these systems, computational fluid dynamics calculations have shown that the membrane surface is effectively refreshed as a result of buoyancy effects [125].

In a kilogram-scale reaction-pervaporation unit, the method has been tested extensively. The applied membranes showed high permeability and selectivity towards water during the whole reaction period. Besides that, the membranes appeared to be thermally and chemically stable for the reaction conditions applied. For this specific application the energy savings as compared to conventional methods are estimated to be more than 40%, and the reactor efficiency can be increased by at least 30% [99, 100].

5.3.2
Large-scale Dehydrogenations with Inorganic Membranes

Over the years, several processes for the catalytic dehydrogenation of propane to propylene have been developed, which can be divided into processes based on an adiabatic or an isothermal reactor concept, respectively. The processes currently applied on an industrial scale are based on adiabatic systems, such as the Catofin (Lummus/Air Products) and the Oleflex (UOP) process. As the dehydrogenation of propane to propylene comprises an equilibrium reaction (11), selective removal of hydrogen from the reaction mixture can shift the reaction towards the product side. At high temperatures, thermal cracking may occur.

$$C_3H_8 \iff C_3H_6 + H_2 \tag{11}$$

Van Veen et al. [126] studied the technical and economic feasibility of the application of ceramic membranes in different dehydrogenation processes. As the Oleflex process uses four reactor beds in series, this process is more suitable

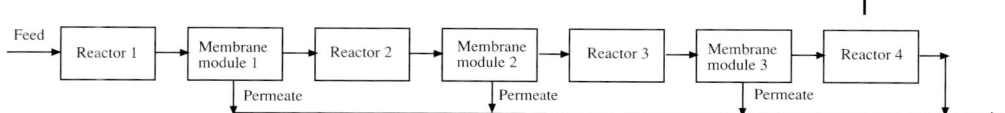

Fig. 5.12 Generalized process flow diagram of the Oleflex process extended with four membrane modules [126].

for implementation of ceramic separation units than the Catofin process, which uses a parallel reactor system. Figure 5.12 shows the configuration of the Oleflex process extended with four membrane modules.

Based on the adiabatic reactor concept, several configurations of the membrane units have been considered as compared to the conventional Oleflex process [126]. Concerning the technical feasibility, the propane dehydrogenation process requires membranes with a selectivity much higher than Knudsen-diffusion-based selectivity, in combination with a reduced permeate pressure. Additionally, the membranes have to be stable at the working conditions ($T=650\,°C$, $p=1.5$ bar) and their performance should be indifferent to coke formation. In an isothermal reactor concept, application of inorganic membranes may lead more easily to a technically feasible process as additional heat for propane conversion is available. However, the difference in price level between feedstock and product is rather small to give an economically viable membrane-assisted dehydrogenation process of propane.

The potential application of ceramic membranes for the dehydrogenation of ethylbenzene to styrene (12) has also been evaluated [126]. In the conventional process, two radial reactors in series are used with one preheater and one interstage heater. Steam acts as an energy carrier and as a diluent.

$$C_6H_5\text{-}CH_2CH_3 <\!-\!> C_6H_5\text{-}CHCH_2 + H_2 \tag{12}$$

For the dehydrogenation of ethylbenzene in a packed-bed ceramic membrane reactor, three configurations are possible using a specific sweep gas in combination with a hydrogen or oxygen selective membrane (see Fig. 5.13).
- The hydrogen permeates through the hydrogen-selective membrane tube under the influence of a pressure difference over the membrane. The hydrogen is removed with an inert sweep gas such as steam (A).
- The hydrogen permeates through the membrane and is removed using air as a sweep gas. Subsequently, the hydrogen is burned by the oxygen (B).
- Oxygen permeates through the oxygen-selective membrane into the reaction mixture. By oxidative dehydrogenation, the oxygen burns the hydrogen that is formed (C).

According to the study by van Veen [126], a ceramic membrane reactor does not lead to a feasible process for the dehydrogenation of ethylbenzene. The profit

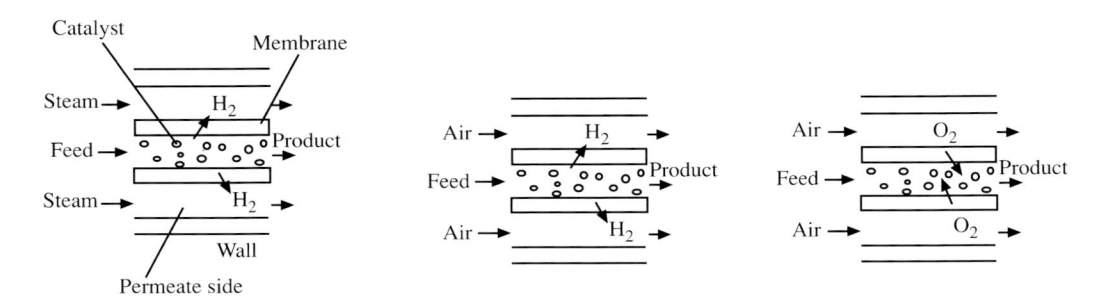

Fig. 5.13 Possible membrane reactor subconfigurations for the dehydrogenation of ethylbenzene [126].

from the higher styrene yield by application of ceramic membranes does not compensate for the expensive membranes. A viable membrane-assisted dehydrogenation of ethylbenzene asks for cheaper membranes, being highly selective with a higher permeability than the membranes currently available.

5.3.3
OTM Syngas Process

An alliance of five international companies including Amoco, BP Chemicals, Praxair, Sasol and Statoil, has put significant effort into the development and commercialization of a novel technology to address overall cost reduction in the production of synthesis gas (13) [127].

$$CH_4 + 1/2\, O_2 \rightarrow CO + 2H_2 \tag{13}$$

Conventional processes for the production of syngas involve partial oxidation or steam reforming. In the process with oxygen, an expensive air-separation plant is required, whereas in the case of steam reforming high-temperature heat additions are necessary.

The alliance is developing the OTM (Oxygen-Transport-Membrane) Syngas process, which integrates the separation from air, steam reforming and natural gas oxidation (see Fig. 5.14). Air is introduced on one side of the membrane, whereas natural gas is added on the other side. Oxygen is separated from air by adsorption on the surface of the membrane where it is subsequently dissociated and ionized. The oxygen ions diffuse through the membrane. The subsequent reaction with natural gas to syngas takes place in the presence of a reforming catalyst at the permeate side of the membrane. High selectivity and high yields are obtained by controlling the oxygen flux through the membrane. For the OTM process, dense ceramic materials are used, which are related to inorganic

OTM Syngas Generation

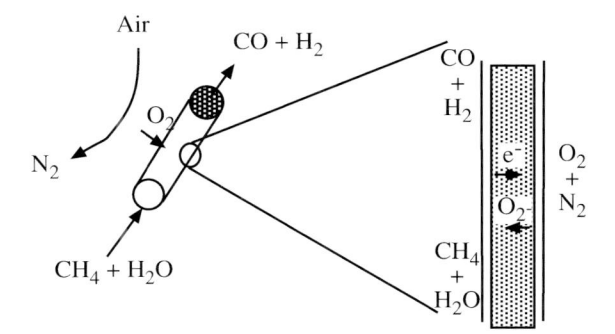

Fig. 5.14 Schematic setup of the OTM syngas process [127].

perovskite structures, although no exact details are given on the membranes applied.

The technical challenges of the OTM process comprise amongst others the material performance, fabrication processing reliability, process integration, engineering-scale-up and cost competitiveness. The alliance participants expect to move the OTM program towards commercialization over the next several years.

5.3.4
Membrane Recycle Reactor for the Acylase Process

A particular example of a membrane-assisted process applied on a large scale is the acylase-catalyzed resolution of N-acetyl-D,L-amino acid, as developed by Degussa AG [128, 129]. Annually, the industrial plant produces several hundreds of tonnes of enantiomerically pure L-amino acid. D,L-amino acid is acetylated in a Schotten-Baumann reaction to N-acetyl-D,L-amino acid. Subsequently, the L-amino acid enantiomer is obtained via an acylase reaction. Figure 5.15 shows the reaction scheme.

Regarding the economical viability of the plant, the retention and stability of acylase are essential features for the process. An ultrafiltration unit retains acylase as the mobile catalyst in the reactor. Alternatively, acylase can be immobilized in a fixed or fluidized bed. A mobile catalyst system is preferred compared to the immobilized form, as the mobile catalyst system avoids mass-transfer limitations. Additionally, regeneration of the catalyst and scale-up of the reactor are much easier as compared to the process with the immobilized acylase. With respect to the deactivation of the catalyst, the thermal as well as the operational stability of acylase has been evaluated extensively [128, 129]. At a pH of 7, acylase appears to be sufficiently stable for L-amino acid manufacture.

Figure 5.16 shows the process scheme of the commercial acylase process. In a separation unit of several ultrafiltration modules in parallel, acylase is removed from the reactor outlet stream and recycled to the reactor. Subsequently,

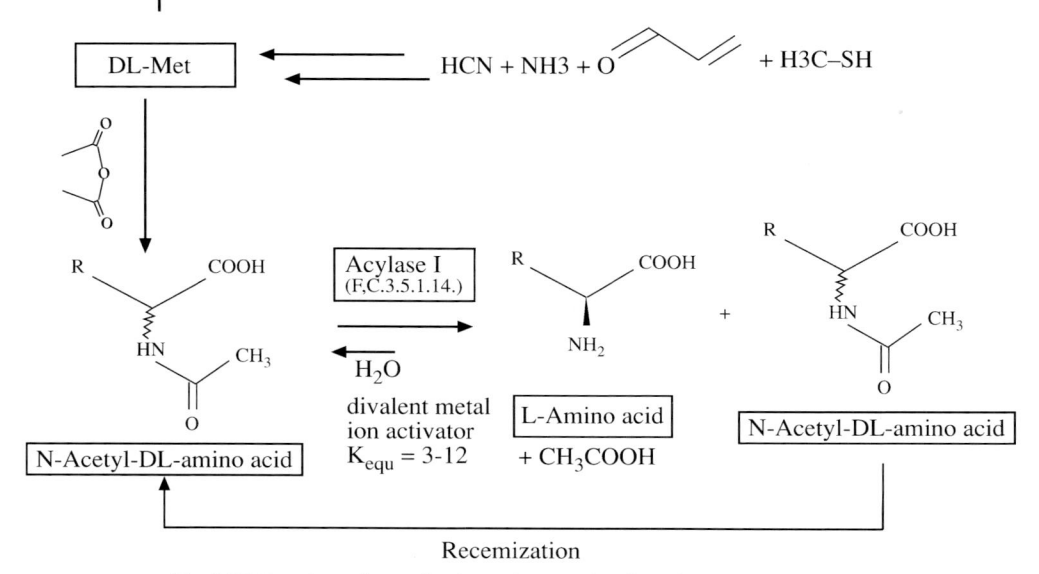

Fig. 5.15 Reaction scheme for the acylase-catalyzed resolution to L-amino acid

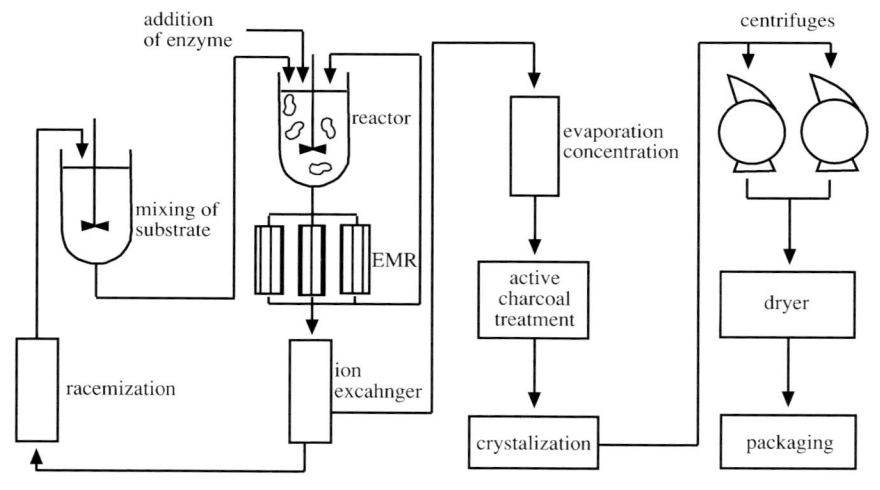

Fig. 5.16 Process scheme of the Degussa acylase process [128, 129].

the ion exchanger separates the L-amino acid product from the non-converted N-acetyl-D-amino acid, which is racemized and recycled to the reactor.

The acylase process is a typical example of reaction and separation by membranes in a sequential mode. The process is fully developed up to industrial scale, yielding high-quality products at good cost effectiveness.

5.3.5
Membrane Extraction Integrated Systems

An example of an industrial membrane bioreactor is the hollow-fiber membrane system for the production of (–)-MPGM (1), which is an important intermediate for the production of diltiazem hydrochloride [130, 131]. For the enantiospecific hydrolysis of MPGM a hollow-fiber ultrafiltration membrane with immobilized lipase from Serratia marcescens is used. (+)-MPGM is selectively converted into (2S,3R)-(+)-3-(4-methoxy-phenyl)glycidic acid and methanol. The reactant is dissolved in toluene, whereas the hydrophilic product is removed via the aqueous phase at the permeate side of the membrane (see Fig. 5.17). Enantiomerically pure (–)-MPGM is obtained from the toluene phase by a crystallization step. In cooperation with Sepracor Inc., a pilot-plant membrane reactor has been developed, which produces annually about 40 kg (–)-MPGM per m^2 of membrane surface.

In a comparable system, (R,S)-ibuprofen can be separated by a membrane reactor [132] (see Fig. 5.18). The technique comprises a stereospecific hydrolysis by an enzyme. Subsequently, the enantiomeric ester is extracted into the organic phase on the other side of the membrane. In the system developed by Sepracor Inc., (R)-ibuprofen is selectively hydrolyzed by proteases in a hollow-fibre unit and the (S)-ibuprofen ester can be isolated at 100% yield. This configuration also applies for enantioseparation of other acids such as naproxen and 2-chloropropionic acid.

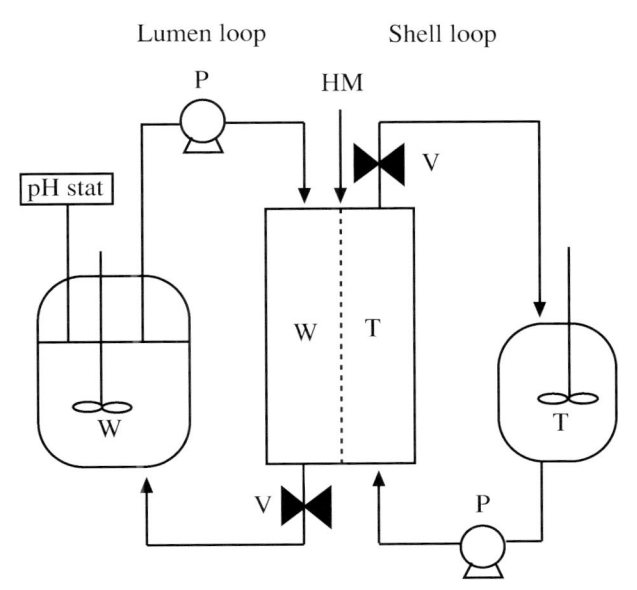

Fig. 5.17 Flow diagram of the membrane reactor for the production of (–)-MPGM. W: water, T: toluene, HM: hydrophilic membrane, P: recycle pump; V: throttle valve.

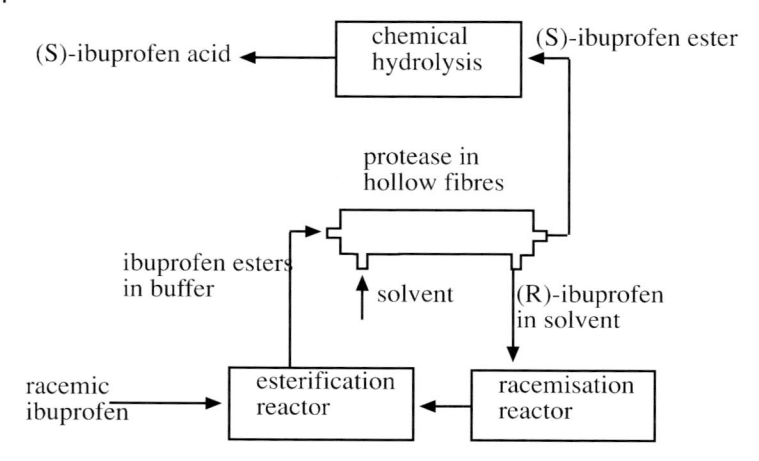

Fig. 5.18 Configuration of the Sepracor membrane bioreactor

Although at the moment no large-scale method exists for the production of enantiomerically pure components, it can be foreseen that both the MPGM and Sepracor reactor system have potential for application on a larger scale due to their ease of scale-up.

5.4
Concluding Remarks and Outlook to the Future

For the industrial application of membrane reactors, it can be concluded that these are accepted as proven technology for many biotechnological applications. The membranes used in this area can operate under relatively mild conditions (low temperature and aqueous systems). However, there is a tremendous potential for membrane reactors in the (petro)chemical industry, requiring application at elevated temperatures in nonaqueous systems. Often, this will make the use of inorganic membranes mandatory. Especially with respect to the application of inorganic membranes, several key issues need to be addressed in the near future. One of them is the development of high-surface-area-per-volume systems. Potential solutions are the use of ceramic hollow fibers or membranes fabricated using microsystem technology. Long-term stability of the membrane materials is a second important issue that will require an ongoing development from materials scientists. A third issue relates to scale up. First, most investigations described in the literature use small (typically 1–100 cm^2) membranes. From past experience in the production of inorganic ultrafiltration and microfiltration membranes it can be expected that the development of a production technology for large amounts of membrane area will require a substantial effort and will probably take several years. Finally, once large amounts of membrane area are available, they will have to be placed in the appropriate modules. The development of these modules will have to comprise several issues. These include high-temperature sealing, heat transfer and the development of flow patterns avoiding polarization effects, thus allowing for an effective use of the intrinsic membrane properties and the required (not necessarily being the intrinsic) reaction kinetics.

The main part of the developments on membrane reactors for chemical processes has focused on very large scale processes, with emphasis on applications in the petrochemical industry and in the production of bulk chemicals. As described above, many hurdles will have to be taken before implementation on this scale will take place. Although most of the earlier work on membrane reactors has focused on shifting the reaction equilibrium, currently a shift can be observed towards systems aiming on selectivity increase and controlled reactant dosage. As reaction selectivity is of major importance in the production of fine chemicals and pharmaceutical products, it seems plausible to expect that membrane reactors will find their way in the production of chemicals through applications in these areas.

References

1 A. G. Dixon, *Catalysis,* **14** (1999), 40.
2 G. Saracco, V. Specchia, *Catal. Rev. – Sci. Eng.,* **36** (1994), 305.
3 G. Saracco, H. W. J. P. Neomagus, G. F. Versteeg, W. P. M. van Swaaij, *Chem. Eng. Sci.,* **54** (1999), 1997.
4 K. K. Sirkar, P. V. Shanbhag, A. S. Kovvali, *Ind. Eng. Chem. Res,* **38**, (1999), 3715.
5 T. T. Tsotsis, A. M. Champagnie, R. G. Minet, P. K. T. Liu, Catalytic membrane reactors, in Computer-aided design of catalysts; E. R. Becker, E. R. Pereira eds., Chemical Industries, Marcel Dekker New York, **51** (1993).
6 M. Cheryan, M. A. Mehaia, Membrane Bioreactors. In Membrane Separations in Biotechnology; W. C. McGregor ed.; *Bioprocess Technology,* Marcel Dekker, New York, **vol. 1** (1991).
7 G. Belfort, *Biotechnol. Bioeng.,* **33**, (1989), 1047.
8 J. N. Armor, *J. Membr. Sci.,* **147** (1998), 217.
9 W. J. Koros, *Chem. Eng. Progr.*, October, (1995), 68.
10 C. W. Jones, W. J. Koros, *Ind. Eng. Chem. Res.,* **34** (1995), 158.
11 C. W. Jones, W. J. Koros, *Ind. Eng. Chem. Res.,* **34** (1995), 164.
12 A. Singh, W. J. Koros, *Ind. Eng. Chem. Res.,* **35** (1996), 1231.
13 J. Smid, C. G. Avci, V. Günay, R. A. Terpstra, J. P. G. M. van Eijk, *J. Membr. Sci.,* **112** (1996), 85.
14 R. M. Waldburger, F. Widmer, *Chem. Eng. Technol.,* **1** (1996), 117.
15 A. L. Y. Tonkovich, R. Secker, E. Reed, E. Roberts, J. Cox, *Sep. Sci. Technol.,* **30** (1995), 397.
16 A. M. Ramachandra, Y. Lu, Y. H. Ma, W. R. Moser, A. G. Dixon, *J. Membr. Sci.,* **116** (1996), 253.
17 V. M. Gryaznov, *Platinum Metals Rev.,* **30** (1986), 68.
18 H. Nagamoto, H. Inoue, *Chem. Eng. Commun.,* **34** (1985), 315.
19 S. J. Xu, W. J. Thomson, *AIChE J.,* **43** (1997), 2731.
20 T. Hibino, T. Sato, K. Ushiki, Y. Kuwahara, *J. Chem. Soc. Faraday Trans.,* **91** (1995), 4419.
21 J. E. ten Elshof, H. J. M. Bouwmeester, H. Verweij, *Appl. Catal. A,* **130** (1995), 195.
22 Y. Lu, A. G. Dixon, W. R. Moser, Y. H. Ma, *Catal. Today,* **56** (2000) 297.

23 W. Wang, Y. S. Lin, *J. Membr. Sci.*, **103** (1995), 219.
24 D. Lafarga, J. Santamaria, M. Menendez, *Chem. Eng. Sci*, **49** (1994), 2005.
25 J. Coronas, M. Menendez, J. Santamaria, *Chem. Eng. Sci.*, **49** (1994), 2015.
26 A. L. Y. Tonkovich, D. M. Jimenez, J. L. Zilka, G. L. Robberts, *Chem. Eng. Sci.*, **51**, (1996), 3051.
27 A. L. Y. Tonkovich, R. B. Secker, E. L. Reed, G. L. Roberts, *Sep. Sci. Technol.*, **30** (7–9), (1995), 1609.
28 A. L. Y. Tonkovich, D. M. Jimenez, J. L. Zilka, G. L. Roberts, *Chem. Eng. Sci.*, **51** (1996), 3051.
29 G. Capinelli, E. Carosini, F. Cavani, O. Monticelli, F. Trifiro, *Chem. Eng. Sci.*, **51** (1996), 1817.
30 C. Tellez, M. Menendez, J. Santamaria, *AIChE J.*, **43** (1997), 777.
31 V. T. Zaspalis, W. van Praag, K. Keizer, J. G. van Ommen, J. R. H. Ross, A. J. Burggraaf, *Appl. Catal.*, **74** (1991), 205.
32 V. T. Zaspalis, W. van Praag, K. Keizer, J. G. van Ommen, J. R. H. Ross, A. J. Burggraaf, *Appl. Catal.*, **74** (1991), 223.
33 V. T. Zaspalis, W. van Praag, K. Keizer, J. G. van Ommen, J. R. H. Ross, A. J. Burggraaf, *Appl. Catal.*, **74** (1991), 235.
34 V. T. Zaspalis, W. van Praag, K. Keizer, J. G. van Ommen, J. R. H. Ross, A. J. Burggraaf, *Appl. Catal.*, **74** (1991), 249.
35 H. J. Sloot, G. F. Versteeg, W. P. M. van Swaaij, *Key Eng. Mater.*, **61/2** (1991), 261.
36 V. T. Zaspalis, W. van Praag, K. Keizer, J. G. van Ommen, J. R. H. Ross, A. J. Burggraaf, *Appl. Catal.*, **74** (1991), 249.
37 G. Saracco, S. Specchia, V. Specchia, *Chem. Eng. Sci.*, **51**, (1996), 5289.
38 H. J. Sloot, G. F. Versteeg, W. P. M. van Swaaij, *Chem. Eng. Sci.*, **45** (1990), 2415.
39 H. J. Sloot, C. A. Smolders, W. P. M. van Swaaij, G. F. Versteeg, *AIChE J*, **38** (1992), 887.
40 J. W. Veldsink, R. M. J. van Damme, G. F. Versteeg, W. P. M. van Swaaij, *Chem. Eng. Sci.*, **47** (1992), 2939.
41 R. Prasad, K. K. Sirkar, *AIChE J.*, **33** (1987), 1057.
42 R. Prasad, K. K. Sirkar, *AIChE J.*, **34** (1988), 177.
43 M. C. Yang, E. L. Cussler, *AIChE J.*, **32** (1986), 1910.
44 G. T. Frank, K. K. Sirkar, *Biotechnol. Bioeng. Symp. Ser.*, **15** (1985), 621.
45 W. Kang, R. Shukla, K. K. Sirkar, *Biotechnol. Bioeng.*, **34** (1990), 826.
46 G. T. Frank, K. K. Sirkar, *Biotechnol. Bioeng. Symp. Ser.*, **17** (1986), 303.
47 R. Shukla, W. K. Kang, K. K. Sirkar, *Biotechnol. Bioeng.*, **34** (1989), 1158.
48 M. M. Hoq, T. Yamane, S. Shimizu, T. Funada, S. Ishida, *J. Am. Oil Chem. Soc.*, **62** (1985), 1016.
49 W. Pronk, P. J. A. M. Kerkhof, C. van Helden, K. van 't Riet, *Biotechnol. Bioeng.*, **32** (1988), 512.
50 R. Molinari, M. E. Santoro, E. Drioli, *Ind. Eng. Chem. Res.*, **33** (1994), 2591.
51 A. van der Padt, M. J. Edema, J. J. W. Sewalt, K. van 't Riet, *J. Am. Oil Chem. Soc.*, **67** (1990), 347.
52 T. J. Stanley, J. A. Quinn, *Chem. Eng. Sci.*, **42**, (1987), 2313.
53 J. C. Charpentier, *Chem. Eng. J.*, **11** (1976), 161.
54 M. Herskowitz, J. M. Smith, *AIChE J.*, **29** (1983), 1.
55 P. Cini, M. P. Harold, *AIChE J.*, **37**, (1991), 997.
56 M. Torres, J. Sanchez, J. A. Dalmon, B. Bernauer, J. Lieto, *Ind. Eng. Chem. Res.*, **33** (1994), 2421.
57 M. W. Reij, J. T. F. Keurentjes, S. Hartmans, *J. Biotechnol.*, **59** (1998), 155.
58 U. Bäuerle, K. Fischer, D. Bardtke, *STAUB Reinhaltung der Luft*, **46** (1986), 233.
59 S. Hartmans, E. J. T. M. Leenen, G. T. H. Voskuilen, In Biotechniques for air pollution abatement and odour control policies, A. J. Dragt, J. Ham eds., Elsevier, Amsterdam, (1992).
60 M. W. Reij, G. T. H. Voskuilen, S. Hartmans, in Biofilms-Science and Technology, L. F. Melo, T. R. Bott, M. Fletcher, B. Capdeville, eds., Kluwer, Dordrecht, (1992).
61 M. Reiser, K. Fischer, K. H. Engesser, *VDI Berichte*, **1104** (1994), 103.
62 M. Hinz, F. Sattler, T. Gehrke, E. Bock, *VDI Berichte*, **1104**, (1994), 113.

63 R. A. Binot, P. Paul, S. Keuning, S. Hartmans, D. de Hoop, *ESA Technol. Prog. Quart.*, **4** (1994), 14.
64 L. M. Freitas dos Santos, U. Hömmerich, A. G. Livingston, *Biotechnol. Rog.*, **11** (1995), 194.
65 M. W. Reij, C. D. de Gooijer, J. A. M. de Bont, S. Hartmans, *Biotechnol. Bioeng.*, **45** (1995), 107.
66 M. G. Parvatiyar, R. Govind, D. F. Bishop, *Biotechnol. Bioeng.*, **50** (1996), 57.
67 L. A. Bernstein, C. R. F. Lund, *J. Membr. Sci.*, **77** (1993), 155.
68 M. P. Harold, C. Lee, A. J. Burggraaf, K. Keizer, V. T. Zaspalis, R. S. A. de Lange, *MRS Bulletin*, April, (1994), 34.
69 J. Yehenskel, D. Leger, F. Courvoisier, *Adv. Hydrogen Energy Prog.*, **2** (1979), 569.
70 D. Friesen, Euromembrane '92, (1992), 357.
71 D. J. Edlund, W. A. Pledger, *J. Membr. Sci.*, **77** (1993), 255.
72 A. M. Champagnie, T. T. Tsotsis, R. G. Minet, E. Wagner, *J. Catal.*, **134** (1992), 713.
73 N. Itoh, Y. Shindo, K. Haraya, K. Obata, T. Hakuta, H. Oshitome, *Int. Chem. Eng.*, **125** (1985), 138.
74 N. Itoh, *AIChE J.*, **33**, (1987), 1576.
75 J. C. S. Wu, P. K. T. Liu, *Ind. Eng. Chem. Res.*, **31** (1992), 322.
76 Y. L. Becker, A. G. Dixon, W. R. Moser, Y. H. Ma, *J. Membr. Sci.*, **77** (1993), 233.
77 A. G. Dixon, Y. L. Becker, Y. H. Ma, W. R. Moser, Preprints – Separations Topical Conference, AIChE Annual Meeting, Miami Beach, (1992).
78 E. Gobina, K. Hou, R. Hughes, *J. Membr. Sci.*, **105** (1995), 163.
79 Z. D. Ziaka, R. G. Minet, T. T. Tsotsis, *AIChE J.*, **39** (1993), 526.
80 J. P. Collins, R. W. Schwartz, R. Sehgal, T. L. Ward, C. J. Brinker, G. P. Hagen, C. A. Udovich, *Ind. Eng. Chem. Res.*, **35** (1996), 4398.
81 Y. Yildirim, E. Gobina, R. Hughes, *J. Membr. Sci.*, **135** (1997), 107.
82 H. Weyten, K. Keizer, A. Kinoo, J. Luyten, R. Leysen, *AIChE J.*, **43** (1997) 1819.
83 T. Ioannides, G. R. Gavalas, *J. Membr. Sci.*, **77** (1993), 207.
84 Y. V. Gokhale, R. D. Noble, J. L. Falconer, *J. Membr. Sci.*, **105** (1995), 63.
85 M. E. Rezac, W. J. Koros, S. J. Miller, *Ind. Eng. Chem. Res.*, **34** (1995), 862.
86 T. T. Tsotsis, A. M. Champagnie, S. P. Vasileiadis, Z. D. Ziaka, R. G. Minet, *Chem. Eng. Sci.*, **47** (1992), 2903.
87 G. Barbieri, F. P. Di Maio, *Ind. Eng. Chem. Res.*, **36** (1997), 2121.
88 A. M. Adris, C. J. Lim, J. R. Grace, *Chem. Eng. Sci.*, **52** (1997), 1609.
89 H. V. Jansen, M. de Boer, R. Legtenberg, M. Elwenspoek, *J. Micromech. Microeng.*, **9** (1995), 115.
90 S. Pancharatnam, R. A. Huggins, D. M. Mason, *J. Electrochem. Soc.*, **122** (1975), 869.
91 Y. Nigara, B. Cales, *Bull. Chem. Soc. Jpn.*, **59** (1986), 1997.
92 A. G. Dixon, W. R. Moser, Y. H. Ma, *Ind. Eng. Chem. Res.*, **33** (1994), 3015.
93 M. Vasudevan, T. Matsuura, G. K. Chotani, W. R. Vieth, *Ann. N.Y. Acad. Sci.*, **506** (1987), 345.
94 C. Heath, G. Belfort, *Int. J. Biochem.*, **22**, (1990), 823.
95 T. K. Ghose, J. T. A. Kostick, *Biotechnol. Bioeng.*, **12** (1970), 921.
96 M. O. David, R. Gref, T. Q. Nguyen, J. Neel, *Trans IchemE*, **69** A, (1991), 341.
97 K. Okamoto, M. Yamamoto, Y. Otoshi, T. Semoto, M. Yano, K. Tanaka, H. Kita, *J. Chem. Eng. Japan*, **26** (1993), 475.
98 J. T. F. Keurentjes, G. H. R. Janssen, J. J. Gorissen, *Chem. Eng. Sci.*, **49** 24A, (1994), 4681.
99 W. J. W. Bakker, I. A. A. C. M. Bos, W. L. P. Rutten, J. T. F. Keurentjes, M. Wessling, Int. Conf. Inorganic Membranes (1998), Nagano, Japan, 448.
100 W. J. W. Bakker, W. L. P. Rutten, J. T. F. Keurentjes, M. Wessling, PCT/EP99/ 03435 (1999) Akzo Nobel nv.
101 U. Kragl, C. Dreisbach, C. Wandrey, *Appl. Homogeneoous Catal. Organomet. Comp.*, **2** (1996), 832.
102 A. S. Bommarius, in Biotechnology, Bioprocessing, H. J. Rehm, G. Reed, A.

Pühler, P. Stadler, G. Stephanopoulos, eds., **3** (1993), 427.
103 D. M. F. Prazeres, J. M. S. Cabral, *Enzyme Microb. Technol.*, **16** (1994), 738.
104 U. Kragl, in Industrial Enzymology, T. Godfrey, S. West, eds., 2nd ed., Macmillan Press Ltd., London, (1996), 271.
105 G. Giffels, J. Beliczey, M. Felder, U. Kragl, Tetrahedron Asymmetry, **9** (1998), 691.
106 K. Soai, S. Niwa, *Chem. Rev.*, **92** (1992), 833.
107 U. Kragl, C. Dreisbach, *Angew. Chem.*, **108** (1996), 684.
108 F. Lipnizki, R. W. Field, P. Ten, *J. Membr. Sci.*, **153** (1999), 183.
109 R. Atra, G. Vatai, E. Bekassy-Molnar, *Chem. Eng. Process.*, **38** (1999), 149.
110 S. Y. Nam, H. J. Chun, Y. M. Lee, *J. Appl. Polym. Sci.*, **72** (1999), 241.
111 M. Ghazali, M. Nawawi, R. Y. M. Huang, *J. Membrane Sci.*, **124** (1997), 53.
112 R. Y. M. Huanga, R. Pal, G. Y. Moon, *J. Membrane Sci.*, **160** (1999), 17.
113 R. Y. M Huangb, R. Pal, G. Y. Moon, *J. Membrane Sci.*, **160** (1999), 101.
114 R. W. van Gemert, F. P. Cuperus, *J. Membrane Sci.*, **105** (1995), 287.
115 A. W. Verkerk, P. van Male, M. A. G. Vorstman, J. T. F. Keurentjes, *Sep. Pur. Technol.* (1999) to be published.
116 L. Bagnell, K. Cavell, A. M. Hodges, A. W. Mau, A. J. Seen, *J. Membrane Sci.*, **85** (1993), 291.
117 G. K. Pearce, EU Pat. 0210055 A1, (1987), Bp Chemicals, Ltd.
118 R. M. Waldburger, PhD thesis, Swiss Federal Institute of Technology, Zurich (1993).
119 R. M. Waldburger, F. Widmer, W. Heinzelmann, *Chem. Ing. Tech.*, **66** (1994), 850.
120 H. Kita, S. Sasaki, K. Tanaka, Okamoto, K. M. Yamamoto, *Chem. Lett.*, **10** (1988), 2025.
121 R. Gref, M. O. David, Q. T. Nguyen, J. Néel, Fourth Int. Conf. Pervaporation Processes in the Chemical Industry, Ft Lauderdale, (1989), 344.
122 M. O. David, PhD thesis, Institut National Polytechnique de Loraine, Nancy, (1991).
123 J. F. Jennings, R. C. Benning, US Pat. 2956070 (1960), American Oil Comp.
124 H. M. van Veen, Y. C. van Delft, C. W. R. Engelen, P. P. A. C. Pex, Book of Abstracts, volume 2, Euromembrane 99, Leuven, Belgium, 209.
125 G. J. S. van der Gulik, R. E. G. Janssen, J. G. Wijers, J. T. F. Keurentjes, to be published.
126 H. M. van Veen, M. Bracht, E. Hamoen, P. T. Alderliesten, Fundamentals of Inorganic membrane science and Technology, (A. J. Burggraaf, L. Cot eds.) 1996, Elsevier Science, 641.
127 C. A. Udovich, Natural Gas Conversion V, Studies in Surface Science and Catalysis (eds. A. Parmaliana et al.) 119 (1998) 417.
128 A. S. Bommarius, M. Schwarm, K. Drauz, *Chimica Oggi*, (1996), 61.
129 A. S. Bommarius, K. Drauz, U. Groeger, C. Wandrey, Membrane bioreactors for the production of enantiomerically pure α-amino acids, Chirality in Industry, A. N. Collins, G. N. Sheldrake, J. Crosby, eds. (1992).
130 H. Matsumae, H. Furui, T. Shibatani, *J. Ferment. Bioeng.*, **41** (1993), 979.
131 H. Matsumae, M. Furui, T. Shibatani, T. Tosa, *J. Ferment. Bioeng.*, **75**, (1993), 93.
132 S. L. Matson, S. A. Wald, C. M. Zepp, D. Dodds, PCT Int. Pat. Appl., WO 8909765 (1991).

6
Electromembrane Processes

T. A. Davis, V. D. Grebenyuk and O. Grebenyuk

6.1
Ion-exchange Membranes

Ion-exchange membranes are synthetic membranes permeable to either positively or negatively charged ions in aqueous solution. This unique property makes ion-exchange membrane applications very attractive for chemical industry, because it allows for the removal, addition, substitution, depletion, or concentration of ions in process solutions.

The membranes that are selectively permeable to positively charged ions are usually named cation-exchange membranes, or simply cation membranes, and membranes selectively permeable for negatively charged ions are called anion membranes. The selectivity occurs due to high concentration of immobile (fixed) ions within the membrane body. Cation membranes have negatively charged fixed ions usually sulfonic or carboxyl groups chemically bound with the membrane's matrix. Their charge is neutralized by positively charged ions (counterions). An anion membrane would have positively charge fixed ions, usually quaternary ammonium groups, and negatively charged counterions. Fixed ions and counterions are connected by ionic bounds in a dry membrane. In the swollen membrane this bond would be dissociated. Therefore, the counterion is mobile and can be replaced by another ion. Hence the membrane would be permeable to ions with the charge sign opposite to the charge of the fixed ion.

An important feature of ion-exchange membranes is their permeability to counterions and their impermeability to co-ions (ions with a charge like that of the fixed charge). But they are not perfectly impermeable to co-ions. As solvent penetrates ion-exchange material some co-ions penetrate the membrane as well, driven by the difference of chemical potentials of solution and membrane. To follow the law of electric neutrality, an additional amount of counterions equivalent to the amount of co-ions must move into the ion-exchange membrane. This quantity of electrolyte absorbed by this mechanism may be estimated by the Donnan equation, which is based on the law of equity of chemical potentials in both liquid and solid phases. In the case of an electrolyte in which both the an-

Membrane Technology in the Chemical Industry.
Edited by Suzana Pereira Nunes and Klaus-Viktor Peinemann
Copyright © 2006 WILEY-VCH Verlag GmbH & Co. KGaA, Weinheim
ISBN: 3-527-31316-8

ion and the cation have a single charge, and assuming that the membrane's ion-exchange capacity is much higher than the equilibrium solution's concentration, the Donnan equation can be written as:

$$\bar{m} \sim m^2 \theta / E \bar{\theta}$$

The electric field applied across a membrane would determine the direction of ion movement. Cations move toward the cathode and anions move toward the anode. With the membrane selectively permeable for only cations or only anions, a separation process would take place.

The chemical structures of ion-permeable membranes as well as ion-exchange resins are three-dimensionally crosslinked polymers with ionic groups attached. The structural units of the most common ion-exchange membrane are shown in Fig. 6.1.

Bipolar membranes have two layers: cation (C) and anion (A) films. Boundary C–A have specific electrochemical property. At current direction C → A electrolyte concentration increases at the boundary. But at the other current direction A → C electrolyte concentration decreases at the boundary to the level at which current transference by hydrogen and hydroxyl ions generated by water dissociation occurs. Low dissociated groups of iron and some other element presence is important for inhabitation recombination of hydrogen and hydroxyl ions. A potential drop of only 0.8 V is necessary for modern bipolar membrane performance [22].

Singly charge selective membranes (also called monovalent selective, but "valence" should be only for elements) are penetrated most easily by singly charged cations or singly charged anions. These membranes may be synthesized by introducing a barrier layer of oppositely charged fixed groups on the ion-perme-

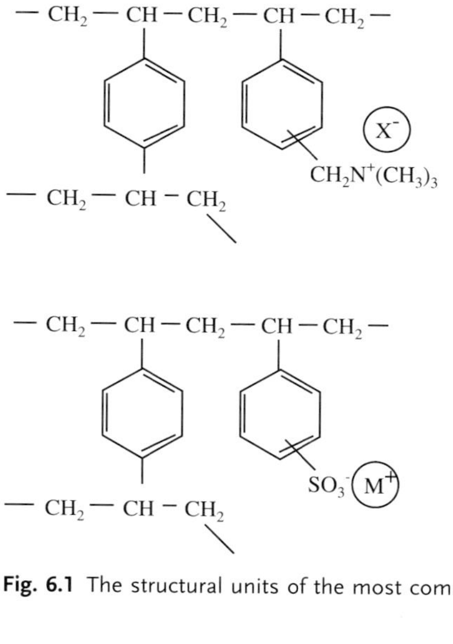

Fig. 6.1 The structural units of the most common ion-exchange membrane.

able membrane surface. They resemble bipolar membranes, but the density of the oppositely charged fixed ions (by absolute value) on the ion-permeable membrane surface is much less than the density of the fixed ions in the bulk membrane. All counterions are affected by repulsive forces in this barrier layer, but counterions with multiple charges are repelled more strongly. It should be noted that this repulsion is not absolute, and counterions with multiple charge will carry an increasing fraction of the current through the barrier layer as singly charged ions are depleted from the solution.

Membranes for electrodialysis are typically hydrocarbon films with ion-exchange functional groups attached to the polymer chains. Hydrocarbon membranes are usually categorized as homogeneous or heterogeneous. The heterogeneous membranes are made simply by grinding ion-exchange resins to a powder, dispersing that powder in a powder of thermoplastic film-forming polymer, applying the powder mixture to a reinforcing fabric, and hot pressing to form a reinforced polymer film with imbedded ion-exchange particles. The film-forming polymer is usually polyethylene or poly(vinylidene fluoride), but other polymers could also be used. Typically heterogeneous membranes are thick, opaque and mechanically strong, but they tend to have higher resistance and lower permselectivity than homogeneous membranes, because their ability to transfer ions relies on the continuity of particle-to-particle contact of the ion-exchange powder, which is the discontinuous phase in the fused polymer mixture.

The term "homogeneous" is used loosely to describe membranes that are not classified as heterogeneous. On a molecular scale no membrane is truly homogeneous, because the unlike parts of the polymer, hydrophilic ion-exchange functional groups and hydrophobic hydrocarbon chains, tend to become segregated into clusters that are much smaller than the dimensions of the resin particles in the heterogeneous membranes.

Typical homogeneous membranes have a polymer matrix of styrene crosslinked with divinylbenzene (DVB) and ion-exchange functional groups of sulfonic acid or quaternary amines. Manufacture of DVB yields an impure product containing nearly 50% of ethylstyrene, which participates in the polymerization but does not accept functional groups readily. In some cases the film of styrene-DVB copolymer is made in one step, and the functional groups are added in subsequent steps.

Membranes made with only styrene and DVB tend to have poor physical properties after they are functionalized, so other monomers or solvents are usually added to the formulation before polymerization. Ionics used the approach of adding a nonreactive, high-boiling, water-soluble solvent to the monomers prior to polymerization [27]. Tokuyama mixes the monomers with a plasticizer and PVC powder. This paste is applied to a reinforcing fabric and cured while being held between two release films [28].

Cation-exchange membranes are made by adding a sulfonic acid functional group to the benzene ring of the styrene group, usually by treatment with concentrated sulfuric acid, sulfur trioxide or chlorosulfonic acid. Anion-exchange groups can also be added to the benzene ring, but a key reagent for that procedure, chloromethyl methyl ether, is a dangerous carcinogen. That danger is avoided by replacement of styrene with chloromethylstyrene and treatment of the polymer with trimethylamine to form a quaternary amine functional group. Alternative monomers for anion membranes include vinylpyridine or methylvinylpyridine, both of which are quaternized with methyl iodide after polymerization [29].

Ion-permeable membranes are also made by swelling existing films with styrene and DVB, which can then be post-treated to add functional groups, or by grafting of ion-exchange functional groups directly onto the polymer matrix of existing films. For example, free radicals formed by radiation of polyethylene or fluoropolymer films become sites for addition of vinyl sulfonic acid, acrylic acid or vinyl amines [30].

The ion-permeable membrane properties vital for the electrodialysis efficiency are discussed below.

6.2
Ion-exchange Membrane Properties

6.2.1
Swelling

Ion-exchange membrane swelling is determined by the presence of hydrophilic groups in the membrane structure. These groups are fixed ions, counterions, and co-ions. Water can be found in two states – bound and free. There is no clear distinction between these two states, because the molecules are constantly changing their state. And the time a molecule is standing in the same state is far less then 0.001 s. The most common ion-exchange membranes based on styrene and divinylbenzene have a hydrophobic matrix. The sulfonic groups serving as fixed ions in strong cation-exchange membranes are hydrated with one water molecule. The most common single-charge cations are hydrated with 3–4 water molecules. Double-charge cations have from 5 to 8 molecules of hydrate water. Anions are hydrated a little. For example, halogen ions are surrounded by 1.5–2.4 water molecules (Tab. 6.1).

The ions' hydration increase as crystallographic radius decrease and the ions' charge increase. This effect is very clearly seen for s-elements. The interaction between these elements and water in ion-exchange material has electrostatic nature. The metal's cations with filled d-level are capable of forming complexes with water molecules.

The different character of interactions between water molecules on the one hand and cations and anions on the other hand has to be considered. Cations orientate water molecules in order to allow oxygen atoms to form a covalent bond with cations. In the case of anions, water molecules will face the anion with protons. It is possible to form a hydrogen-type bond in this case.

All kinds of membranes have higher moisture content if the counterion has higher hydration. The only exception is hydrogen ion, which has a special type

Table 6.1 Counterion hydration [3].

Cation-exchange membrane					
Counterion	H^+	Li^+	Na^+	K^+	Rb^+
Water content, eq/eq of resin	12.4	12.4	10.6	9.0	8.5
Hydration number	2.7	3.7	3.1	3.0	2.6
Cation-exchange membrane					
Counterion	Cs^+	NH_4^+	Ca^{2+}	Mg^{2+}	Zn^{2+}
Water content, eq/eq of resin	8.45	11.6	9.7	11.6	12.2
Hydration number	2.0	3.5	4.9	8.0	6.4
Anion-exchange membrane					
Counterion	I^-	Br^-	Cl^-	F^-	OH^-
Water content, eq/eq of resin	8.4	11.5	13.5	16.0	17.7
Hydration number	1.4	1.7	1.7	2.4	3.0

of hydration. The hydration numbers for counterions vary widely depending on the method of analysis, because different methods have different sensitivity for the interaction between an ion and its hydration shell. Some membranes can be damaged if they are allowed to dry out, but variation in swelling related to changing the ionic form has little effect on membrane properties.

6.2.2
Electrical Conductivity

The electrical conductivity of ion-exchange membranes depends on their chemical structure, ionic form, temperature, pH, and solution concentration. Multiple measurements have shown that the electrical conductivity $\bar{\chi}_m$ of ion-exchange materials increases as swelling, temperature, and counterions mobility increase and the ion-exchange constant and counterion charge decrease. The most influential is the nature of the counterion. The following rule is always true for a strong ion-exchange membrane [1]:

$$\bar{\chi}_m^I > \bar{\chi}_m^{II} > \bar{\chi}_m^{III}$$

where I, II and III are the charges of the ion.

There are no known exceptions to this rule for cation membranes. The few exemptions for anions can be explained by higher hydration and by smaller ion-exchange constants for the corresponded counterions.

Multicharged ions are more tricky for electrodialysis. They have less mobility and can be accumulated in an ion-permeable membrane reducing its electric conductivity. Multicharged ions are shielding fixed charges more than single-charged ones. This shielding decreases the transport numbers and reduces the current efficiency.

Ion-permeable membranes are changing ionic forms during electrodialysis. The electrical conductivity of ion-permeable membrane ($\bar{\chi}_m$) containing two ion species 1 and 2 can be calculated as the average between two extreme values. One of them is based on the model of parallel independent movement of two types of ions ($\bar{\chi}_m^{\parallel}$) and another one is based on the model of successive ion movement from one fixed charge to another ($\bar{\chi}_m^{\perp}$) [2]:

$$\bar{\chi}_m = a\bar{\chi}_m^{\parallel} + (1-a)\bar{\chi}_m^{\perp}$$

$$\bar{\chi}_m^{\parallel} = \bar{\gamma}_1 \bar{\chi}_{1m} + \bar{\gamma}_2 \bar{\chi}_{2m}$$

$$1/\bar{\chi}_m^{\perp} = \bar{\gamma}_1/\bar{\chi}_{1m} + \bar{\gamma}_2/\bar{\chi}_{2m}$$

$$\bar{\gamma}_1 + \bar{\gamma}_2 = 1$$

$$a = 2\bar{\chi}_{2m}/(\bar{\chi}_{1m} + 2\bar{\chi}_{2m}) + 2\bar{\gamma}_1(\bar{\chi}_{1m}^2 - \bar{\chi}_{2m}^2)/((\bar{\chi}_{2m} + 2\bar{\chi}_{1m})(\bar{\chi}_{1m} + 2\bar{\chi}_{2m}))$$

The membranes in different ionic forms may have very different degrees of swelling. In this case the values $\bar{\chi}_{1m}$ and $\bar{\chi}_{2m}$ have to be adjusted to accommodate a decrease in fast-moving ion mobility and an increase in slow-moving ion mobility.

Ion-permeable membranes are complicated heterogeneous systems. Even ion-exchange beads and so-called homogeneous membranes in fact behave as heterogeneous ones. The term "homogeneous" is more likely a historical tradition rather then a term that strictly satisfies the thermodynamic meaning of this term. However, the term "homogeneous membrane" is generally accepted. We shall use this terminology keeping in mind its conventionality.

For the purpose of simplicity let us hypothesize an ion-exchange membrane as a two-phase system. Each phase is continuous and randomly positioned. Assume that all fixed ions are concentrated in homogeneous (in the thermodynamic meaning) gel areas. The gel areas are partially in contact with each other and partially separated by intergel areas. Assume also that intergel areas do not contain any fixed ions and are filled by equilibrium solution. Electric current can pass by one of three ways: through connected gel areas only (column II), through intergel areas only (column III) or through both (column I). Figure 6.2 illustrates this model. Three columns on this picture illustrate three possible current passages. The shaded area would correspond to a gel area. The column width a, b, and c would be proportional to the current fraction passing through a specific passage.

The electric conductivity of this system can be described by the following formula [3]:

$$K_m = aK_d/(e + dK_d) + bK_d + c$$

$$K_m \equiv \chi_m/\chi$$

$$K_d \equiv \bar{\chi}/\chi$$

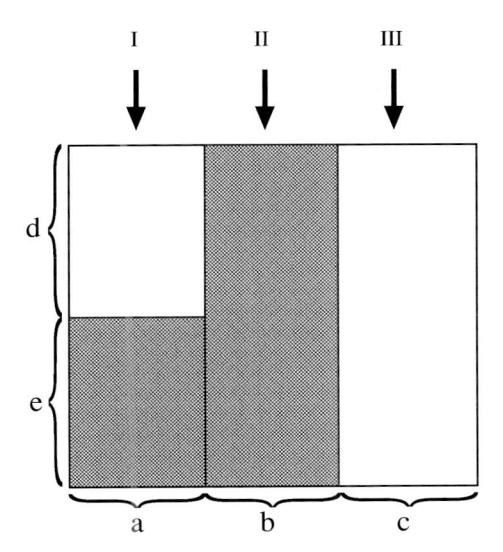

Fig. 6.2 Three-conductor model of ion-exchange membrane conductivity.

The total sum of all fractions should be equal to unity:

$a + b + c = 1$

$e + d = 1$

The total fraction of gel area (*f*) in the ion-exchange membrane can be expressed through the model's parameters:

$ae + b = f$

The set of equations above are used for calculation of ion-exchange membrane conductivity and selectivity versus external solution conductivity.

The counterion concentration is usually high and reaches up to 3 N in common ion-exchange materials. This is the reason for ion-exchange membrane's high electric conductance. Donnan exclusion of electrolyte in gel areas determines their weak dependence from equilibrium electrolyte concentration. At the same time electric conductivity of a solution of strong electrolyte is approximately in linear proportion to the concentration up to 1 normal solution. We have to take into account the lower mobility of ions in ion-exchange membranes compared to ion mobility in solution due to a slowdown by the polymer matrix. Therefore, the conductivity of an ion-exchange membrane is lower than the conductivity of an electrolyte at moderately high concentrations because of high ion mobility in solution. But ion-exchange membrane conductivity is higher than electrolyte conductivity at low concentrations because of the larger number of current carriers in ion-exchange membranes. Thus at certain concentration the

Table 6.2 Parameters of the three-conductor model.

Parameters	Resin	Membrane	Ion-exchange column
a	0.65	0.18	0.25
b	0.35	0.72	0.83 f
c	0.00	0.10	0.75–0.83 f
d	0.17	0.15	1–0.71 f
e	0.83	0.85	0.71 f

electric conductivity of ion-exchange membrane and equilibrium electrolyte will be equal. This concentration is called the isoconductivity point. The intergel areas in our model do not contain fixed ions or polymer chains. Their electric conductivity must be equal the solution conductivity. So electric conductivity of both gel areas and intergel areas must be equal at the isoconductivity point. The fraction of gel areas f in ion-exchange membranes would fluctuate in a narrow range for various ionic forms of the same ion-exchange membrane. However, we need to note that f will increase as swelling is increasing. For the sodium form of sulfonated styrene-divinylbenzene cation-exchange resin and the same type cation-exchange membrane the parameters of the three-conductor model can be used (Tab. 6.2) [1].

These parameters can be used for the approximate electric conductivity calculation of various ionic forms and various concentrations of equilibrium solution. The conductivity at the isoconductivity point has to be used for both ion-exchange resins and membranes.

The electric conductivity of ion-exchange membranes increases as temperature rises. This change may be characterized with an equation similar to the Arrhenius equation for viscosity:

$$\bar{\chi} = \chi_0 \exp(-E/RT)$$

The activation energy E is estimated at 21±3 kJ/mol for sulfonated styrene-divinylbenzene cation-permeable membranes in equilibrium with one of the following solution: 0.1–1 N NaCl; 0.001–1 N $MgCl_2$; 0.01–0.5 N $CdCl_2$ and 0.1–2.5 N $FeCl_3$. The energy E would increase if weakly ionized groups were present in the membrane.

A change in pH has little effect on the electrical conductivity of ion-permeable membranes containing strong fixed ions. Membranes containing weak ionic groups may change conductivity sharply with pH change, because the ionic groups in a weak cation-exchange membrane have a high degree of dissociation in alkaline solution and the groups in a weak anion-exchange membrane are highly dissociated in acidic solution.

Bifunctional ion-exchange membranes show two bends on the conductivity versus pH curve. Each bend corresponds to one of two types of functional groups with different dissociation degree at low and high pH values.

Structural factors have large impact on the ion-exchange membrane's conductivity. For example, increasing of the crosslinking would induce linear decrease of the membrane's conductivity.

6.2.3
Electrochemical Performance

Electrochemical performance of ion-exchange membranes is characterized by transport numbers, selectivity and specific selectivity. The ion transport number (t_i) is the fraction of the electric current carried by specific ion type:

$$t_i = I_i/I_0$$

For all ions participating in current transport the following expression will be true:

$$\Sigma \bar{t}_g + \Sigma \bar{t}_c = 1$$

The selectivity P can be represented by the expression:

$$P = (\bar{t}_g - t_g)/(1 - t_g)$$

Selectivity demonstrates the relative transport property divergence between real and ideal membranes. Commercial ion-permeable membranes typically have selectivity from 0.93 to 0.99.

The specific selectivity of a membrane for ion type A in the presence of ion type B (P_B^A) can be expressed as:

$$P_B^A = \bar{t}_A C_B / \bar{t}_B C_A$$

In the case of a two-component electrolyte the counterion transport number would be expressed as:

$$\bar{t}_g = 1/(1 + \bar{u}_c \bar{a}_c / \bar{u}_g \bar{a}_g)$$

The ion activity can be replaced by concentration to a first approximation. The co-ion concentration can be calculated with the Donnan equation based on the capacity of an ion-exchange membrane. It is more difficult to determine the ions' mobility ratio. The counterion mobility can be calculated based on isoconductivity point for the appropriate ionic form. The mobility of co-ions can be determined based on a self-diffusion coefficient measurement.

The transport numbers can be calculated using the three-conductor model. The co-ion concentration in gel areas of ion-exchange membrane is far less than in intergel areas, so it would be a valid assumption that co-ions are transported through intergel areas only. As can be seen from the picture of the three-conductor model the current would be carried only by counterions in parts a and b. Both counterions and co-ions would carry the current in part c. Therefore the following equation would describe the transport number of the counterion in the membrane:

$$\bar{t}_g = 1 - t_c c / K_m$$

$$K_m \equiv \bar{\chi}_m / \chi$$

For practical purposes the most common characteristic of membrane electrochemical performance is current efficiency, which is defined as current fraction η carried by a specific ion type i.

$$\eta_i = q_i / Q$$

The major difference between transport number and current efficiency is that the latter characterizes total transport through the membrane including diffusion. (Diffusion is suppressed during experimental determination of transport numbers.)

6.2.4
Diffusion Permeability

Electrical conductance of an ion-exchange membrane $\bar{\chi}_m$ in various ionic forms can be used for an approximate estimation of the ions' self-diffusion coefficients $\bar{D}_i$ using the Nernst-Einstein equation:

$$\bar{D}_i = \frac{RT}{z_i F} \frac{\bar{\chi}_m \bar{t}_i}{E}$$

The diffusion coefficient calculated in this way is usually higher than values measured by independent methods. This can be explained by the fact that during the self-diffusion measurements ions are moving toward each other and the water present in the ion-exchange resin is motionless. At the time of electric conductance measurements all counterions are moving in one direction and create electroosmotic flow. It is obvious that in this case the ion's movement would have less resistance as they move along with the liquid flow.

The counterion diffusion coefficients obtained by a kinetic study or by conductivity measurements cannot be used as membrane diffusion coefficients, because counterions and co-ions are moving in opposite directions in electrodialysis. However co-ions are motionless at the time of ion exchange between equal activity solutions. Therefore the ion-migration conditions are totally different in these cases.

It should be noted that ion-exchange membrane swelling depends on the external solution concentration. This fact has a significant impact on the membranes' diffusion permeability. Moreover, the real-life situation often includes an ion-exchange membrane facing a concentrated solution on one side and a dilute solution on the other side. A concentration gradient occurs within the membrane and affects its diffusion and osmotic permeability.

By order of magnitude the diffusion permeability of commercial membranes is about 10^{-6} cm^2 s^{-1}.

6.2.5
Hydraulic Permeability

Hydraulic permeability is measured as the volume of liquid passed through one unit of a membrane surface for one time unit at one unit of pressure difference. For most commercial membranes this parameter has a value of $10^{-10} - 10^{-11}$ cm^3 g^{-1}s^{-1}.

Water can penetrate through either gel areas or intergel areas in a homogenous membrane. The intergel areas are less crosslinked, and the water flow would meet less resistance in these areas. However, the volumetric fraction of intergel areas is small; therefore, it is not possible to predict the contribution of gel areas and intergel areas in the total hydraulic permeability.

In a heterogeneous membrane water can penetrate through ion-exchange particles, through gaps between them, and through the binding polymer. The pore-radius estimation and comparison of water transfer coefficients with water self-diffusion coefficients suggests that all three transports take place.

The external solution concentration has an opposite effect on the membranes hydraulic permeability for homo- and heterogeneous membranes. As the solution concentration is increased the hydraulic permeability of a homogenous membrane decreases because water penetrates mostly through the gel of homogeneous membranes, because the gel area progressively dehydrates as the solution concentration increases. In the case of a heterogeneous membrane the water passes mostly through the cracks between the gel areas and the bounding polymer, and these cracks enlarge as the solution concentration increases, because of shrinkage of the resin particles.

6.2.6
Osmotic Permeability

The osmotic permeability of ion-exchange membranes D_{osm} can be estimated by comparing the self-diffusion coefficient of water in the membrane $\bar{D}_w$ for various degrees of hydration of the membrane. As the water content increases (up to 4 mole of water per equivalent of fixed ion) the water molecule mobility increases, rising steeply at the beginning and more gradually later. It also rises with increasing crystallographic radius of the counterion.

The nature of the membrane matrix has significant influence on water mobility in the membrane. Water diffusion coefficients are high in perfluorinated membranes with weak hydrophilic interaction [2]. The coefficients are lower for membranes with hydrocarbon matrix with stronger hydrophilic interaction. A correlation exists between the self-diffusion coefficients of water and co-ions: the water self-diffusion coefficient increases as the co-ion self-diffusion coefficients increase. It is possible that co-ions migrate in swelling water along the pore axis.

As the solution concentration increases, the osmotic permeability of heterogeneous membranes increases linearly and the osmotic permeability of a homogeneous membrane decreases logarithmically. A membrane in single-charge ionic form would have an osmotic permeability approximately twice as high as the same membrane in double-charge ionic form. The order of magnitude of D_{osm} would be about 10^{-5} cm^2 s^{-1}.

6.2.7
Electroosmotic Permeability

As transport numbers can characterize ion transport through a membrane, a water transport number $\bar{t}_w$ can be assigned to characterize solvent carried through a membrane. This number would express a number of moles of water carried through the membrane with one Faraday of electrical charge.

$$\bar{t}_w = n_g \bar{t}_g - n_c \bar{t}_c$$

The solvent transport number is not a characteristic of a membrane or a specific pair of ions. It can have either linear or nonlinear dependency on $\bar{t}_g$ because this characteristic would depend on the degree of membrane's heterogeneity, current density, the nature of the polymer matrix and the nature of the ion. It was determined for cation membranes that the solvent transport is higher when a counterion's hydration number is higher.

Higher solution concentration leads to lower electroosmotic flow through a homogeneous membrane with water content over 19%. This effect is more pronounced if the counterion is more hydrated, and it can be explained as increasing electrostatic interaction in the gel phase of the ion-exchange membrane concurrent with decreasing pore radius. In the case of heterogeneous membranes and homogeneous membranes with water content less then 19%, the electroosmotic flow is independent of the solution concentration. The explanation of this effect is connected with the inner structure of the ion-exchange membrane in a way similar to the hydraulic permeability.

The electroosmotic permeability increases with the increased swelling. The nature of fixed groups and ion-exchange capacity is negligible. Also, the membrane nature has little influence in the case of highly crosslinked structures with equal hydration numbers. However, for low crosslinked membranes the characteristic $\bar{t}_w$ would have a higher value in the case of a hydrophobic matrix than in the case of hydrophilic ones. For example, perfluorinated membranes have higher electroosmotic coefficients then membranes based on a hydrocarbon matrix. Not all water in a membrane moves with counterions. It was determined that only 75% of total water content moves in a perfluorinated membrane, 60% – in a polystyrene based membrane, and 33% – in a polyacrylic-based membrane. The membrane electroosmotic permeability has a magnitude of 10^{-3} cm^3 A^{-1} s^{-1}.

6.2.8
Polarization

The electric current is carried predominantly by only one kind of ion in an ion-exchange membrane – by cations in a cation membrane and by anions in an anion membrane – in contrast to the case in free solution where both kinds of ions carry current. Therefore concentration changes take place in the solution close to the membrane surface. These changes are called concentration polarization. Let us assume that there is a thin nonmixed solution layer near the membrane with thickness δ. The concentration in this layer would change linearly from the bulk concentration C_0 to the concentration C_1 on the membrane surface. Let us disregard the electrolyte diffusion on the opposite side of the membrane. Then it is obvious that at certain current density the concentration would approach zero on the receiving membrane surface. This current density is named the limiting current density $i_{\lim}$ and it can be described by the equation:

$$i_{\lim} = DC_0 F/\delta(\bar{t}_g - t_g)$$

Taking into account the electrolyte diffusion on the other side of the membrane, the equation should be [1]:

$$i_{\lim} = (D/\delta + 4\bar{D}C/Xd_m)zFC/(\bar{t}_g - t_g)$$

In case of free convection:

$$\delta \sim [Dv/(C_0 - C')]^{1/4}$$

In the case of laminar flow:

$$\delta \sim v^{-1/2} v^{1/6} D^{1/3}$$

In the case of turbulent flow passing through a chamber with a mixing screen the thickness of the diffusion layer can be described [1]:

$$\delta = 0.0685 K_s [D_0 T[1 + 0.0158(T - 298)]]^{1/3} (ld_0)^{0.25} (1 - h/d_0)^{0.5} / v^{0.5}$$

For practical purposes the relation between the limiting current density and the flow velocity can be approximated:

$$i_{\lim} = Av^n$$

The empirical coefficients A and n can be determined from current–voltage curves. Several researchers reported $n = 1/2$ independent of the flow profile. It is possible to calculate that the coefficient $n = 1/2$ in the case of a uniform flow profile. The theoretical value for the case of parabolic flow distribution is one third. It is in a good agreement with experiment:

Coefficient	Theoretical value	Experimental value
A	1.05	0.96
n	0.33	0.34

We have to note that the calculated electric resistance of an electrodialysis compartment based on polarization curves is somewhat higher than experimental ones across all flow rates. This may be explained by influence of volume charge, concentration, heat, and natural convection on the polarization characteristic.

The method of laser interferometry allows measurement of concentration in the narrow boundary layer with resolution up to 2 micrometer. The results of these measurements are expressed as an empirical equation linking diffusion layer thickness and flow velocity:

$$\delta = 2.1 \times 10^{-2} v^{-0.5}$$

These measurements demonstrated that as the current approaches the limiting current, the concentration is decreasing but it never descends below one half of the bulk concentration. These measurements also reveal a nonlinear concentration profile in the diffusion layer at a current density below the limiting one. Convection occurs near a cation membrane surface at current densities higher than the limiting value. This convection levels the concentration at the thin boundary layer. Adsorption of some surfactants on a membrane surface may cause microconvective flow in the diffusion layer, and in some cases that may cause the limiting current to increase. The adsorption effect may be used for synthesis of new membrane materials and optimization of electromembrane purification processes [49]. However, there is no convection flow near the anion membrane at currents exceeding the limiting one. Therefore the concentration in the diffusion layer decreases to very low values. Hydrolysis of weak-electrolyte ions (like phosphoric, carbonic, etc. acids) and homogeneity/heterogeneity of the membrane surface should be considered for ion transference through anion membranes [50–52]. Applying an electrical field with special pulse characteristics can diminish the effect of concentration polarization and save power for desalination processes [54].

An increase in membrane uniformity leads to a higher limiting current. The membrane surface condition has a noticeable influence on the limiting current value. The polarization characteristic may deteriorate as a result of fixed ion concentration reduction due to ion-exchange membrane degradation at the time of manufacturing or at the time of operation, or adsorption of low-mobility organic substances present in the solution or leached from the membranes.

Desalination of very diluted solutions is done in an electrodeionization apparatus (please refer to Section 6.3.2). The desalination compartment of an electrodeionization stack is filled with bead-type or fibrous ion-exchange material. Such packing not only lowers the electrical resistance of desalination chambers, but also improves the current efficiency. The conductivity of the system of ion-exchange resins–solution can be calculated on the equation of three-conductor model using volume-average conductivity of the cation and anion resin mixture. The course of an electrical current through the system of mixed ion-exchange resin–solution is associated with some features that are not observed in the case of individual resins [3]. Assuming random distribution of anion (A) and cation (C) exchange beads four types of bead connections can be distinguished along the electric currents direction ($\rightarrow$): two cation-exchange granules (C $\rightarrow$ C), two anion-exchange beads (A $\rightarrow$ A), and contacts between two unlike beads (C $\rightarrow$ A, A $\rightarrow$ C). Let us examine processes on each type of contacts during the current passing. Each pair of granules has both direct contact and contact through thin layer of electrolyte between them. In the case of the two cation-exchange beads C $\rightarrow$ C, cations are crossing over from the left bead to the right one either through the area of direct contact or the solution between the beads. The electric current running through this pair does not generate any concentration change neither on the boundary itself nor in the surrounding area. The situation is similar near the A $\rightarrow$ A type contact. The only difference is the opposite direction of the current-carrying ion movement. On the boundaries C $\rightarrow$ A and A $\rightarrow$ C the swapping of current-carrying ions takes place. The swapping from cation conductivity to anion conductivity on the C $\rightarrow$ A type contact is associated with ion accumulation, which means increasing solution concentration. On the contrary, swapping from anion conductivity to cation conductivity on the A $\rightarrow$ C type contact leads to ion withdraw into the beads and reduction of solution concentrations. As concentration decreases to a very low level the "water-splitting" phenomenon occurs. This phenomenon represents splitting a water molecule into hydrogen and hydroxyl ions. Sometimes this phenomenon is also called acid-alkali generation, and its use decreases power-consumption and increases stack capacity [53].

For more deteiles of ion-transport and polarization phenomena the monograph [15] is recommend.

6.2.9
Chemical and Radiation Stability

Mostly, ion-exchange membranes are composite materials. Because of this, their chemical, thermal and radiation stability is determined not only by the stability of the separated materials but also by the changes that occur at the phase boundaries. The latter is especially important for heterogeneous membranes because of damage to the bonding between high molecular polyelectrolyte and the film-forming polymer as well as delaminating of the reinforcing fabric.

There are notions that ion-exchange membranes based on polystyrene and polyethylene are stable in 30% HCl and 25% NaOH as well. But it has been reported that membrane selectivity decreased monotonically during heating in water at 80 °C, which may be explained by expansion of membrane pores filled with external solution.

During exposure of ion-permeable membranes to radiation, their mechanical firmness and flexibility decreases. Ion-exchange capacity, crosslinking, conductivity and selectivity decrease as well.

Application of the three-conductor model allows the estimation of the scale of porosity increase during exposure of ion-permeable membranes to radiation. The first portion of radiation increases porosity significantly; subsequently porosity changes a little. But changes in swelling indicate macromolecule destruction with higher radiation doses. Reinforcing ion-permeable membranes does not improve their radiation stability significantly. For example, mechanical firmness of swelled ion-exchange membranes based on polystyrene and polyethylene decrease to half their original value at 80 Mrad. At 200 Mrad they cannot be used at all because they may be destroyed by slight bending [3].

Interestingly, perfluorinated membranes have superior chemical stability, but their radiation stability is worse than that of membranes with a hydrocarbon matrix.

6.3
Electromembrane Process Application

6.3.1
Electrodialysis

Electrodialysis (ED) is used to remove ionized substance from liquids through selective ion-permeable membranes. ED is the most widely commercialized electromembrane technology. Desalination of brackish water is the area of electrodialysis application with the largest number of installations. This chemical-free technology competes with reverse osmosis. Electrodialysis shows better resistance to fouling and scaling. It also has an economical advantage in desalination of low-salinity solutions [13]. Also, it should be kept in mind that because of small material consumption ED is the most environmental friendly process for solution desalination [14].

Electrodialysis has the ability to concentrate salts to high levels with much less energy consumption than evaporation would require. This capability has been utilized in Japan to make edible salt by recovering NaCl from seawater and concentrating it to 20% before evaporation. The plants there are huge; some have greater than 100 000 square meters of membrane. Salt recovered by electrodialysis in Kuwait is the raw material for a chlor-alkali plant there. Electrodialysis has also been used to concentrate salts in reverse osmosis brines [32].

Electrodialysis is used in a wide variety of food applications. Throughout the world it is used to remove salt from cheese whey so that the other components of whey can be used as food for humans and animals. In Japan, the mineral composition of cows milk intended for infant formula is altered by electrodialysis to more closely resemble the composition of mother's milk. In France, potassium tartrate is removed from wine to prevent its precipitation. In Japan, salt is removed from soy sauce to allow its use by people with hypertension. Salts of organic acids, e.g. lactic and succinic, produced by fermentation are recovered from the fermentation broth and concentrated by electrodialysis [31]. Amino acids made by synthesis or by hydrolysis of proteins are desalted at their isoelectric pH.

In the food applications mentioned above it is impractical to remove components that could foul the membranes, because these are necessary constituents of the product. In such cases the process is operated under conditions that minimize fouling, and then the fouling that does occur is handled by cleaning in place (CIP). The CIP procedures can include soaking in brine, current reversal and washing with acid, base and nonionic surfactants.

The mandatory condition for an electrodialysis process to be executed is an alternating order of cation and anion membranes and electric field applied across the entire assembly (Fig. 6.3). Between the alternating membranes are two types of compartments – desalination and concentrating. Ions will migrate from the compartments where electric current is passing from an anion membrane to a cation one (the even compartments in Fig. 6.3). They will be transferred to the successive compartments (the odd compartments in Fig. 6.3). These compartments will accumulate the ions because the ion-exchange membrane between them would prevent ions from moving further. Therefore the solution in the even compartments will be demineralized and solution in the odd compartments will be concentrated.

The solution is alkalized in the cathode compartment and it is acidified in the anode compartment. As a result, the entire flow incoming to the electrodialysis apparatus can be separated into desalinated and concentrated streams. Co-ions absorbed by the membrane reduce the efficiency of this process. The highest possible degree of concentration can be achieved if no incoming solution is supplied to the concentrated compartments. In this case, water would be delivered to the concentrate compartments by osmosis and electroosmosis and in the hydration shells of the transported ions.

The electroosmotic phenomenon makes it possible to use electrodialysis for concentrating uncharged substances, which would not normally migrate in an

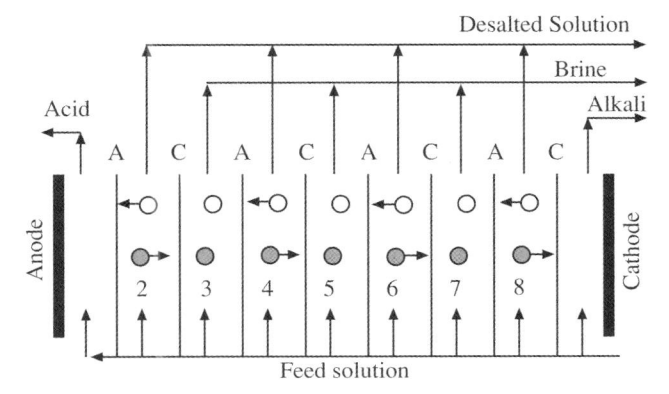

Fig. 6.3 Arrangement of membranes for electrodialysis.

electric field. For example, by adding a dry electrolyte to a nonelectrolyte solution and then desalting it in an electrodialysis apparatus it is possible to remove a considerable portion of water from the original solution. Water will be moved by electroosmosis from desalination compartments to concentrating ones.

There are several common elements in the design of any electrodialysis stack: end blocks, end frames with electrodes, membranes, spacers between membranes, and manifolds for inlet and outlet of fluids. The electrodes used in an electrodialysis stack must withstand the electrochemical reactions and the solutions that circulate within the electrode chambers, as well as to the electrolytes, which are carried there due to electrodialysis and form as a result of electrolysis. Residue must not be formed inside the electrode chambers, while the chamber itself must be thin enough to not create too much resistance to the electric current.

Titanium covered by platinum or by dioxide of manganese, ruthenium, iridium or other substances is most commonly used as an anode. Graphite and graphite covered with lead dioxide have also been used as anodes. Under some conditions, high pH and absence of salts in the anolyte, nickel can be used as the anode. Stainless steel is commonly used as the cathode. If current reversal is employed, the same material, platinized titanium or graphite, is used for both electrodes. Electrode chambers should be flushed with a large flow of rinse solution in order to remove the electrode reaction products.

Special spacers are used for membrane separation. Along with membranes, spacers are important structural parts of the electrodialysis stack. In fact, the type of spacer used usually dictates the design of the electrodialysis stack. In electrodialysis, the spacer regulates the distance between the membranes and guides the flow of fluid in a certain way. The spacer is usually made of nonconducting and hydrophobic materials, elastic enough to conform to and form a seal with the membrane and yet rigid enough not to be pushed out from the stack at the time of stack tightening. Materials such as polyethylene, polypropylene, polyvinyl chloride, and various elastomers are used.

In normal electrodialysis the spacers are made of nonconductive material. However, ion-conducting spacers are used to receive extremely desalinated water. Depending on design, the fluid between the membranes can move through channels, which form either a tortuous path or a set of parallel paths.

In a tortuous-path spacer, the flow of fluid changes direction several times and gains large speeds due to lathes attached to the walls. In order to prevent the displacement of these lathes, they are fastened by crosspieces, which are thinner than the spacer. The use of the crosspieces also allows whirling of the flow of fluid and lowering the concentration polarization. For better whirling, crosspieces should be located at angles of 15–75 degrees to the flow. Tortuous-path spacers are made by die cutting sheets of polymer to form lathes, ribs and a peripheral gasket. Two pieces of slightly different configuration are glued together to form a single unit. These spacers are easy to manufacture, but they significantly obscure the surface of the membrane.

Spacers for parallel path flow usually consist of two elements – a peripheral forming the outside walls of the chamber, and a separator, which is nested inside the gasket. The spacers for the desalination and concentration chambers are often identical. Potential drawbacks of such a design are poor flow distribution and stack deformation. To increase the stability of the design plastic guide rods can be put through the gaskets and membranes.

Some separators are made of crimped perforated sheets and sheets modified by the break-and-stretch method. These separators are easy to make, good at whirling the fluid, but significantly mask the membrane.

Netting separators, which appear as two layers of parallel filaments fused at the crossover points, have a smaller footprint on the membrane surface. Vexar nonwoven netting, developed by DuPont but now made by many others, is the most commonly used separator in parallel path electrodialysis stacks. These separators can be made by extrusion of a variety of polymers.

The electrodialysis reversal systems with netting-based spacers (Fig. 6.4) have higher efficiency and lower energy consumption. Nonwoven netting is a better turbulence promoter. This leads to higher limiting current density and better ion transport through the membranes. However, it also results in higher hydraulic resistance of the desalination channel [17].

Depending on the hydraulic scheme of connection all electrodialysis apparatuses can be separated into single-pass and circulating. Circulating apparatuses can be of periodical and continuous types. A single-pass apparatus can be single-stage or multistage. In each stack the feed can be parallel or successive. As the fluid is desalted, the critical speed of the flow should be increased. Therefore, multistage ar-

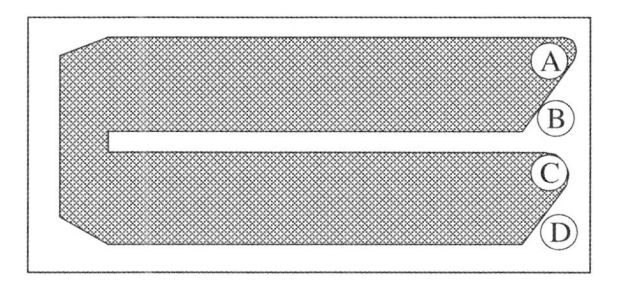

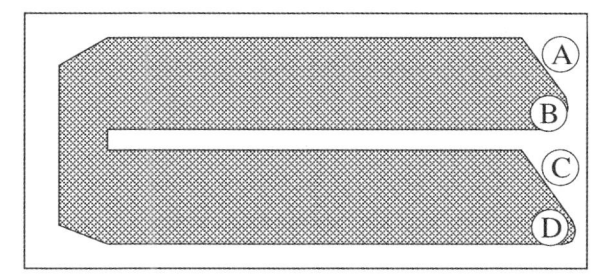

Fig. 6.4 Demineralizing (top) and concentrating (bottom) spacers. A, B, C, D are the openings in the spacer forming the flow passage through the EDR stack.

rangements have advantages, because the number of chambers can be decreased at each subsequent stage. Desalted solution goes through the compartments successively, and its speed increases. Thus, all compartments provide optimal conditions for desalination. In some cases, it is most rational to connect desalination chambers or packets in sequence and concentration chambers in parallel. The common disadvantage of all devices with a successive distribution of flows is the significant change in pressure from one chamber to the next, which can cause chambers to collapse with contact between or even tears in the membranes.

Counterflow feed of desalination and concentration compartments might be considered as a way to decrease the diffusion transport of salt and increases the output of electrical current. Electrodialysis efficiency would increase, but pressure difference between chambers would be higher than with parallel flow. That is why, in practice, the parallel system of solution flows in neighboring chambers is preferred, disregarding its higher energy demands. A parallel flow within a single stack and counterflow within the stack sequence is a practical solution for a large-scale commercial system.

Also done on a small scale is the flow arrangement in crisscrossing directions with a parallel connection of the same kind of compartments.

The most commonly used method of solution feed to a filter-press type electrodialysis stack is through matching openings in spacers and membranes, which in an assembled state form manifolds along the length of the stack. The channels are connected with the insides of the appropriate compartments. Common disadvantages of all filter-press type stack designs are the passage of electric current (called shunt currents) through the manifolds and the possibility of crossleak of liquid between the manifolds of desalination and concentration solutions.

Scaling and fouling can cause serious operational problems in electrodialysis. Precipitates in an electrodialysis apparatus have low specific conductivity that can substantially increase the electric resistance of the membrane and the whole apparatus. Consequently, it leads to an increase in specific energy consumption per unit of product and to local heating. Sometimes the heating may be so significant that it results in membrane melting, cell leakage and crossleaks. Because they cause local distortion of current and solution flow, precipitates tend to spread from the point of their origin. The precipitate's origin can be one of the following: electrophoretic deposit of solid particles suspended in solution; increasing ion concentration above the solubility limit, and pH change in the boundary layer. The precipitation occurs more often in concentration cells where the concentration on ions capable of forming low-solubility compounds reaches the maximum.

The maximum possible degree of concentration K can be calculated with the following formula [4]. This formula takes into account the selectivity of ion transport through the membranes for the concentrate's ion strength μ from 2 to 6.5.

Also, it is very convenient to use the Langelier index [5].

$$K \leq \left(\frac{2.9 \times 10^{-3} \mu}{C_{NaCl} K_{Ca} [Ca^{2+}] K_{SO_4} [SO_4^{2-}]} \right)^{0.5}$$

The danger of precipitate formation is especially high if the current density is above the limiting value. At this condition, current is carried by ions from the dissociation of the solvent. In water solution they would be the hydrogen and hydroxyl ions. The pH will rise in the boundary layer near the cation membrane on the side facing the desalination cell and near the anion-exchange membrane on the side facing the concentration cell. The rising pH can reduce the solubility of some solutes. Practically the most common precipitate is calcium carbonate:

$$HCO_3^- + OH^- \rightarrow CO_3^{2-} + H_2O$$

$$Ca^{2+} + CO_3^{2-} \rightarrow CaCO_3$$

The scale usually is confined on ion-exchange membrane surfaces. The reason for the local sediment growth is the concentration fluctuations. The concentration fluctuations are the results of nonuniform flow distribution in the cell.

More often the sediment contains carbonate and hydroxide. Therefore, the most common method to remove the sediment is concentrate acidification or current reversal [18]. Electrodialysis reversal (EDR) is an electrodialysis process with periodic reversal of the electric field. As the direction of electric current is switched desalination chambers become concentrating ones and vice versa. Colloidal and organic precipitate accumulates in the desalination chamber. As this chamber becomes a concentration one after switching the current direction, the organic or colloidal precipitate is washed away with concentrate flow.

Fouling of ion-permeable membranes is often connected to the presence of weakly ionizable organic substances in the process solution. The organic ions produced by these substances have low mobility in the membrane phase, so they concentrate at the membrane/solution interface. Eventually these ions are absorbed by the membrane, which leads to membrane poisoning [7].

Current reverse and regular acid/base treatment will not completely restore the original polarization characteristics. Visual observation with a microscope may reveal "pickling" on the membrane surface. Mechanical renewal of the membrane surface would help to restore the original polarization characteristics. However, the properties are changed irreversibly if degradation penetrates deep into the membrane. Cation membranes are usually more stable then anion membranes. An alkaline environment accelerates membrane degradation. A process of "self-poisoning" takes place. For example, an alkaline extract from anion membrane added to the solution treated by electrodialysis would cause 30–40% reduction in the limiting current, and a 7–8 fold increase in polarization, reaching 8–9 V at a current density of twice the limiting value.

Deteriorating strongly ionized groups can transform into weakly ionized groups. Weakly acidic and weakly basic groups would have a low degree of dissociation in neutral solutions or at a current density close to the limiting one.

Microflora can form an inhibiting film on the membrane surface. This film increases the membrane's surface resistance and decreases the limiting current density.

Electrodialysis of a solution containing organic matter has additional complications. Organic substances may decrease the limiting current density due to concentration-profile changes in the diffusion layer as well as due to forming a fouling film on membrane's surface. For example the presence of polystyrene sulfonate in potassium chloride solution at the time of electrodialytic treatment leads to a decrease of potassium and chloride ion concentration and an increase of polystyrene sulfonate ion concentration in the diffusion layer. In the presence of the negatively charged organic ions the chloride transport in a solution decreases and limiting current density falls. The same effect causes reduction of membrane selectivity due to ionic group blocking by organic ions.

A significant reduction of limiting current density can be caused by formation of a surface polarized membrane-fouling film of macromolecules or colloidal substances. The fouling film can be electrical neutral, or with the fixed charge like the fixed charge of an ion-permeable membrane. The film may comprise unreacted intermediates and byproducts of membrane synthesis or their decomposition products. The presence of a porous film is an explanation for different transport numbers of hydrogen and hydroxyl ions through the membranes at current densities above the limiting value.

Weakly ionized organic substances have very low mobility in the membrane. Therefore they accumulate near the membrane/solution interface where they are absorbed by the membrane and result in membrane fouling. The large intensity of an electrical field at the membrane/solution interface, a consequence of the second Wien effect, increases the degree of dissociation of weakly ionizable organic substances, which also promotes fouling of membranes.

The most widespread methods of prevention of a poisoning of membranes are the pretreatment of solutions and periodical polarity reversing. The most recent method of protection is to deposit on the membrane surface a thin film with the same charge as the fouling substances. The charge density in the modifying film should be much less than the charge density of the membrane [6,1].

6.3.2
Electrodeionization

Electrodeionization (EDI) is a process widely employed in chemical industry for ultrapure solution preparation. EDI is a membrane-based process driven by electrical current. The EDI process resembles ED. However, the EDI stack has ion-exchange material added to the desalination compartments. The filler can be in form of beads or fibers [1]. The major advantages of EDI over ion exchange is consistent part-per-billion and part-per-trillion level of contaminants; continuous high recovery operation; no chemical consumption for regeneration; no chemical waste; safety and reliability. The feed water requires pretreatment before entering the EDI process in order to meet requirements of low scaling potential and low level of fouling organic and colloidal particles. Commonly reverse osmotic permeates serve as a feed stream for EDI.

Fig. 6.5 General view of a commercial EDI installation produced by Ionics Inc.

In some instances, water produced by EDI with 10–18 megohm-cm resistivity can be used directly. Commonly, an EDI system is followed by a mixed-bed polishing filter in order to assure high purity. For customers EDI demineralization means lower water-treatment cost and less risk to downstream equipment and products [8, 9].

The EDI stack consists of alternating pairs of strong cation and anion membranes, high-quality ion-exchange resins, polymeric flow channel spacers and a pair of electrodes. Electric current passing through the stack not only removes the ions present in the solution, but also splits water into hydrogen and hydroxide ions. These ions are continuously produced at points of contact between anion and cation exchange surfaces in the EDI stack (see Section 6.2.8). Constant flow of hydrogen and hydroxyl ions regenerates ion-exchange beads and membranes inside the diluting compartment. Therefore there are favorable conditions for adsorption of weakly ionized substances. EDI has gained its popularity for very high removal rates of silica, carbon dioxide, and boron [21]. Figure 6.5 shows the general view of a commercial EDI installation produced by Ionics Inc.

In order to avoid secondary contamination at the C → A contacts, the number of these contacts should be minimized by using uniform-size beads in desalination compartment or beads may be arrangement in such a way that the C → A contacts never occur in dilute cell (sparse media [53]). This allows decreased power consumption and increased stack capacity. Or, the beads can be arranged in form of alternating layers where each layer contains only one type of ion-exchange resin.

6.3.3
Electrochemical Regeneration of Ion-exchange Resin

Electrochemical regeneration of ion-exchange resin is a process of electromigration by which previously absorbed counterions are replaced by hydrogen and hydroxyl ions. The hydrogen and hydroxyl ions are generated either by the bipolar membrane or on electrodes separated from the resin by ion-exchange membranes. The advantage of electrochemical regeneration over conventional chemical regeneration of an ion-exchange resin is reduction in chemical consumption and wastewater generation.

The kinetic study of this process reveals that in the case of individual resins two different situations occur. The first situation takes place when the solution convection levels the concentration in all points, and the second one happens when there is no convection in the system.

If electric current is passing through ion-exchange beads only, the degree of their regeneration (ω, fraction) can be characterized by the following equation derived from Faraday's law and mass balance [3]:

$$\eta Q = \frac{1}{\beta}\left(\ln\frac{1}{1-\omega} - \omega\right) + \omega$$

β is the ratio of the ion mobility for ions coming out over the ions coming in.

$$Q = i\tau/Fq_0$$

τ – time of current flow
q_0 – ion-exchange resin quantity
η – fraction of current carried by counterions on the discharge side of the membrane.

If there is no ion exchange between the different points of the system, then the ion-exchange column regeneration can be described by the equation in following form:

$$\omega = \begin{cases} Q & Q \leq \beta \\ \dfrac{2\sqrt{\beta Q} - \beta(1+Q)}{1-\beta} & \beta \leq Q \leq 1/\beta \\ 1 & 1/\beta \leq Q \end{cases}$$

It is necessary to use conductivity values determined at isoelectroconductivity conditions. Because the value of β characterizes the ratio of ion mobility in an ion-exchange resin or more precisely – in its gel portion. Analysis of the equations above shows that the rate of electrochemical regeneration should be higher if there is no ion exchange between different parts of the system. The source of ions for regeneration makes no difference for the regeneration rate.

Research on influences such as the fixed ion type, temperature, nature and concentration of a solution on kinetic and energy characteristics of the process of electrochemical regeneration has been reported [3]. The analysis of the reported equations shows that there is an optimum concentration of a solution at which the electric power consumption on regeneration is minimal. For sulfostyrene cation-exchange resin this minimum is about 0.01 to 0.03 eq/l. The power consumption would be 0.36, 0.95, and 2.3 kWh/eq for 30, 60, and 85% regeneration, respectively.

Electrochemical regeneration of the mixed-bed ion-exchange resin can take place in both desalination and concentration chambers of electrodialysis apparatus. Withdrawal of ions from desalination chambers would be responsible for the resin regeneration in this chamber. Ion desorption on the C→A contact points and water splitting on the A→C contact points would carry out the regeneration in the concentration compartment. In all cases the current efficiency will increase as the current density grows.

6.3.4
Synthesis of New Substances without Electrode Reaction Participation: Bipolar-membrane Applications

The unique ability of ion-exchange membranes to conduct ions selectively is a foundation for technological processes based on the swapping ion mechanism. Double-exchange reaction by normal chemical methods is rather difficult, but it can be done easily with high efficiency in one stage in an electrodialyzer shown in Fig. 6.6.

A bipolar membrane in an electric field generates hydrogen ions on one side and hydroxyl ions on the other side. Therefore electrodialysis with a bipolar membrane can be used to produce acid and base from salt. For splitting 0.1 N NaCl at a current density $i=20$–150 A/sq. m concentrations of NaOH and HCl (C_{NaOH}, C_{HCl}, equiv./l) may be expressed as:

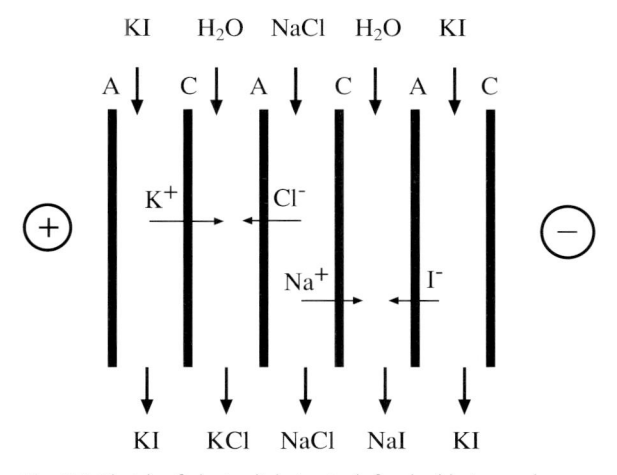

Fig. 6.6 Sketch of electrodialysis stack for double ion-exchange reaction.

$C_{NaOH} = 1.46 + 0.16\,i$

$C_{HCl} = 0.67 + 0.20\,i$

Power consumption for generation of NaOH and HCl is 77.5–108 and 172–303 W h per equivalent at a current density $i = 40$–100 A/sq. m [23].

This method would not require hazardous raw materials. For example, the electrodialyzers with bipolar membranes are applied for processing of spent solutions of HNO_3 and HF after stainless steel pickling. The solution is neutralized by KOH followed by removal of heavy-metal hydroxide precipitates. Then the $KNO_3 + KF$ solution is treated in the electrodialyzers with bipolar membranes. The solutions $HNO_3 + HF$ and KOH are recovered and the cycle is repeated [16]. Bipolar membranes are also used for cation- and anion-exchanger regeneration [24]. Bipolar-membrane application in biochemical reactors allow very soft pH regulation without a large pH change in the place of acid and alkali introduction [10], recycling of dimethyl isopropyl amines from a waste air stream, itaconic acid and sodium methoxide production [25]. A major commercial application is the conversion of organic acids and amino acids from their salts to the protonated form.

Bipolar membranes may be used to avoid the acid rain problem [11]. From 65 to 90% of the acid rain problem is due to sulfur-dioxide pollution from the smoke of power stations. A process was developed to sorb sulfur dioxide from flue gases by sodium sulfite solution. Sodium sulfite was converted to sodium

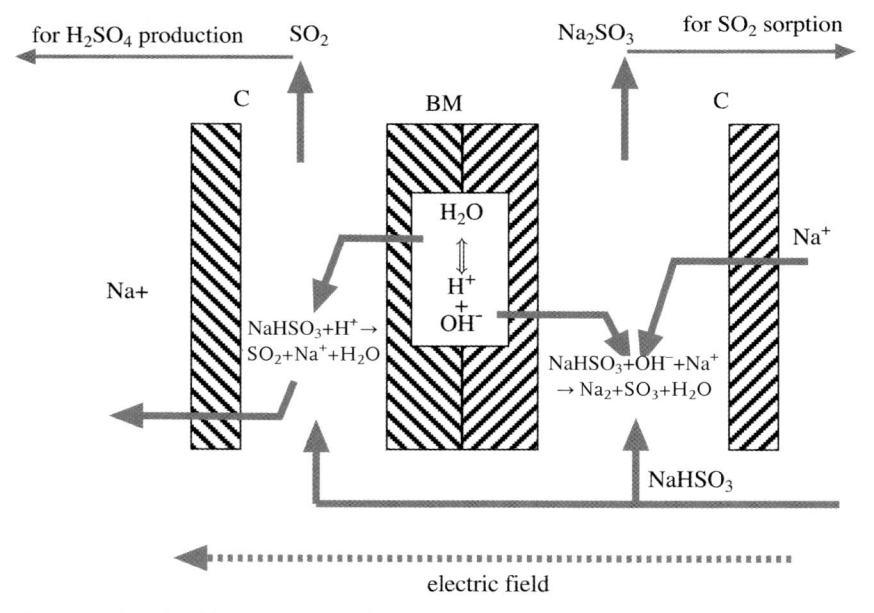

Fig. 6.7 Sodium bisulfite to sodium sulfite conversion by bipolar electrodialysis.

bisulfite [26]. Then in bipolar electrodialyzers sodium bisulfite is converted back to sodium sulfite and pure sulfur dioxide (Fig. 6.7). Sodium sulfite is used for sulfur-dioxide sorption again and pure sulfur dioxide may be used for sulfuric acid production. This process was demonstrated on a pilot scale in power plants in North America.

6.3.5
Isolation of Chemical Substances from Dilute Solutions

Ion-exchange membranes may be used in a process aimed at isolation of a valuable chemical substance from dilute solutions. For example, a dye can be extracted from a dilute solution [2]. The process is based on the reversible precipitation of the chemical compound on the surface of a polarized ion-exchange membrane. Electromembrane apparatus with only the one type of ion-exchange membrane (anion – for anion dyes and cation – for cation dyes) are used for these processes. The ions are delivered to the membrane by diffusion and electromigration. Because the membrane is not permeable for large dye molecules the concentration of dye increases near the membrane surface and reaches the limit of its solubility. Then, dye crystallization occurs and microcrystals get induced dipole moments. The first crystal layer is held by electrostatic interaction. The second and following layers are held by dipole–dipole interaction. After the electric current is turned off dipole–dipole interaction disappears, the precipitate is resuspended and it can been removed with a small volume of water. This method allows not only dye concentration but also dye purification from substances nonperceptible on membranes (for example, readily soluble nonorganic and organic compounds) occurs. The following parameters of the process are described: flow velocity 0.6 cm/s, C.I. Direct Black 3 concentration –29 mg/l, energy consumption 0.6–0.9 kWh/g for complete dye extraction, current density 50–100 mA/sq. cm, and capacity 430 mg/sq. m membrane surface. The dye concentration after desorption was 1–2 g/l. The highest degree of purity is confirmed by the fact that the actual specific conductivity of dye produced is only a fraction of that reported in the literature. Purification of colloids and ionized substances nonpermeable for an ion-exchanger membrane may be done using this method.

6.3.6
Electrodialysis Applications for Chemical-solution Desalination

Desalination of aqueous solutions of valuable chemical products is an effective technology for purification of products recovered from waste streams in the chemical industry.

Of special note is the electrodialysis application for solution desalination in organic synthesis. For example, electrodialysis with ion-exchange membranes based on polystyrene and polyethylene for HCl removal from solution in chlorhydrin synthesis is reported [2]. It is possible to get 1 N HCl from 1–3.5 g/l solutions with a current efficiency of 60%. Transport coefficients for chlorhydrin derivatives do not exceeds 10^{-7} cm^2/s.

Desalination of 70–75% diethyleneglycol solution by electrodialysis requires only 0.7–1.3 kWh/kg [2].

High water consumption by the chemical industry makes very important usage of electrodialysis for wastewater recovery [19]. For example, electrodialysis reversal demonstrated a stable performance and good fouling resistance in desalting wastewater from a paper mill [20]. The estimated cost of EDR treatment with 70% TDS removal and 85% recovery rate is $ 8.52/m^3.

6.4
Electrochemical Processing with Membranes

6.4.1
Electrochemistry

Electrochemical processing is a major part of the chemical industry. Primarily, electrochemistry is used to achieve oxidation or reduction of chemicals. These redox reactions are achieved by using electrodes to add (or remove) electrons to a reactant. The reactant to be oxidized or reduced can be a gas, liquid, solid or solute, and often the redox reaction converts the reactant from one state to another. For example, Cl$^-$ ions in solution are oxidized to gaseous Cl$_2$ in a chlor-alkali cell, gaseous H$_2$ and O$_2$ are combined to produce electrical energy and liquid H$_2$O in a fuel cell, and metallic copper from an anode is converted to Cu$^+$ ions in solution.

Electrochemistry accounts for about 12% of the product value in the chemical and primary metal industries. Aluminum and magnesium are major contributors. Why are they produced electrolytically? It might be better to ask what other process would allow the production of such reactive metals. Indeed, the reason for using electrochemistry for a specific chemical transformation is probably because there are problems in achieving that transformation by more conventional means. These problems might include unwanted side reactions or subsequent purifications to make an acceptable product. Electrochemistry requires the input of electrical energy, which is considerably more expensive than thermal energy. The efficiency of converting fossil fuel energy to electrical energy is only about 30%. If the energy of the fossil fuel could be used directly to produce the material in the desired form and purity, electrochemistry would probably receive no consideration.

The benefit of using a membrane in an electrolytic cell can be illustrated in the electrolysis of NaCl. If an aqueous solution of NaCl flows between two electrodes, NaOH forms at the cathode, and Cl$^-$ ions are oxidized to Cl$_2$ at the anode by the reaction

$$2Cl^- - 2e^- \rightarrow Cl_2 \tag{1}$$

In an undivided electrolytic cell these two products will react quickly to form NaOCl. However, if an ion-permeable barrier is positioned between the electrodes, the gaseous Cl_2 and the NaOH can be recovered separately. The ion-permeable barrier in the divided cell can be made of a variety of materials. Prior to development of ion-exchange membranes, the barrier was an asbestos diaphragm. Modern chlor-alkali cells utilize perfluorinated sulfonic acid membranes based on poly(tetrafluoroethylene) (PTFE), which is commonly known as Teflon®. DuPont developed the first commercial perfluorinated membrane and called it Nafion®[33]. Nafion® has been improved over the years and is the most commonly used membrane in divided electrolytic cells.

In contrast to the electrodialysis stacks discussed elsewhere in this chapter, the cells for electrochemical processing typically have one membrane, or at most a few membranes, between a pair of electrodes. A simple membrane cell is illustrated in Fig. 6.8. The components from left to right are anode, anode compartment containing the anolyte solution, cation-exchange membrane, cathode compartment containing the catholyte solution, and cathode. In a typical cell the electrodes are metal plates, but they can also be in the form of screens, fabrics, felts, porous sheets, or carbon blocks. The materials of the electrodes are varied to meet the needs of a particular situation. Except in cases where the electrode material is intentionally sacrificial, the electrode surface contacting the solution must be inert to avoid deterioration. Often the electrode surface is chosen to be catalytic in order to enhance the rate of a desired reaction and suppress an undesired reaction. When the electrode reactions are sufficiently selective, high product yields and substantial savings in product purification can accrue.

In some applications the electrolysis of water is the primary electrode reaction. This is the case at the cathode in a chlor-alkali cell.

$$2H_2O + 2e^- \rightarrow H_2 + 2OH^- \tag{2}$$

When NaCl is fed to the anode compartment, Na^+ ions carry electric charge through the membrane. However, the membrane is not perfectly selective, and some OH^- ions leak through into the anolyte. A small amount of HCl is added to the anolyte to neutralize the OH^- ions that leak through to avoid formation of $HClO^-$.

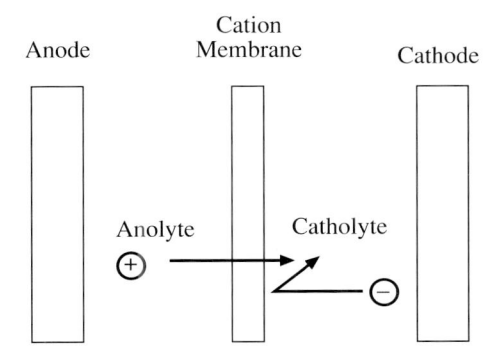

Fig. 6.8 Simple membrane cell for electrolysis.

If the anolyte does not contain Cl⁻ or some other anion that is subject to oxidation at the anode, then O_2 will be generated at the anode by the reaction

$$2H_2O - 4e^- \rightarrow O_2 + 4H^+ \tag{3}$$

Thus the electrolysis of water produces one mole of OH^- ion, one mole of H^+ ion, one-half of a mole of H_2, and one-quarter of a mole of O_2. In a 2-compartment cell the H^+ ions would carry electric charge through the cation-exchange membrane and neutralize the OH^- ions in the catholyte. Figure 6.9 shows a 2-compartment cell used to generate NaOH from Na_2SO_4. As long as the H^+ ions generated at the anode are consumed by the conversion of sulfate to bisulfate, current efficiencies are high and NaOH concentrations up to 15% are achievable. But Fig. 6.10 [34] shows the rapid decline in current efficiency as the H_2SO_4 concentration of the anolyte increases in a 2-compartment cell. Current efficiency can be substantially increased if a 3-compartment cell is used.

Figure 6.11 shows a 3-compartment cell with an anion-exchange membrane next to the anode and a cation-exchange membrane next to the cathode. A solution of Na_2SO_4 is flowing between the two membranes. H_2SO_4 is formed in the anode compartment by the transport of $SO_4^=$ through the anion-exchange membrane and the generation of H^+ ions at the anode. NaOH is formed in the cathode compartment just as in the chlor-alkali cell. Although the anion-exchange membrane does eliminate the Na^+ in the H_2SO_4 produced by this process, there is still some loss in current efficiency due to backmigration of protons through the anion-exchange membrane. One successful approach to eliminate the loss in current efficiency was to inject ammonia into the anolyte [35]. Ammonia neutralizes the acid in the anolyte, and the $(NH_4)_2SO_4$ produced in the process is used as fertilizer. High-quality NaOH is still produced in the catholyte.

The 3-compartment cell shown in Fig. 6.11 has the limitation that the anion-exchange membrane is vulnerable to attack by the oxidizing conditions of the anode. There have been several variations of the 3-compartment cell that deal

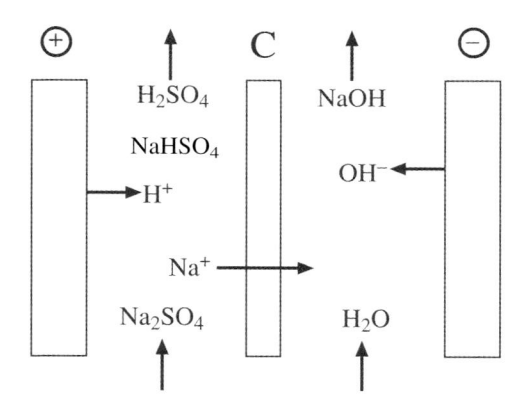

Fig. 6.9 Splitting Na_2SO_4 in 2-component cell.

Performance of 2-Compartment Cell

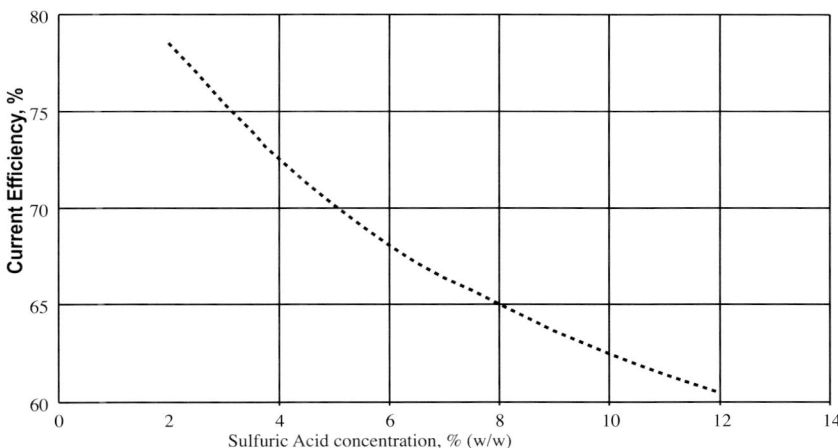

Fig. 6.10 Current efficiency as a function of sulfuric acid concentration in a 2-component cell.

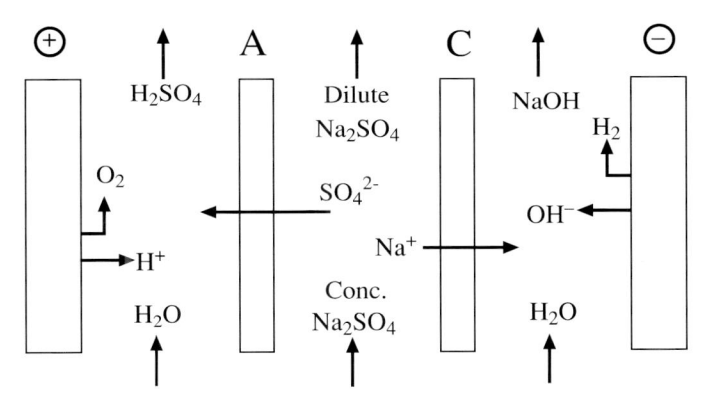

Fig. 6.11 Splitting Na_2SO_4 in 3-component cell.

with this problem. The HYDRINA® process developed by De Nora overcame the problem by introducing H_2 gas at the anode as shown in Fig. 6.12 to eliminate the oxidizing conditions at the anode [36]. In fact, the H_2 gas generated at the cathode can be used as the reducing gas at the anode. An added benefit is a lower cell voltage. Another approach to dealing with the anion-exchange membrane in the 3-compartment cell was to replace it with a perfluorinated cation-exchange membrane. In Fig. 6.13 the Na_2SO_4 solution is again fed to the center compartment where it is converted to $NaHSO_4$ by H^+ ions from the anode compartment. The solution then moves to the anode compartment where more of the Na^+ ions are replaced with H^+ ions to form H_2SO_4. But the conversion is not complete because the more mobile H^+ ions begin to carry more of the current as their relative concentrations builds up.

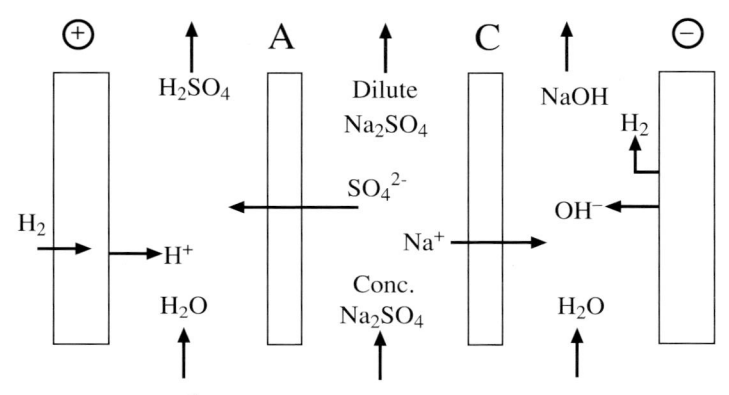

Fig. 6.12 HYDRINA® process for splitting Na_2SO_4.

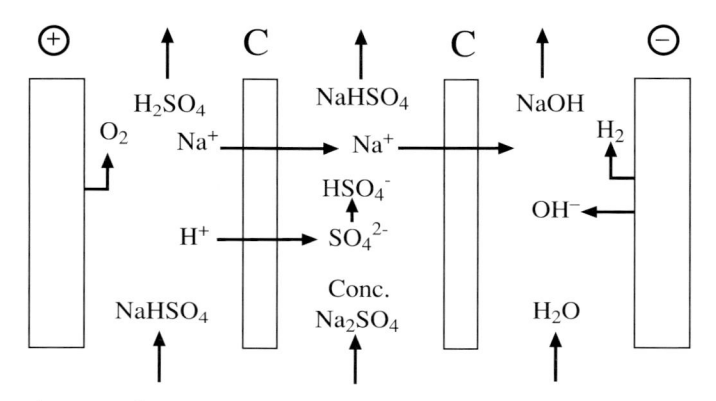

Fig. 6.13 Splitting Na_2SO_4 with two cation membranes.

The 3-compartment cell is satisfactory for acids that do not decompose at the anode, but it would be unsatisfactory for splitting NaCl, because Cl_2 would form at the anode. To alleviate this problem a cation-exchange membrane could be added next to the anode, and H_2SO_4 could be circulated as the anolyte. The additional membrane could be a perfluorinated membrane that resists oxidation at the anode, and that addition would improve the chemical stability of the cell.

The reader might reasonably want to compare at this point the relative merits of splitting a salt by electrolysis or with bipolar membranes discussed elsewhere in this chapter. Electrolysis has the advantages that the acid and base can be made more pure, at higher concentrations and with higher operating current density. Bipolar membranes have the advantage of substantially lower energy consumption and probably lower capital cost.

6.4.2
Chlor-alkali Industry

Production of Cl_2 and NaOH by electrolysis of NaCl is a huge industry with annual production capacity in excess of 50 million tons of NaOH per year. Membrane cells are the state-of-the-art technology, but mercury and diaphragm cells are still used because the capital cost for their replacement is substantial. The mercury cell technology is more than a century old and still accounts for nearly half of the world's production capacity. Chlorine evolves from a DSA (dimensionally stable anode) situated above a pool of mercury with NaCl brine in between. Mercury reacts with sodium to form sodium amalgam, which is removed and hydrolyzed in a separate reactor.

$$Na^+ + Hg + e^- \rightarrow NaHg$$
$$2NaHg + 2H_2O \rightarrow H_2 + 2NaOH + 2Hg \tag{4}$$

The reconstituted metallic mercury is returned to the cell. The major advantage of the mercury cell is that it produces very pure NaOH at 50% concentration. But electrical energy consumption is high, and the threat of mercury pollution is a major concern. Traces of mercury appear in the NaOH and in air emissions from the cells. Because of their design it is impractical to modify mercury cells into membrane cells.

The diaphragm cell has a membrane of sorts, with a sheet of asbestos felt separating the two compartments. The diaphragm keeps the chlorine gas out of the cathode compartment, but it offers no selectivity for diffusion and migration of ions. Therefore, the NaOH produced is dilute (12% maximum) and loaded with NaCl. Evaporation of the catholyte to concentrate the NaOH to 50% causes most of the NaCl to precipitate, but the chloride level of the product remains about 1%. In some cases the asbestos diaphragms have been replaced with other porous membranes (principally PVC) to improve membrane life and avoid environmental criticism. In other cases diaphragm cells have been converted to membrane cells.

Membrane cells are acknowledged as the most efficient for chlor-alkali. Since the process is so energy intensive and the market is so large, there has been considerable competition to improve membranes, electrodes, cell design and operating conditions. Moreover, the technology developed for the chlor-alkali industry has been beneficial to the electromembrane processing in general. Therefore, it would be useful to describe some of the technology developments here.

6.4.3
Perfluorinated Membranes

The development of perfluorinated membranes has been most important to the progress in membrane cells. The first perfluorinated membranes were made by DuPont, and they were followed by products from Asahi Glass [37] and Asahi

Chemical in Japan where membrane cells are now dominant. Because of their Teflon®-like chemical composition, perfluorinated membranes resist chemical and thermal degradation better than any of the hydrocarbon ion-exchange membranes that preceded them. For most of them the starting materials are perfluorinated monomers such as tetrafluoroethylene (TFE) $CF_2=CF_2$ and hexafluoropropylene oxide (HFPO)

$$CF_3CF\overset{O}{\overset{/\,\backslash}{-}}CF_2$$

The preparation of perfluorinated membranes from these monomers is rather complex, and several research groups developed different routes to similar end points. Typically they utilized the monomers TFE and HFPO and a variety of other reagents to synthesize a complex perfluorinated monomer (PFM) containing ether linkages. The general structure of PFM is

$$CF_2=CF-(OCF_2-CFCF_3)_p-O-(CF_2)_q-X \quad X=SO_2F \text{ or } COOCH_3; \; p=0-2; \; q=1-4$$

The TFE and PFM monomers are then copolymerized by radical-initiated polymerization to make a polymer of the structure

$$[(CF_2\text{-}CF_2)_m-CF_2-CF-(OCF_2-CFCF_3)_p-O-(CF_2)_q-X]_n \quad m=6-8; \; n=600-1500$$

Calculation of the formula weight for $n=1$ gives the "equivalent weight" for the polymer, which is the average polymer weight per ionic charge. (Membranes with higher equivalent weight have higher permselectivity but also higher electrical resistance.) For DuPont's Nafion R1100 resin used to make most of its membranes, the values of variables in its PFM are $p=1$, $q=2$. Knowing that the equivalent weight is 1100 allows one to calculate a value of $m=6.56$. This indicates that well over half of the membrane is comprised of a Teflon®-like structure.

Before fabrication of the membranes, the terminal group is converted to a more stable form, sulfonyl fluoride or methyl carboxylate. Then the copolymer is extruded to form a membrane. The extruded films can be further processed to achieve a variety of membrane properties. They can be laminated with PTFE fabric to produce a reinforced membrane, and membranes made of polymers with different values of m, n, p or q can be laminated to form composite membranes with desired properties. After fabrication is complete, the X group is converted by hydrolysis to SO_3^- or COO^-.

Perfluorinated membranes used in chlor-alkali cells normally have a thin layer of carboxylate on the cathode-facing surface of a sulfonate membrane. Nafion 901 was introduced as such a membrane [38]. It achieved 33% NaOH concentration with 95% current efficiency in cells operating at 3 kA/m^2 and 3.3 to 3.9 V. The carboxylate layer can be prepared by lamination, but the layer can be thinner if it is made by surface treatment. This modification is made by treating a surface of the sulfonate membrane successively with phosphorus pentachloride, hydrazine and mild oxidation. The carboxylic acid layer on the membrane is beneficial because it helps in the rejection of hydroxyl ions, the transport of which reduces the current efficiency of the process. Sulfonate membranes tend to be leaky to bases, and any base that leaks through to the anolyte must be neutralized by addition of HCl to the brine so that its pH can be maintained between 2 and 4. A higher pH promotes NaOCl and O_2 formation. A lower pH reduces current efficiency, because H^+ ions compete with Na^+ ions in transporting current through the membrane. If H^+ ions reach the carboxylate layer of the membrane and convert it to the carboxylic acid, the membrane resistance increases dramatically.

6.4.4
Process Conditions

With the carboxylate layer on the membrane, the concentration of NaOH in the catholyte can be maintained at about 32%, which compares favorably with the 12% product from diaphragm cells. Consequently, considerably less thermal energy is needed for evaporation to the commercial 50% product.

Voltage requirements vary inversely with temperature, so it is advantageous to operate at elevated temperatures. The perfluorinated membranes can tolerate sustained temperatures of 90 °C.

Because the anolyte and catholyte are both quite concentrated, there are osmotic forces at work to dehydrate the membrane. To some extent dehydration is beneficial, because it increases the ratio of sulfonate to water in the membrane, which in turn improves permselectivity. But eventually the dehydration leads to an increase in membrane resistance. Therefore, raising the NaOH concentration to reduce evaporation costs would likely increase electrical power costs.

Calcium and magnesium in the brines can cause major problems in chlor-alkali cells, because these ions will migrate through the membrane along with the Na^+ ions. But the pH gradient across the membrane prevents the completion of their passage due to the insolubility of their hydroxides. Build-up of $Ca(OH)_2$ and $Mg(OH)_2$ causes an increase in membrane resistance, and even worse, disruption of the membrane structure. To avoid these problems, calcium and magnesium must be removed from the brine before it is introduced to the electrolytic cell. Chelating ion-exchange resins in the Na^+-ionic form are typically used to remove these ions to concentrations of a few parts per billion.

Electrode materials in chlor-alkali cells are typically nickel for the cathode and RuO_2-coated titanium for the anode. The composition of the coating can have a dramatic effect on the cell voltage. The RuO_2 coating protects the titanium from oxidation and catalyzes the preferred gas evolution reaction to generate Cl_2. Some chlor-alkali cells have a noble-metal coating on the nickel cathode to reduce the cell voltage.

6.4.5
Zero-gap Electrode Configurations

In the diagrams of electrolytic cells shown above to illustrate the electrode and membrane configurations, the anolyte and catholyte are shown to flow between the membrane and the electrode surface. The gap between the membrane and the electrode allows space for the gas bubbles to disengage and allows for flow of solution delivering the ions, such as Cl^-, to the electrode surface. However, the gap is not always necessary. Often the molecules that react at the electrode surface are the solvent, typically water dissociating to form H^+ and O_2 or OH^- and H_2. In such cases the solvent is already present at the electrode surface. But there is still a need to remove the gas so that it does not block the flow of electric current between the electrode and the membrane. In "zero-gap" configurations, the gas is allowed to escape through the electrode. This can be done by utilizing a screen, grid or perforated metal electrode or by making the electrode porous. In some cases the membrane is held firmly against the electrode by applying a pressure differential across the membrane. In other cases the membrane is formed directly on the surface of the porous electrode.

The zero-gap design is also utilized for processes in which gases are consumed at the electrodes. The most noteworthy example is in fuel cells where H_2 and O_2 react at the electrode surfaces to form H_2O and release electrical energy.

Porous electrodes must be carefully designed to ensure that four processes can occur:
- gas transport to or from the electrode surface
- flow of electrons via a continuous path of electrode material
- transport of ions to and through the membrane material
- transport of solvent to the reaction site (except in fuel cells)

Ideally four media, a hydrophobic pore connected to a gas manifold, a protrusion of the electrode, a protrusion of the membrane, and a hydrophilic pore connected to a solvent manifold, would converge at a point that is geometrically optimized so that the resistance to transfer is the same in all four media. Further, there would need to be an abundance of these convergence points packed into a small volume with very short transport distances to minimize resistances. Practically it is very difficult to achieve the convergence of four separate media and maintain their connectivity with short transport distances. Therefore, it is usually necessary for one or more of the media to perform double or triple duty.

In the conventional design where a solution flows between the membrane and the electrode surface, the solution must perform three of the functions listed above. That solution must contain an electrolyte so that ions can transport the current through the gap. The formation and growth of gas bubbles blocks the current flow, so the bubbles must be swept away by swift solution flow.

In zero-gap membrane–electrode assemblies the membrane performs the double duty of ion removal and delivery of solvent molecules to the electrode surface. Since the water flux is proportional to the membrane thickness, it is desirable to have very thin membranes, which is much easier when the membrane and electrode are manufactured as one piece.

By their very nature porous zero-gap systems are three-dimensional. They are comprised of interpenetrating media for conduction of electrons, ions, solvent and gas. Often they are made as composites of fibers, powder, polymers, metals, etc., that are mixed, pressed and cured to form felt-like structures.

6.4.6
Other Electrolytic Processes

As mentioned above, the electrolysis of NaCl to from Cl_2 and NaOH is the largest application of membranes in electrolytic cells. There are also other brine electrolyzes of commercial importance. NaBr brine is electrolyzed to form Br_2. (Another method of Br_2 formation is to treat bromide brines with Cl_2 derived from electrolysis.) Electrolysis of KCl brines is the preferred process for making KOH. Because K^+ ions are less hydrated than Na^+ ions, the membrane is more effective at blocking the backdiffusion of KOH, which allows production of KOH in concentrations as high as 47%.

Chlorine is used on a large scale for production of chlorinated hydrocarbons with vinyl chloride monomer being the largest single product. Until recently, Cl_2 had been a mainstay oxidizing agent for many industries, the two major applications being disinfection of drinking water and bleaching of the brown fibers from wood to produce white paper. Then, pressure from environmentalists forced the industry to seek oxidizing agents that were perceived to be more acceptable. The major objection to Cl_2 seems to be the unwanted chlorinated byproducts that result from reactions with organics. Chlorine dioxide ClO_2 overcomes that objection, but it has the disadvantage that it cannot be stored and shipped safely. Instead, it must be produced on site from another oxidizing agent, sodium chlorite, $NaClO_2$. Some chemical routes to this conversion are

$$2NaClO_2 + Cl_2 \rightarrow 2ClO_2 + 2NaCl \tag{5}$$

$$2NaClO_2 + NaOCl + H_2SO_4 \rightarrow 2ClO_2 + NaCl + Na_2SO_4 + H_2O \tag{6}$$

which have the unwanted byproducts NaCl and Na_2SO_4. An electrochemical alternative is to convert $NaClO_2$ to ClO_2 at the anode by the reaction

$$2NaClO_2 \rightarrow 2ClO_2 + Na^+ + 2e^- \tag{7}$$

This process has the advantage of no unwanted byproducts. Moreover, the yield of ClO_2 can exceed 90%.

Another precursor for ClO_2 is sodium chlorate $NaClO_3$. The pulp and paper industry consumes 95% of global chlorate production. The $NaClO_3$ can be reduced to ClO_2 by several different agents, which are listed below along with their reactions:

$$SO_2+H_2SO_4+2NaClO_3 \rightarrow 2ClO_2+2NaHSO_4 \tag{8}$$

$$2HCl+2NaClO_3 \rightarrow 2ClO_2+NaCl+0.5\ Cl_2+H_2O \tag{9}$$

$$2CH_3OH+6H_2SO_4+9NaClO_3$$
$$\rightarrow 9ClO_2+3Na_3H(SO_4)_2+0.5CO_2+1.5HCOOH+7H_2O \tag{10}$$

$$3H_2O_2+4H_2SO_4+6NaClO_3 \rightarrow 6ClO_2+2Na_3H(SO_4)_2+6H_2O+3O_2 \tag{11}$$

All of these reactions produce some unwanted byproducts, which can be avoided by splitting $NaClO_3$ in a 3-compartment cell to chloric acid, $HClO_3$, which can be stored and shipped safely [39]. The $NaClO_3$ is fed between two cation-exchange membranes, and H^+ ions from the anolyte replace Na^+ ions that migrate to the catholyte to form NaOH coproduct. The $HClO_3$ can be catalytically converted to ClO_2 by the reaction

$$4HClO_3 \rightarrow 4ClO_2+2H_2O+O_2 \tag{12}$$

Chloric acid is also produced directly from HOCl in the anode compartment of a 2-compartment electrolytic cell [40].

$$4HOCl+2H_2O-6e^- \rightarrow 2HClO_3+Cl_2+6H^+ \tag{13}$$

Sodium hydrosulfite, $Na_2S_2O_4$, also known as sodium dithionite, is a reducing agent used industrially in the processing of pulp and paper, textiles and clay. It is made in a 2-compartment cell with a nickel anode and a special flow-through cathode made of stainless-steel fibers [41]. The starting material is SO_2 that reacts with water to form H_2SO_3. The reduction takes place in the cathode compartment.

$$2H_2SO_3+2e^- \rightarrow S_2O_4^= +2H_2O \tag{14}$$

The reader has probably noticed by now that in many of the processes described the right-hand side of the cell has the same electrolytic reaction

$$2H_2O+2e^- \rightarrow H_2+2OH^- \tag{15}$$

Indeed, with a few exceptions such as the production of $Na_2S_2O_4$, the majority of electrochemical processing seems to be oxidation at the anode and/or removal of a cation through a membrane. This means that electrochemical processes produce large volumes of H_2. In fact, some plants are designed to produce H_2 by electrolysis of water. In this process the anode reaction is

$$2H_2O-4e^- \rightarrow O_2+4H^+ \tag{16}$$

Thus, a Faraday of electric charge produces $^1/_2$ mole of H_2 and $^1/_4$ mole of O_2, and the gases are very pure. The major criterion for keeping the gases pure is avoidance of mixing. This can be done in the laboratory by simply performing the electrolysis in a U-shaped tube containing H_2SO_4 with the electrodes positioned high enough so that the gases move upward to collection points, but this arrangement requires excessive energy to push the current through the large gap between the electrodes. A more energy-efficient apparatus utilizes a membrane to separate the electrodes, one of which can have zero-gap. With an adequate supply of water to the electrodes, both can be zero-gap.

6.4.7
Fuel Cells

Chemists usually consider a gas-producing reaction to be irreversible, because one of the reaction products, in this case the gas, is removed from the reaction site. But if the gas is retained at or introduced to the reaction site, the process can be reversible. H_2 and O_2 can be brought together and burned to form water and release energy, or they can be combined in the reverse of the electrolysis reaction to form water and electrical energy. The apparatus for conversion of H_2 and O_2 to electrical energy is the fuel cell, and those containing ion-exchange membranes are often called PE (polymer electrolyte), PEM (proton-exchange membrane) or SPE (solid polymer electrolyte) fuel cells. Of course there are several different types of fuel cells that do not utilize membranes, but they are outside the scope of this chapter.

Fuel cells can be used to recover energy from the H_2 gas generated by an electrolytic cell. Since both processes operate with direct current, the cost of rectifiers is avoided. Return on investment for a fuel cell in such an application is reported to be about 5 years. In some situations it might be advisable to consider reusing the H_2 directly in the electrolytic cell to depolarize the anode. Such schemes have been proposed for cells where O_2 is usually evolved at the anode [42]. Depolarization reduces the over voltage at the anode and allows about the same amount of energy recovery with just a modification to the existing electrolytic cells instead of the expenditure for fuel cells.

Figure 6.14 shows the structure and operation of the membrane fuel cell. The membrane is sandwiched between two porous electrodes coated with a thin layer of platinum catalyst. H_2 gas enters through the porous anode and reacts on the catalyst surface to form protons and release electrons. The electrons move through the external electrical circuit. The protons move through the membrane, which is impermeable to electrons and the gases. O_2 adsorbed on the cathode catalyst where it picks up electrons to form oxygen anions, which react with protons to form water. The fuel cell can produce up to 1 V, but generally operates at 0.5 to 0.8 V depending upon the current, and its energy efficiency approaches 50%. Stacking individual fuel cells with their electrodes connected in series allows output voltages to be increased to practical levels. The main incentive for developing membrane fuel cells is the prospect that they can

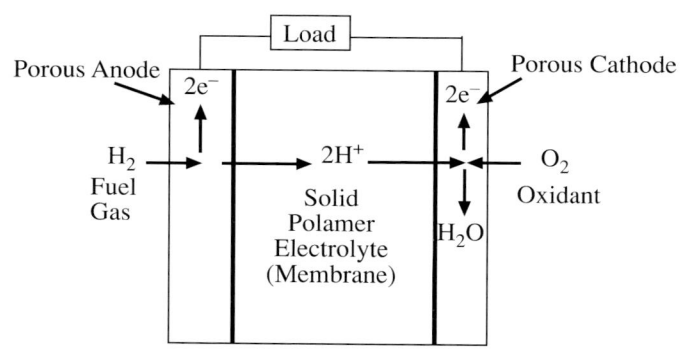

Fig. 6.14 Membrane or SPE fuel cell.

be used to power automobiles. A major drawback is the high cost of platinum catalyst and membranes, but these costs are coming down with design improvements that require lower catalyst loadings and thinner membranes.

Perfluorinated membranes have the physical characteristics necessary for fabrication and operation in a fuel cell, but their cost is prohibitive. Alternative materials are being developed that could provide cheaper membranes that equal or even exceed the performance of Nafion. One example is a membrane formed by grafting onto poly(vinylidine fluoride) (PVDF) film [43]. An 80-µm PVDF film is irradiated under nitrogen in an electron accelerator and immediately soaked in a solution of 80% styrene monomer and 20% tetrahydrofuran under reflux at 70 °C. The degree of grafting is determined by the time of contact (0.4 to 4 h), and at least 20% grafting is needed to ensure penetration to the center of the film. After chloroform extraction to remove unreacted monomer and ungrafted homopolymer, the film is treated with 0.5 M chlorosulfonic acid in 1,2-dichloroethane at ambient temperature for 24 h to obtain 95% sulfonation of the grafted polystyrene. The grafted PVDF membrane was reported to perform as well as Nafion in fuel-cell tests.

Membrane fuel cells have zero-gap electrodes on both sides of the membrane. Typically the electrodes are made of a carbon fiber mat impregnated with platinum on carbon (Pt/C) catalyst. To achieve an extended surface for the gas to be adsorbed and react and to maintain continuity for ionic transport, interpenetration of electrode and membrane is necessary. This is usually accomplished by impregnating the porous electrode with Nafion solution. One assembly technique is to suspend the Pt/C in Nafion solution by sonication and spray it onto carbon paper. Then the membrane is hot pressed between the two impregnated electrodes [44]. Another approach is to make an aqueous suspension of three powders – Pt/C, carbon black, and PTFE – and spray it onto the carbon paper. Then 5% Nafion solution is applied by spraying or by floating the electrode on the Nafion solution, after which the membrane is pressed between the electrodes [45]. The Nafion solution serves as an adhesive as well as a means of extending the electrolyte into the structure of the porous electrode.

Gore-Select membrane is a recent innovation that is useful for fuel cells. The membrane is made by filling the pores of Gore-Tex (expanded PTFE) with a perfluorinated ionomer like Nafion. The resulting membrane is about 20 μm thick, about one tenth the thickness of a Nafion membrane, and it has about half the resistance of Nafion. Membrane–electrode assemblies are made by applying platinum black coatings (0.3 mg/cm^2) to both surfaces. They are reported to perform well in fuel cells [46].

6.4.8
Electroorganic Synthesis

Many examples of electroorganic synthesis have been reported and a few are commercial. The first significant commercialization of electroorganic synthesis was the electrohydrodimerization (EHD) of acrylonitrile to adiponitrile, an intermediate for hexamethylenediamine, which is a monomer for nylon.

$$2CH_2=CHCN+2H_2O+2e^- \rightarrow NC(CH_2)_4CN+2OH^- \tag{17}$$

The reduction takes place at the cathode with a catholyte composed of aqueous quaternary ammonium salt. The anolyte is sulfuric acid. H$^+$ ions generated at the anode pass through the membrane and neutralize the OH$^-$ ions generated in the catholyte. Commercial installations are operated by Solutia (spun off from Monsanto), BASF, Asahi Chemical, and Rhodia. Some of these manufacturers have developed undivided cells for this process.

In some syntheses, including the EDH described above, the organic to be modified is brought directly into contact with the electrode. In other cases indirect electrosynthesis is accomplished with an intermediate redox agent such as cerium, which can have its oxidation state changed in an electrochemical cell. There are benefits to using an intermediate redox agent. An inorganic agent is typically more soluble in aqueous systems, and it can be treated at high current densities without unwanted side reactions. Since the electrochemistry can be separated from the chemical step, both processes can be optimized.

An example is the oxidation of naphthalene with cerium, which has been demonstrated on a pilot scale [47]. Ce(III) in and aqueous solution of methanesulfonic acid is oxidized to Ce(IV) at the anode. The Ce(IV) reacts with naphthalene in a separate vessel to form naphthoquinone and Ce(III). The aqueous phase containing the methanesulfonic acid and Ce(III) is separated and returned to the electrolytic cell. The naphthoquinone is an intermediate for dyes, agricultural chemicals, and anthraquinone, which is made by a Diels-Alder reaction with butadiene.

6.4.9
Electrochemical Oxidation of Organic Wastes

Widely practiced disposal methods for organic wastes, incineration, biological degradation, and landfill are unacceptable for some wastes because of the presence of hazardous materials in the waste. A case in point is the accumulation of plutonium-contaminated tissue paper in a nuclear laboratory, which led to the development of an oxidation process called "Silver Bullet" based on Ag(II) generation at an anode [48]. A mixture of the organic waste and $AgNO_3$ is circulated through the anode compartment where Ag(I) is oxidized to Ag(II), which then reacts with the organic material to convert it to CO_2. Side reactions with water and nitrate form OH and NO_3 free radicals that also aid in the destruction of the organics. Heteroatoms are also oxidized, phosphorus to phosphate and sulfur to sulfate, and these must be removed by a bleed of the anolyte. Chloride is partially oxidized to chlorine gas, but some AgCl also forms. The process is effective on a variety of wastes including pesticides.

Acknowledgments

The authors would like to express their gratitude to W. W. Carson and A. von Gotteberg for reading the draft and making their comments.

List of Symbols

m	solution molality
$\bar{m}$	molality of Donnan electrolyte
E	molality of counterions
$\theta, \bar{\theta}$	average molal activity coefficient in solution phase and in solid phase, respectively
$\bar{\chi}_m$	electric conductivity of ion-permeable membrane
$\bar{\chi}_m^{\|}$	theoretical electric conductivity of ion-permeable membrane based on parallel ion-movement model
$\bar{\chi}_m^{\perp}$	theoretical electric conductivity of ion-permeable membrane based on successive ion-movement model
$a, b, c, e,$ and d	the parameters of the three-conductor model
f	is the fraction of gel areas in the ion-exchange membrane
χ_0	constant
R	molar gas constant
T	absolute temperature
I_0	total electric current through the membrane
I_i	the fraction of electric current carried by i-type ions
$\bar{t}_g$	counterions transport number
$\bar{t}_c$	co-ions transport number

t_g	counterions transport number in solution
C_A, C_B	ion concentration for ion type A and ion type B, respectively, in the releasing solution
$\bar{u}_g, \bar{u}_c$	counter and co-ion mobility respectively
$\bar{a}_g, \bar{a}_c$	counter and co-ion activity respectively
χ_m, χ	specific resistance of membrane and solution, respectively
q_i	quantity of type i ion, gram-equivalent
Q	electric current passed through the membrane, Faraday
K	the maximum possible degree of concentration
μ	solutions' ion strength
K_{Ca}, K_{SO4}	ion ratio in concentrate and dilute flows
$[Ca^{2+}], [SO_4^{2-}]$	the ion concentrations in the feed solution
C_{NaCl}	sodium chloride concentration in the concentrate
n_g	quantity of water carried with one equivalent of counterions through a membrane
n_c	quantity of water carried with one equivalent of co-ions through a membrane
X	fixed ion concentration in the membrane
v	kinematic viscosity, sq. cm/s
l	distance between the mixing-screen elements
d_0	the thickness of the chamber, cm
h	the height of the mixing-screen elements, cm
v_0	kinematic viscosity at 298 K
D, D_0	diffusion coefficient at given temperature and at 298 K
d_m	membrane thickness
a	proportionality factor
γ_i	fraction of membrane capacity occupied by ion specie i
C_0	bulk solution concentration
i_{lim}	limiting current density
z	ion charge
C'	solution concentration at the receiving surface of the membrane
K_s	spacer factor
δ	diffusion layer thickness

References

1 GREBENYUK VD (1976) Electrodialysis Technic Kiev, p 160 (in Russian).
2 GREBENYUK VD, PONOMAROV MI (1992) Electromembrane separation of mixtures Naukova Dumka, Kiev, p 184 (in Russian).
3 GNUSIN NP, GREBENYUK VD (1972) Electrochemistry of granulated ion exchangers, Naukova Dumka, Kiev, 180 (in Russian).
4 GREBENYUK VD, PENKALO II, LUKACHINA VV (1995) Acceptable ratio of salt concentration in electrolyte NaCl–Na$_2$SO$_4$, J. Water Chemistry & Technology 17: 3: 225–229.
5 MELLER FH (1984) Electrodialysis and Electrodialysis Reversal Technology, IONICS, Inc. p66

6 C, CHEBOTAREVA RD, PETERS S, LINKOV V (1998) Surface modification of anion-exchange membrane to enhance antifouling characteristics, *Desalination*, **115**, p 313–329.

7 SCHOEMAN JJ, HILL E, STEYN AA (1993) An investigation into the organic fouling of ion- exchange membranes, Report to the Water Research Commission by the Division of Water Technology, CSIR, WRC Report # 396193.

8 HERNON BP, ZANAPALIDOU RH, ZHANG L, SIMMS KJ, SIWAK LR (1994) Electrodeionization in Power Plant Application, Ultrapure Water, p33, July/August.

9 TERADA T, TODA H, IWAMOTO J, UMEMURA K, KOMATSU K, HUEHNERGARD MP, TESSIER DF, TOWE IG. (1995) US Patent 35961805, Method and Apparatus for Producing Deionized water.

10 BERSIER P (1999) At The 13th International forum Electrochemistry in the Chemical Industry. Clear Water Beach, Florida, November.

11 GREBENYUK VD, SOBOLEVSKAYA TT, GREBENYUK OV, KONOVALOVA ID, VISOTSKII SP (1992) *Russian J. Applied Chemistry*, **65**, #5, p 1059–1065.

12 VERBICH SV, GREBENYUK OV, Problem of Electrofiltration of Disperse System (1991), *J Water Chemistry & Technology*, **13**:12: p 1059–1076

13 NIKONENKO VV, ISTOSHIN AG, ZABOLOTSKY VI, LARCHET C, BENZARIA (1999) Analysis of electrodialysis desalination costs by convective-diffusion model. *Desalination*, **126**, 207–211.

14 GREBENYUK VD, MAZO AA, LINKOV VM (1996) New ecological problems of desalting and water re-use. *Desalination*, **105**, 175–183.

15 ZABOLOTSKY VI, NIKONENKO VV (1996) Ion transport in membranes. 'Nauka', Moscow 392 p (Rus).

16 DAVIS TA, GENDERS JD, PLETCHER DA (1997) First Course in Ion-permeable membranes, The Electrochemical Consultancy, 255 pages.

17 VON GOTTBERG, AJM (1998) New High-Performance Spacers in Electrodialysis Reversal (EDR) system, Proceedings, American Water Works Association Annual Conference, Dallas, TX.

18 ALLISON RP (1991) Surface and wastewater desalination by electrodialysis reversal, Proceedings, American Water Works Association Conference, Orlando, FL.

19 WONG KHM (1993) Wastewater desalination by electrodialysis reversal, Proceedings, Water Technology Seminar, Singapore, October.

20 CHEN W, HORAN NJ (1998) The Treatment of High Strength Pulp and Paper Mill Effluent for Wastewater Reuse, Environmental Technology, Vol.19, pp 861–871.

21 HERNON BP, ZANAPALIDOU RH, ZHANG L, SIWAK LR, SCHOEPKE EJ (1994) Application of Electrodeionization in Ultrapure Water Production: Performance and Theory, Proceedings, 55th Annual Meeting International Water Conference, Pittsburg, PA.

22 STRATHMANN H (1998) At The 12th International forum Electrochemistry in the Chemical Industry. Clear Water Beach, Florida.

23 GREBENYUK VD, PENKALO II, CHALAYA LM (1986) Water desalination with simultaneous acid and alkali generation. *Water Chemistry and Technology* **8**, #2, pp 76–78.

24 DAVIS TA, LATERRA T (1987) On-site Generation of Acid and Base with Bipolar Membranes: A New Alternative to Purchasing and Storing Regenerants, Proceedings, 48th Annual Meeting International Water Conference, Pittsburg, PA.

25 STRATHMANN H, BAUER B, RAPP HJ (1993) Better Bipolar Membranes, *Chemtech*, **June** 1993, 17–24.

26 LIU K-J, CHLANDA FP, NAGASUBRAMANIAN K (1978) Application of bipolar membrane technology: A novel process for control of sulfur dioxide from flue gases. *J. Membrane Sci.* **3**, 57–70.

27 MAC DONALD RJ, HODGDON RB AND ALEXANDER SS (1992) Process for manufactureing continuous supported ion selective membranes using non-polymerizable high boiling point solvents, US Patent 5,145,618.

28 MIZUTANI Y, YAMANE R, IHARA H, MOTOMURA H (1963) Studies of ion-exchange membranes. XVI. The preparation of ion-exchange membranes by the

"Paste Method". *Bull. Chem. Soc. Japan*, **36**:4, 361–366.

29 Mizutani Y (1990) Structure of ion-exchange membranes, *J. Membrane Sci.* **49**, 121–144.

30 Gupta BD and Chapiro A (1989) Preparation of ion-exchange membranes by grafting acrylic acid into pre-irradiated polymer films. *Eur Polym J* **11**, 1137–1148.

31 Davis TA, Glassner DA (1997) Electrodialysis. in Handbook of Downstream Processing. Goldberg E (ed.), Chapman and Hall, London, 140–165.

32 Davis TA (1990) Scale-up of ED/RO system for waste recovery. In: Electrosynthesis from Laboratory, to Pilot, to Production. Genders JD and Pletcher D (eds.), The Electrosynthesis Company, E. Amherst, NY, 239–242.

33 Wolfe WR (1970) British Patent 1181231 DuPont.

34 Martin AD (1992) Sodium hydroxide production by the electrohydrolysis of aqueous effluent stream containing sodium salts, *ICHEME Symposium Series* **127**, 153–162.

35 Thompson J, Genders JD (1992) Process for producing sodium hydroxide and ammonium sulfate, US Patent 5,098,532.

36 Burton LL (1996) Electrochemistry with DuPont's Nafion membranes, Tenth Int'l Symp on Electrosynthesis in the Chem Industry.

37 Ukihashi H (1980) A membrane for electrolysis, *Chemtech* **February** 118.

38 Ibrahim SM, Price EH, Smith RA (1983) High-performance membrane for the chlor-alkali industry, *Proc Electrochem Soc* **83**:206–215.

39 Cawlfield DW, Mindiratta SK (1992) Chloric acid electrochemically for chlorine dioxide, Seventh Int'l Symp on Electrosynthesis in the Chem Industry.

40 Kaczur JJ and Duncan BL (1995) New process applications for electrochemically produced chloric acid, Ninth Int'l Symp on Electrosynthesis in the Chem Industry.

41 Kaczur JJ, Scott LL, Dotson RL (1997) Electrochemical process development at Olin based on advances in electrodes, cell designs, and chemical feedstocks, Eleventh Int'l Symp on Electrosynthesis in the Chem Industry.

42 Pletcher D, Genders JD, Spiegel E, Weinberg NL (1993) Electrochemical methods for the purification of alkali metal hydroxides without the co-production of chlorine, US Patent 5246551.

43 Lehtinen T, Sundholm G, Holmberg S, Sundholm F, Bjornbom P, Bursell M (1998) Electrochemical characterization of PVDF-based proton conducting membranes for fuel cells, *Electrochemica Acta* **43**:1881–1890.

44 Chu D, Jiang R (1999) Comparative studies of polymer electrolyte membrane fuel cell stacks and single cell, *J Power Sources* **80**:226–234.

45 Antolini E, Giori L, Pozio A and Passalacqua E (1999) Influence of Nafion loading in the catalyst layer of gas-diffusion electrodes for PEFC, *J Power Sources* **77**:136–142.

46 Kolde JA, Bahar B, Wilson, MS Zowadzinski TA, Gottesfeld S (1995) Advanced composite polymer electrolyte fuel cell membranes, *Electrochem Soc Proc* **95–23**:193.

47 Harrison S, Labrecque R, Teoret A (1991) Hydro-Quebec's development and demonstration activities in electyroorganic synthesis, Fifth Int'l Symp on Electrosynthesis in the Chem Industry.

48 Steele DF, Richardson D, Craig DR, Quinn JD, Page P (1992) Destruction of industrial organic wastes using electrochemical oxidation, in Electrochemistry for a Cleaner Environment, Genders JD and Weinberg NL (eds.) The Electrosynthesis Company Inc., Amherst, NY.

49 Kononenko NB, Berezina NP, Loza NV (2004). Iteraction of surfactants with ion-exchange membranes. *Journal of Colloids and Surfaces A: Physicochem. Eng. Aspects* **239**:59–64.

50 Pismenskaya N, Nikonenko V, Auclair B, Pourcelly G, (2001) Transport of weak-electrolyte anions through anion-exchange membranes. Current-voltage characteristics. *Journal of Membrane Science* **189**:129–140.

51 Pismenskaia N, Sistat P, Huguet P, Nikonenko V, Pourcelly G, (2004)

Chronopotentiometry applied to the study of ion transfer through anion-exchange membranes. *Journal of Membrane Science* **228**, 65–76.

52 VOLODINA E, PISMENSKAYA N, NIKONENKO V, LARCHET C, POURCELLY G, (2005) Ion transfer across ion-exchange membranes with homogeneous and heterogeneous surfaces, *Journal of Colloid and Interface Science* **285**:247–258.

53 GREBENYUK V, CARSON W, GREBENYUK O, SIMS K, MACDONALD R (2004) Sparse media EDI apparatus and method. International Publication Number Patent WO 2004/024992 A1.

54 MISHCHUK NA, KOOPAL LK, GONZALEZ-COBALLERO F (2001) Intensification of electrodialysis by a non-stationary electric field. *Journal of Colloids and Surfaces A: Physicochemical. Eng. Aspects* **176**, 195–212.

7
Membrane Technology in the Chemical Industry: Future Directions

R. W. Baker

7.1
The Past: Basis for Current Membrane Technology

Attempts to use membranes for practical separations did not begin until the early 1900s when Bechold, Zigmondy and Bachmann, and Elford and Ferry used nitrocellulose membranes to separate laboratory solutions by dialysis and microfiltration. By the 1930s microporous membranes were commercially produced on a small scale and, at about the same time, Teorell, Meyer, and Sievers made the first practical ion-exchange membranes and developed their theory of ion transport through charged membranes. As a result, the elements of modern membrane science were in place by the late 1960s, but essentially no industrial applications of membranes existed; sales of the total membrane industry were probably less than US $10 million/year at that time (in current dollars). Membranes were used in a few laboratory and analytical applications, but not for industrial applications because membranes were still too slow, too expensive, and too unselective for most commercially important separations. The development of solutions to these problems has led to the current widespread and growing use of membranes in the chemical and refinery industries.

7.1.1
Ultrathin Membranes

The seminal discovery that transformed membrane separation from a laboratory to an industrial process was the development in the 1960s of the Loeb-Sourirajan process to make defect-free ultrathin cellulose acetate membranes [1]. Loeb and Sourirajan were trying to use membranes to desalt water by reverse osmosis (RO). The concept of using a membrane permeable to water and impermeable to salt to remove salt from water had been known for a long time, but the fluxes of all the membranes then available were far too low for a practical process. The Loeb-Sourirajan breakthrough was the development of an anisotropic membrane. The membrane consisted of a thin, dense polymer skin 0.2–0.5 µm thick sup-

Membrane Technology in the Chemical Industry.
Edited by Suzana Pereira Nunes and Klaus-Viktor Peinemann
Copyright © 2006 WILEY-VCH Verlag GmbH & Co. KGaA, Weinheim
ISBN: 3-527-31316-8

ported on a much thicker microporous layer. The thin, dense skin layer performed the separation; the microporous support provided the mechanical strength required. These membranes had 10 times the flux and selectivities equivalent to the best membranes then known. Once the advantages of the anisotropic structure had been demonstrated by Loeb and Sourirajan, many other ways of making anisotropic membranes were developed. Anisotropic membranes with effective thicknesses of 0.05 to 0.1 µm are now produced on a large scale.

7.1.2
Membrane Modules

The second advance that made industrial membrane-separation processes possible was the development of methods to incorporate large membrane areas into economical membrane packets or modules. Even with the best anisotropic membranes, most industrial processes require several hundred, sometimes several thousand, square meters of membrane to perform the separation. The three most important configurations, shown schematically in Fig. 7.1, are hollow-fine-fiber, capillary-fiber and spiral-wound modules. Tubular and plate-and-frame

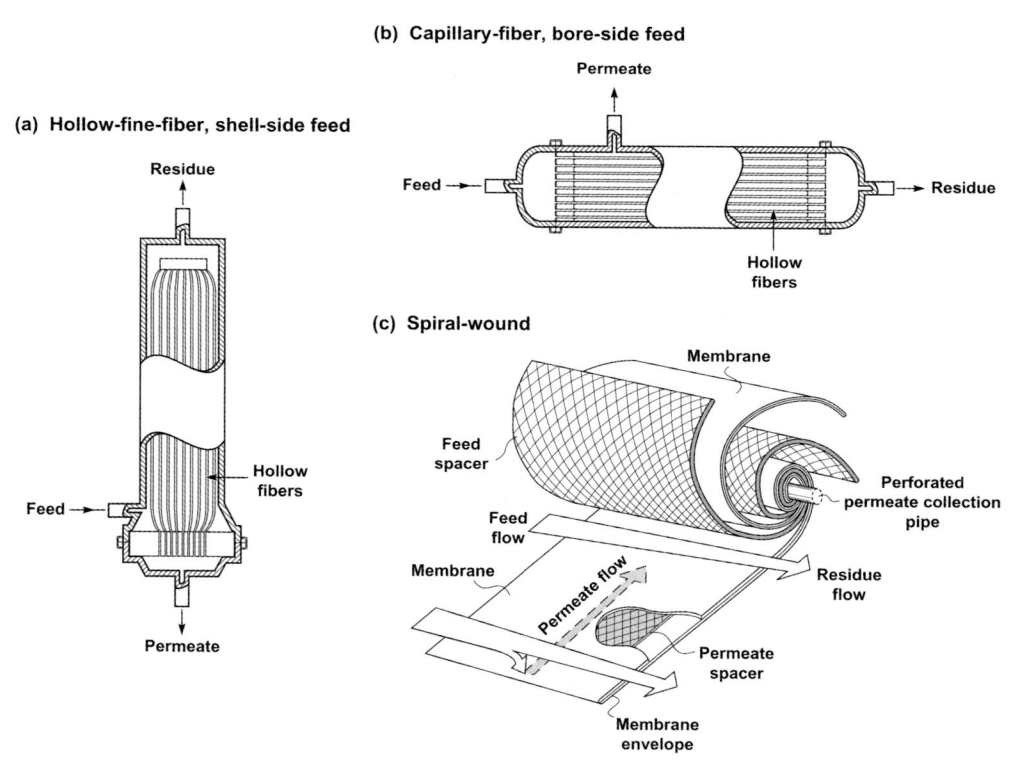

Fig. 7.1 Membrane module configuration.

Table 7.1 Membrane module characteristics.

Module type	Manufacturing cost[a] (US $/m^2)	Area of standard module[b] (m^2)	Characteristics
Hollow-fine-fiber	2–10	100–300	Low cost per m^2 of membrane but modules easily fouled. Only suitable for clean fluids.
Capillary-fiber	5–50	50–100	Limited to low-pressure applications <200 psi; good fouling resistance, can be backflushed. Important in ultrafiltration (UF) and microfiltration (MF) applications.
Spiral-wound	10–50	20–40	The most common RO module. Increasingly used in UF and gas-separation applications.
Tubular, plate-and-frame	50–200	5–10	High cost limits applications.

a) Cost here is early 2000s manufacturing cost; market prices are typically two to five times higher.
b) A standard module is defined as 20 cm (8 in) in diameter and 100 cm (40 in) long.

modules are also used, but their high cost limits them to small-scale or specialized applications. The advantages and disadvantages of the various module types are summarized in Tab. 7.1.

Hollow-fine-fiber membranes are tiny polymeric tubes with diameters of 100 to 250 μm. Between 0.1 and 2 million fibers are packed into an average 20-cm-diameter module. This allows a very large membrane area to be contained in a relatively small volume. Because high-speed automated equipment is used to spin the fibers and prepare the membrane modules, the production cost of these modules on a per-square-meter basis is low, in the range US $2–10/m^2. This cost is the module manufacturing cost; the selling price is higher and reflects the pricing structure adopted by that segment of the industry. Capillary-fiber modules are similar to fine-fiber modules but have larger fiber diameters, generally in the range of 500–2000 μm. In capillary fiber modules, the pressurized feed fluid passes down the bore of the fiber. In hollow-fine-fiber modules, the pressurized feed fluid usually enters the shell side of the fibers, and the permeate fluid passes down the fiber bore.

The relatively low cost of hollow-fiber modules is their principal advantage. A key disadvantage is that the polymer membrane must perform the separation required as well as withstand the pressure driving force across the membrane. Preparing membranes that meet both requirements is difficult. Thus, producing membranes that have high selectivities and high fluxes and that are mechanically stable is more difficult in hollow-fiber form than with the flat sheets used in spiral-wound modules. As a result, membranes used in spiral-wound modules generally have better, sometimes significantly better, performance than their hollow-fiber equivalents. In some cases the best membranes cannot be formed into hollow fibers. Because hollow-fiber membranes must support a relatively high pressure, softening of the microporous support by absorbed materials can lead to catastrophic failure. Plasticization and collapse of the microporous support layer can also be a problem with the flat-sheet membranes used in spiral-wound modules. However, in this case the pressure is supported by the microporous layer and a nonwoven paper support, so the membranes are generally more robust. A final issue with hollow-fiber modules, particularly fine-fiber modules, is their susceptibility to fouling by particulate matter carried into the module with the feed fluid.

For these reasons, despite their apparent cost advantage, hollow-fiber modules are generally limited to separations involving clean particulate-free feed fluids, for example, desalination of seawater, separation of nitrogen from air, and separation of hydrogen from nitrogen, methane and argon in ammonia reactor purge gas.

7.1.3
Membrane Selectivity

The third development of the last 40 years that has made membranes widely applicable to industrial separations is the availability of more-selective, more-permeable membrane materials. In the 1960s and 1970s membranes were generally made from commercially available polymers developed for other purposes. More recently, membrane developers are increasingly using tailor-made polymers. The result is a significant improvement in membrane properties for a number of separations. As an example, Fig. 7.2 is a plot prepared by Robeson showing the oxygen permeability and oxygen/nitrogen selectivities of a large number of membrane materials reported in the literature [2]. There is a very strong inverse relationship between flux and selectivity. This selectivity/permeability tradeoff is apparent in Robeson's plot. Also shown in the figure is a line linking the most permeable polymers at a particular selectivity; this line is called the upper bound, beyond which no better material is currently known. The relative positions of the upper bound in 1991 and in 1980 shows the progress that has been made in tailor-making polymers for this separation. Development of improved membrane materials is a continuing topic of research, so further slow movement of the upper bound is likely.

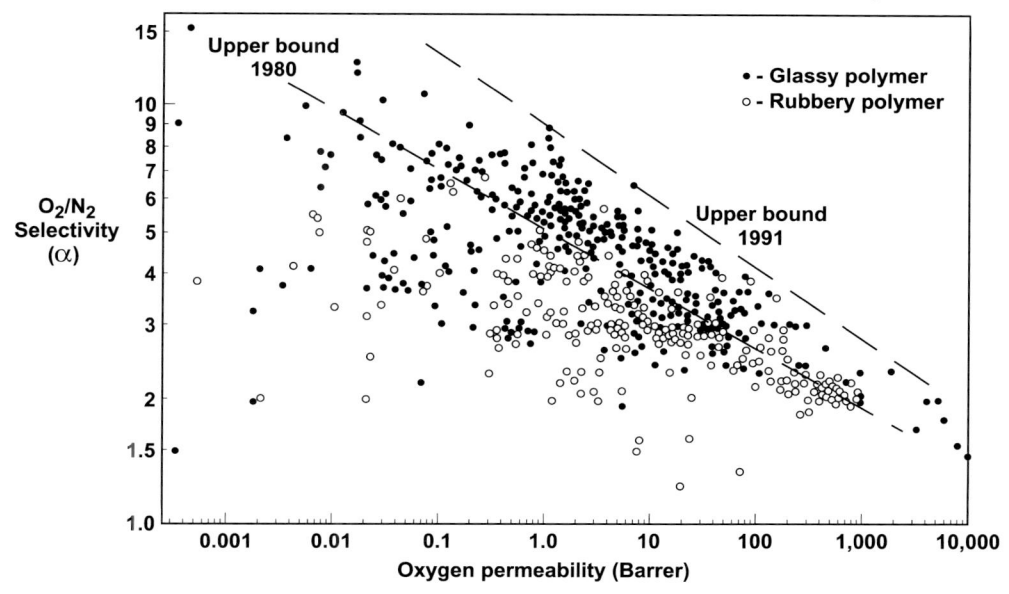

Fig. 7.2 Robeson's plot of polymer membrane oxygen:nitrogen selectivity as a function of oxygen permeability.

7.2
The Present: Current Status and Potential of the Membrane Industry

7.2.1
Reverse Osmosis

Reverse osmosis was the first membrane process to be used on a large commercial scale, following the development of the Loeb-Sourirajan membrane (1962) and the timely infusion of substantial research dollars by the US Office of Saline Water. Membrane and module technology was sufficiently developed for commercial plants to be installed by the late 1960s. The development of interfacial polymerization as a technique to produce composite membranes by Cadotte was another major milestone [3]; currently these membranes have about 90% of the total reverse osmosis market. Hollow-fine-fiber membranes, widely used in the 1970s and 1980s, have been almost completely displaced by the more reliable spiral-wound modules. Currently, approximately 1 billion gallons/day of water are desalted by reverse osmosis. About half of this capacity is installed in the Middle East and other desert regions to provide municipal drinking water. The remainder is used in the industrial world to produce ultrapure water for the electronic and pharmaceutical industries.

The reverse osmosis (RO) industry is now well established; three manufacturers produce 70% of the RO membrane modules. Total RO membrane mod-

Table 7.2 Advances in spiral-wound module reverse osmosis performance (Source: D. Furukawa, 1999).

Year	Cost normalized (1980 US $)	Productivity normalized (to 1980)	Reciprocal salt passage normalized (to 1980)	Figure of merit [a]
1980	1.00	1.00	1.00	1.0
1985	0.65	1.10	1.56	2.6
1990	0.34	1.32	2.01	7.9
1995	0.19	1.66	3.52	30.8
1999	0.14	1.94	7.04	99.3

a) Figure of merit = [(productivity) × (reciprocal salt passage)]/cost

ule sales are currently about US $300 million/year. The industry is extremely competitive, with the manufacturers producing similar products and competing mostly on price. The standard RO membrane is an interfacial composite structure formed from the reaction of trimesoyl chloride with m-phenylenediamine [4,5]. These membranes are typically packaged into 8-inch-diameter spiral-wound modules containing about 40 m^2 of active membrane area. Many incremental improvements have been made to membrane and module performance over the past 15 years, resulting in steadily decreasing water-desalination costs in inflation-adjusted dollars. Some performance values taken from a paper by Dave Furukawa are shown in Tab. 7.2. Since 1980, the cost of spiral-wound membrane modules on a per-square-meter basis has decreased sevenfold. At the same time the water flux has doubled, and the salt permeability has decreased sevenfold. Taking these improvements into account, today's membranes are almost 100 times better than those of the 1980s. This type of incremental improvement is likely to continue for at least a few more years.

Currently, the production of municipal drinking water and ultrapure industrial water accounts for more than 90% of the reverse osmosis market. In the early days of the industry, many people thought that the treatment of industrial process and effluent streams would be a major application. These applications did not develop, however, primarily because of low process reliability due to membrane fouling. A number of improvements that are likely to lead to greater use of reverse osmosis in these areas in the next few years are being made. Fouling-resistant membranes will help, and fouling-resistant module designs may be even more important. Vibrating or rotating modules in which the membrane is moved (rather than the fluid flowing across the surface) have proved able to treat extremely dirty solutions without fouling [6]. The design of one such system is shown in Fig. 7.3. Currently, these modules are extremely expensive – in the range of US $2000–5000/m^2 – compared to alternative designs, which limits their applications. If costs can be reduced, a larger chemical industry market will open up.

A promising new application of reverse osmosis in the chemical industry is the separation of organic/organic mixtures [7, 8]. These separations are difficult,

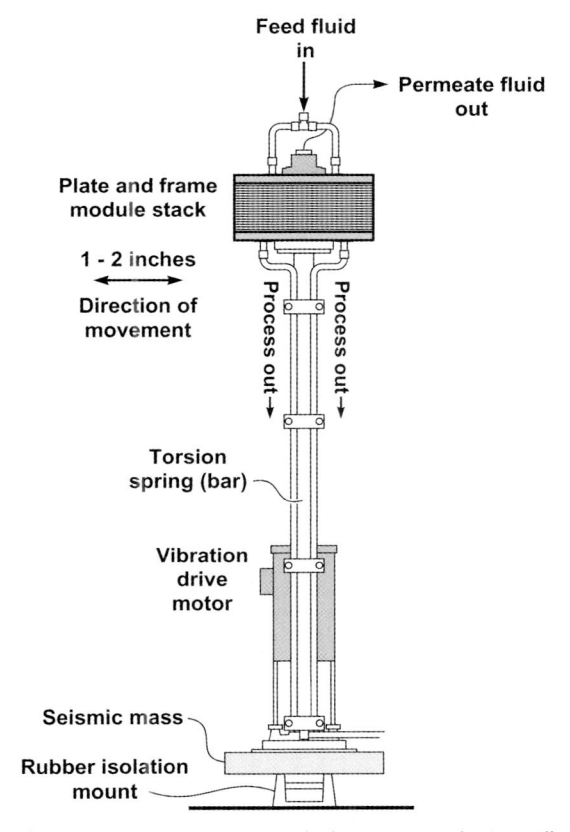

Fig. 7.3 New Logic International vibrating plate-and-frame module design. A motor taps a metal plate (the siesmic mass) supported by a rubber mount at 60 times/s. A bar that acts as a torsion spring connects the vibrating mass to a plate-and-frame membrane module that then vibrates horizontally by 1–2 inches at the same frequencey. By shaking the membrane module in this way, high turbulence is induced in the pressurized feed solution fluid that flows through the module. The turbulence occurs directly at the membrane surface, providing good control of membrane fouling [6].

not only because of the high osmotic pressures that must be overcome, but also because they require membranes that are sufficiently solvent resistant to be mechanically stable, yet sufficiently permeable to provide high fluxes. Nonetheless, this is an area of keen industrial interest, and from 1988 to 2005 more than 40 US patents covering membranes and membrane systems for these applications were issued.

One RO organic/organic separation application that has already reached the commercial stage is the separation of small solvent molecules from larger hydrocarbons in mixtures resulting from the extraction of vacuum residual oil in refineries [9, 10]. Figure 7.4 (a) shows a simplified flow diagram of a refining lube oil separation process. These operations are very large scale. In a typical

(a) Conventional solvent dewaxing process

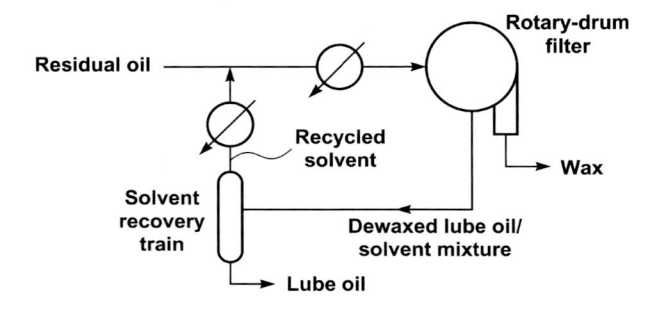

(b) Mobil Oil's solvent dewaxing process

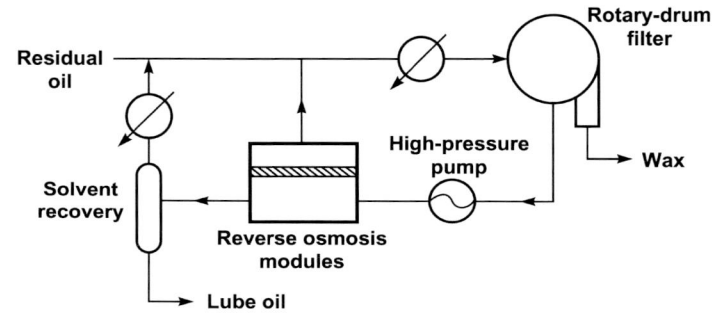

Fig. 7.4 Simplified flow schemes of (a) a conventional and (b) Mobil Oil's membrane solvent dewaxing processes. Refridgeration economizers are not shown. The first 3-million gallon/day commercial unit was installed at Mobil's Beaumont refinery in 1998. Polyimide membranes in spiral-wound modules were used.

100 000 barrel/day refinery, about 15 000 barrels/day of the oil entering the refinery remains as residual oil. A large fraction of this oil is sent to the lube oil plant, where the heavy oil is mixed with 3 to 10 volumes of a solvent such as methyl ethyl ketone/toluene. On cooling the mixture, the heavy wax components precipitate out and are removed by a drum filter. The light solvent is then stripped from the lube oil by vacuum distillation and recycled through the process. The vacuum distillation step is very energy-intensive because of the high solvent-to-oil ratios employed.

A reverse osmosis process developed by Mobil for this separation is illustrated in Fig. 7.4(b). Polyimide membranes formed into spiral-wound modules are used to separate up to 50% of the solvent from the dewaxed oil. The membranes have a flux of 10–20 gal/ft^2 · day at a pressure of 450 to 650 psi. The solvent filtrate bypasses the distillation step and is recycled directly to the incoming oil feed. The net result is a significant reduction in the refrigeration load required to cool the oil and in the size and energy consumption of the solvent recovery vacuum distillation section.

Despite the Mobil success, only one lube oil membrane dewaxing plant has been built to date. However, development of similar applications in other operations is likely, and so the Mobil plant will remain an important technology milestone. Initially, future applications will probably involve relatively easy separations such as the separation of methyl ethyl ketone/toluene from lube oil described above or soybean oil from hexane in food oil production. Livingstone and coworkers in the UK [11] have also described the use of nanofiltration or reverse osmosis to recover and reuse homogeneous organometallic catalysts. Long term, the technology may become sufficiently advanced to be used in more important refining operations, such as fractionation of linear from branched paraffins, or the separation of benzene and other aromatics from paraffins and olefins in the gasoline pool.

7.2.2
Ultrafiltration

Abcor (now a division of Koch Industries) installed the first industrial ultrafiltration plant to recover electrocoat paint from automobile paint shop rinse water in 1969. Shortly afterwards, systems were installed in the food industry for protein separation from milk whey and for apple juice clarification. The separation of oil emulsions from effluent wastewaters has also become a significant application. The current ultrafiltration market is approximately US $ 200 million/year, but because the market is very fragmented no individual end-use segment is more than US $ 10–30 million/year. In the chemical and refining industries, the principal application of ultrafiltration is the treatment of oily wastewater.

The key issue limiting growth of ultrafiltration is its high cost. Because membrane fluxes are modest, large amounts of energy are used to circulate the feed solution to control fouling, membrane modules must be cleaned frequently, and membrane lifetimes are short. These are all different aspects of the same problem – membrane fouling. Unfortunately, membrane fouling is an inherent feature of ultrafiltration. Only limited progress in controlling this problem has been made in the past 20 years and, barring unexpected breakthroughs, progress is likely to remain slow. Better module designs, simple automatic backflushing, and inherently less-fouling membranes are all being developed and used.

Ceramic membranes, which are tougher and longer lasting than polymeric membranes, offer many advantages in ultrafiltration applications but are more than 10 times more costly than equivalent polymer membranes. Thus their use has been limited to small-scale, high-value separations that can bear this cost. One area where ceramic membranes may find a future use is clarification of chemical or refinery process streams, where their solvent resistance is needed. However, it is difficult to see a major business developing from these applications unless costs are reduced significantly.

7.2.3
Microfiltration

The first large use of microfiltration membranes was in laboratory bacteriologic tests; this remained the major use until the 1970s. The introduction of the pleated inline filtration cartridge by Gelman Sciences then led to the development of very large markets for disposable cartridges for sterile filtration in the pharmaceutical industry and particle removal from ultrapure water in semiconductor fabrication processes. In these applications, the membrane cartridge is used as an inline (dead-end) filter and the entire fluid flow is forced through the membrane under pressure. As particles accumulate on the membrane surface or in its interior, the pressure required to maintain the required flow increases until at some point the cartridge must be replaced.

An alternative microfiltration process known as *crossflow* filtration was also developed in the 1970s. In crossflow systems, the feed solution is circulated across the surface of the membrane producing two streams: a clean particle-free permeate and a concentrated retentate containing the particles. The equipment required for crossflow filtration is more complex than for an inline filter, but membrane lifetimes are much longer. Crossflow microfiltration systems have been only a small fraction of the microfiltration market until the last few years. Beginning in the mid-1990s companies such as Memtec, X-Flow, Hydranautics, Kubota, and Zenon [12–16] began to produce various types of crossflow microfiltration systems to treat municipal drinking water and membrane bioreactors to treat municipal and industrial wastewater. The first proof-of-concept plants were installed in 1990–93 [12]. However, the market did not take off until the introduction of rules by the US EPA and European regulators requiring drinking water obtained from surface waters to be treated to control giardia, coliform bacteria and viruses. Many large plants using backflushable, hollow filter membranes have now been built.

The water to be filtered by these water-treatment plants is relatively clean, so a microfiltration system called semi-dead-end filtration is often used. In these systems, the membrane unit is operated as an inline (dead-end) filter until the pressure required to maintain a useful flow across the filter reaches its maximum level. At this point, the filter is operated in a crossflow mode, while concurrently backflushing with air or permeate solution. After a short period of backflushing in crossflow mode to remove material deposited on the membrane, the system is switched back to dead-end operation. This procedure is particularly applicable in microfiltration units used as a final bacterial and virus filter for municipal water treatment plants. A photograph of a municipal microfiltration plant of this type is shown in Fig. 7.5.

Microfiltration plants are also being installed in membrane bioreactors to treat municipal and industrial sewage water. Two types of systems that can be used are illustrated in Fig. 7.6. The design shown in Fig. 7.6(a), using a crossflow filtration module, was developed as early as 1966 by Okey and Stavenger at Dorr-Oliver [17]. The process was not commercialized for another 30 years for lack of suitable membrane technology. In the 1990s, workers at Zenon [15, 16] in Canada and

Fig. 7.5 Photograph of a 25-million gallon/day capillary hollow fiber module plant to produce potable water from a well, installed by Norit (X-Flow) in Keldgate, UK. Courtesy of Norit Membrane Technology B.

at Ebara [13], Kubota [14] and the University of Tokyo [18, 19] in Japan developed submerged hollow-fiber membrane units of the type illustrated in Fig. 7.6(b). The membranes used in this system usually operate with a very small pressure differential of 0.2 to 0.6 bar, provided by a pump that evacuates the permeate side of the membrane. The feed solution is highly fouling, so air is introduced through a sparger placed underneath the membranes. The action of the rising air bubbles scrubs the membranes and allows useful fluxes of 10 to 30 L/m² h to be maintained for 5 to 15 min. The permeate pump is then turned off and the membranes are backflushed for a short period, after which permeation resumes [20, 21]. Plants of this type have been widely installed. The water produced is close to sterile. The manufacturers claim they can "beat" conventional biological treatment plants on the basis of permeate quality, reduced footprint and installed cost.

7.2.4
Gas Separation

The first company to produce a successful membrane-based gas-separation process was Permea, now a division of Air Products, which introduced hollow-fine-fiber polysulfone membranes for the separation of hydrogen from ammonia reactor purge gas in 1980. This application was an immediate success: the feed gas was clean and free of condensable components that might damage the membranes, and the value of the recovered hydrogen provided short payback times. Within a few years, many ammonia plants worldwide had installed these units. Several hundred hydrogen-separating systems have now been installed by Permea and its competitors.

Following this success, Generon, now a division of MG Industries, introduced a membrane process to separate nitrogen from air. The first-generation mem-

(a) Cross-flow filtration bioreactor

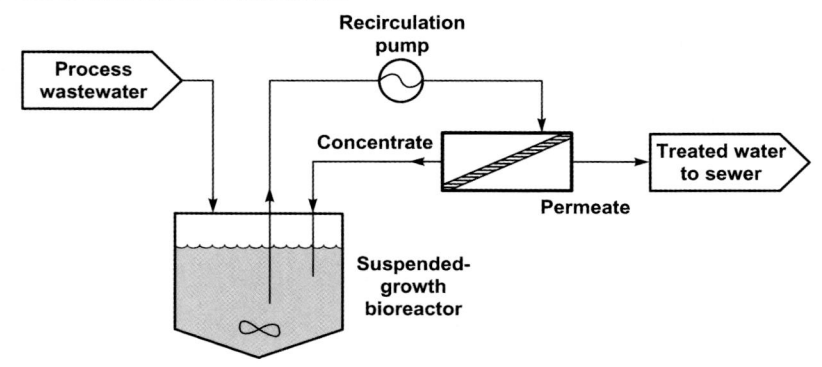

(b) Submerged membrane, air-sparged bioreactor

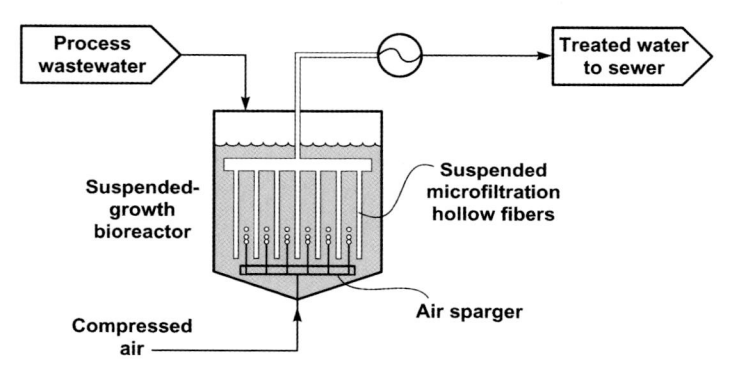

Fig. 7.6 Schematic flow diagram of microfiltration bioreactors operated with (a) crossflow membranes and (b) submerged air-sparged membranes.

branes had modest selectivities and were only able to produce 95% pure nitrogen economically. By the late 1990s, Generon (MG), Medal (Air Liquide), Permea (Air Products) and IMS (Praxair) had all developed tailor-made membrane materials with oxygen/nitrogen selectivities of 7 to 8 for this separation. These membranes produce 97 to 99% nitrogen economically, so a large market developed for systems producing ten thousand to one million standard cubic feet/ day of nitrogen. More than 10 000 of these systems have now been sold.

Other membrane-based gas-separation applications that developed in the late 1980s and the 1990s include the separation of carbon dioxide from natural gas, separation of organic vapors from air and nitrogen, and dehydration of air. Table 7.3 lists the major companies involved in the industry and their principal markets. Currently, total industry sales are estimated to be about US $ 200 million. Of all the industrial membrane-separation processes, gas separation is

Table 7.3 Current membrane gas-separation industry players.

Company	Principal markets/estimated annual sales
Permea (Air Products) Medal (Air Liquide) IMS (Praxair) Generon Ube	The large gas companies are mostly focused on nitrogen/air (US $80 million/yr) and hydrogen separation (US $40 million/yr)
Kvaerner Separex (UOP) Cynara (Narit)	Mostly natural gas separations (US $50 million/yr)
Whatman GKSS Licensees MTR	Vapor/gas separation, air dehydration, other (US $30 million/yr)

These market sizes are informed guesses based on a few industry discussions and should be used with caution.

most closely linked to the chemical and petrochemical-refining industries, so its sales are affected by the chemical and energy industry business cycles. In general, however, the trend for sales of gas-separation membranes is up and this trend will continue, especially if some of the processes now being developed and described below are successfully introduced.

7.2.4.1 Refinery Hydrogen Applications

The separation of hydrogen from nitrogen, argon, and methane in ammonia plant purge streams was the first successful commercial application of membranes for gas separation. This market is now essentially saturated; almost all existing ammonia plants have been retrofitted with membrane units. As a result, the market is limited to replacement units plus new units for the few new ammonia plants that come online each year. Most of the current market for new hydrogen membrane units is in refineries and petrochemical plants, for the recovery of hydrogen from gas streams containing condensable gas mixtures. Figure 7.7 shows a typical application, the recovery and reuse of hydrogen from an oil hydrocracker purge gas [22, 23]. Hydrocrackers and hydrotreaters are used in refineries to break down high molecular weight components, to remove impurities, and to hydrogenate aromatics. Ideally, heavy oil is cracked to C_{5+} hydrocarbons, but inevitably some methane, ethane, and propane are produced as byproducts of the reaction. The oil/gas mixture from the hydrocracker is sent to a lower-pressure separator from which the C_{5+} product is removed. Unreacted hydrogen is recirculated back to the reactor. Methane, ethane, and propane accumulate in the recycle stream and must be removed as an inert gas purge. Typically, 3–4 moles of hydrogen are lost with every mole of light hydrocarbon purged from the reactor.

Typical hydrocracker hydrogen recovery process

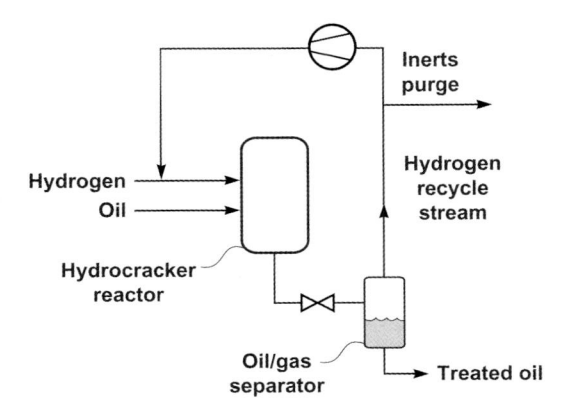

Membrane-based hydrogen recovery

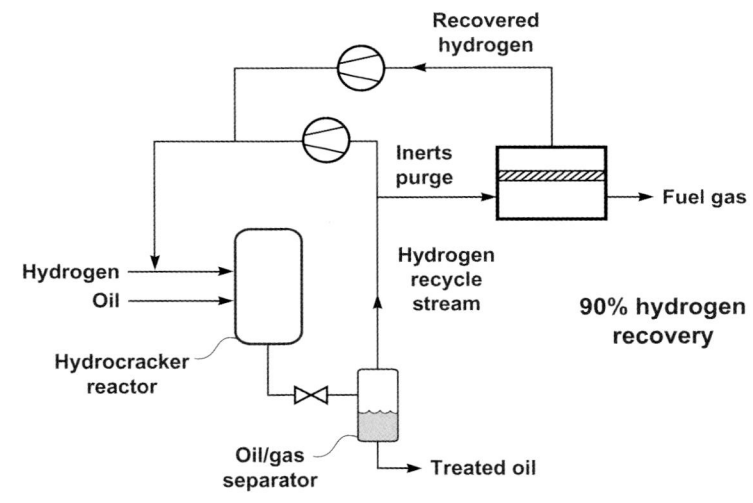

Fig. 7.7 Use of hydrogen-permeable membranes to recover and reuse hydrogen from hydrocracker purge gas streams.

In principle, recovering hydrogen from the inert purge gas is an easy application for membranes. However, as hydrogen is removed through the membrane, the remaining gas becomes enriched in hydrocarbons, and the dew point increases to 60 °C or more. To avoid condensation of hydrocarbons on the membrane, the gas should be heated to at least 60 °C. In practice, to provide a safety margin and to minimize plasticization of the membrane, the gas must be heated to 15–20 °C above the expected dew point, in other words to above 80 °C.

In the past, the available membranes lost a significant fraction of their selectivity when operated at these high temperatures. They also became plasticized by absorbed heavy hydrocarbons in the feed gas. As a consequence, a number of early hydrogen-separation plants installed in refineries had reliability problems. The development of newer polyimide and polyaramide membranes that can safely operate at high temperatures has solved most of these problems and the market for membrane-based hydrogen-recovery processes in refineries is growing.

7.2.4.2 Nitrogen (and Oxygen) Separation from Air

The largest gas-separation process in current use is the production of nitrogen from air. The first membranes used for this process were based on poly(4-methyl-1-pentene) (TPX) and ethyl cellulose. These polymer materials have oxygen/nitrogen selectivities of 4; the economics of the membrane processes were marginal. The second-generation membrane materials now used have selectivities of 6–8, providing very favorable economics, especially for small plants pro-

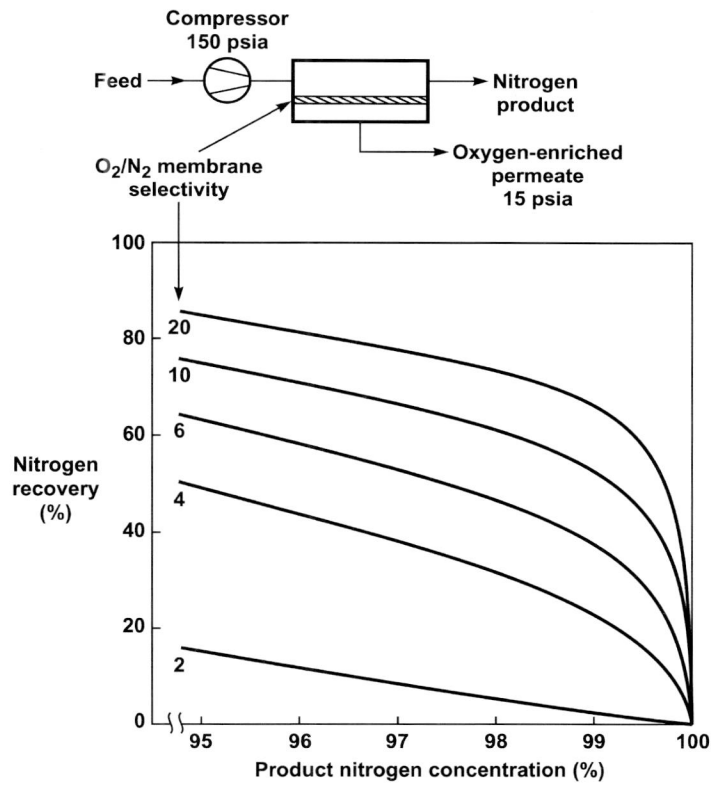

Fig. 7.8 Nitrogen recovery as a function of product nitrogen concentration for membranes with selectivities between 2 and 20.

ducing 5–500 scfm of nitrogen. In this range, membranes are the low-cost process, and most new small nitrogen plants use membrane systems.

A simplified drawing of a nitrogen-from-air separation process and its performance is shown in Fig. 7.8. The feed air is compressed to 8–10 atm with a low-cost screw compressor and then passed through a bore-side hollow-fiber module. The module operates in counterflow mode. The first membranes used for nitrogen separation, with an oxygen/nitrogen selectivity of about 4, could produce 95% nitrogen at a nitrogen recovery of about 50%; unfortunately, the principal market is for 99% nitrogen. At this higher concentration, membranes with a selectivity of 4 achieve only about 25% nitrogen recovery; the other 75% is lost with the permeate stream. The second-generation membranes, with selectivities of 8, could generate a 99% nitrogen product at an overall nitrogen recovery of about 50%. The permeation rates of these membranes are lower than those of earlier low-selectivity membranes, but the cost of the extra membrane modules required to process the same volume of gas is more than offset by the reduced size of the compressor. In a membrane nitrogen-from-air plant, approximately two-thirds of the total component cost of the plant is associated with the air compressor; 20% or less is associated with the membrane modules. The energy used to power the compressor also represents the majority of the operating cost. It follows that reducing the size of the feed-gas compressor is key to lowering the nitrogen production costs for membrane-based separations.

Improvements in membrane selectivity have now reached a point of diminishing returns. A major improvement in selectivity, from 8 to 12, would reduce compressor size by 20%. Unfortunately, based on Robeson's trade-off curve, such an improvement in selectivity is likely to be accompanied by a tenfold reduction in membrane permeability, and a tenfold reduction in permeability means a tenfold increase in membrane area. These effects are shown graphically in Fig. 7.9.

For the reasons outlined above, it is unlikely that significantly improved membranes, or the separation of nitrogen from air, will be introduced in the near future. One place where improved oxygen/nitrogen-selective membranes could make a difference is the production of oxygen from air.

Currently, most oxygen is produced by cryogenic separation of air, at a cost of $40–100/ton. This mature technology has been refined to the point that major cost reductions are unlikely. A new alternative technology is vacuum swing adsorption, which is claimed to produce 90–95% pure oxygen at $40–60/ton. Membranes have the potential to reduce these costs substantially. The target for a membrane process to make a significant impact is an oxygen cost of $15–35/ton EPO_2.[1]

---

1) Cryogenic processes produce essentially pure oxygen, whereas membrane and adsorption processes produce a lower-concentration oxygen product. To compare the economics of different processes producing different concentrations of oxygen-enriched air, oxygen costs in this chapter are reported on an equivalent pure oxygen (EPO_2) basis. EPO_2 is defined as the amount of pure oxygen that must be mixed with normal air to obtain oxygen-enriched air at the specified concentration. For example, in a membrane process producing 40% oxygen-enriched air, only the oxygen added in excess of that contained in air is counted, that is, $40-[(60\times21)/79]$ or 24.1%.

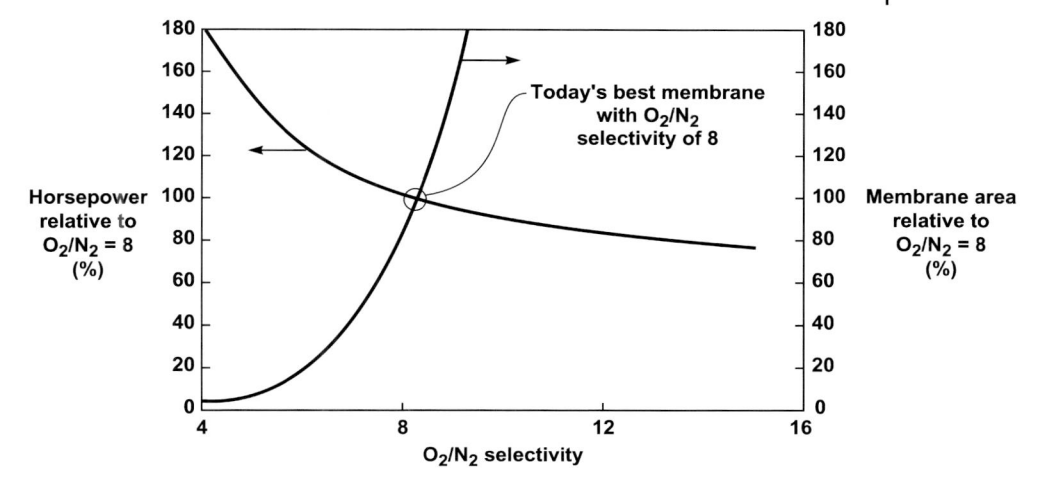

Fig. 7.9 Effect of selectivity on the horsepower and membrane area required for membrane systems producing equivalent amounts of 99% nitrogen from air (countercurrent flow, feed pressure 10 atm, permeate pressure 1 atm). Increasing the membrane selectivity from 8 to 12 reduces the compressor horsepower by 29% but the membrane area increases tenfold.

A simplified flow scheme of a membrane-separation process to produce oxygen-enriched air is shown in Fig. 7.10(a). Feed air containing 21% oxygen is passed across the surface of a membrane that preferentially permeates oxygen. In the scheme shown, the pressure differential across the membrane required to drive the process is maintained by drawing a vacuum on the permeate gas. The alternative is to compress the feed gas.

A few trial calculations show that a process using a feed-gas compressor, even if coupled with an energy-recovery turbine on the residue side, cannot produce low-cost oxygen because of the quantity of electricity consumed. All the feed air must be compressed but only a small portion permeates the membrane. The power consumption of a vacuum pump is less because the only gas evacuated by the vacuum pump is the oxygen-enriched product that permeates the membrane.

Depending on the properties of the membrane and the pressure differential, a permeate gas containing 30–60% oxygen is produced. Such oxygen-enriched air can be used in a number of processes, for example, in Claus plants, in FCC catalyst regeneration for refineries, or to burn methane in high-temperature furnaces or cement kilns. Other processes may require an oxygen content of 95–98%, for example, synthesis gas production. The higher oxygen content can be achieved by adding a second separation stage as shown in Fig. 7.10(b). Because the volume of gas sent to the second-stage separator is one-quarter to one-third the feed to the first-stage unit and the gas is more concentrated, the second stage will be much smaller. This second separation stage could be another membrane unit or, more likely, a vacuum swing adsorption system, which will be more economical in this oxygen concentration range. The size and energy consumption of vacuum swing

(a) One-stage membrane separation process

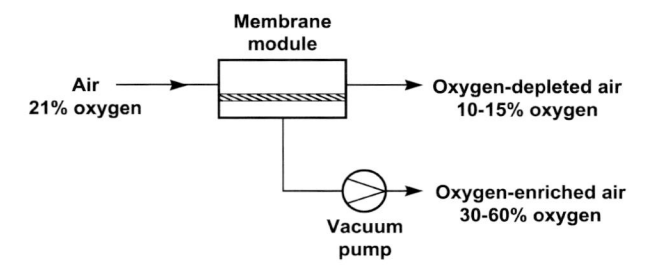

Membrane module

Air 21% oxygen → [Membrane module] → Oxygen-depleted air 10-15% oxygen

↓ Vacuum pump → Oxygen-enriched air 30-60% oxygen

(b) Two-stage separation process

Air 21% oxygen → [Membrane module] → Oxygen-depleted air 10-15% oxygen

↓ Vacuum pump → [Second stage separation unit] → Oxygen-enriched air 95-98% oxygen

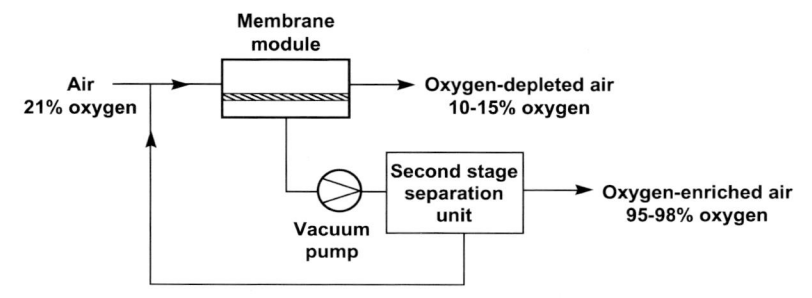

(recycle loop back to feed)

Fig. 7.10 Oxygen/air separation process designs: (a) one-stage membrane-separation process, (b) two-stage separation process.

adsorption systems is almost a direct function of the nitrogen removal required, because it is the nitrogen that adsorbs on the molecular sieve adsorption beds and must be removed with vacuum pumps. Starting from 50% oxygen (1 mole of nitrogen/mole of oxygen) rather than from 21% oxygen (4 moles of nitrogen/mole of oxygen) reduces the mass of nitrogen to be separated per ton of oxygen product fourfold. The resulting cost of upgrading the gas from 50% oxygen to 95–98% oxygen is only $10–15/ton EPO$_2$.

To achieve the process target costs for oxygen-enriched air, very good membranes will be required. Bhide and Stern calculated the cost for oxygen produced by today's best membrane materials [24]. None of these polymers were able to meet the $20/ton EPO$_2$ target. Very substantial improvements in membrane permeances are required. One possible approach is to develop carrier-facilitated transport membranes. The spectacular permeability and selectivity of this type of membrane has maintained interest in facilitated transport membranes despite their many problems, mostly related to membrane stability. The best oxygen-facilitated carrier membrane to date never functioned for more than one month. Nevertheless, this is an area of research where a breakthrough could lead to a very significant commercial result.

7.2.4.3 Natural Gas Separations

The use of membranes to remove impurities from natural gas is probably the fastest growing gas-separation application, and long term, is likely to become the biggest. Raw natural gas varies substantially in composition from source to source. Methane is always the major component, typically 75–90% of the total, but natural gas also contains significant amounts of ethane, some propane and butane, and 1–3% of other higher hydrocarbons. In addition, the gas contains undesirable impurities: water, carbon dioxide, nitrogen, and hydrogen sulfide. Although the composition of raw gas varies widely, the composition of gas delivered to commercial pipeline grids is tightly controlled. Consequently, all natural gas requires some treatment, and about 20% requires extensive treatment, before delivery to the pipeline. Typical US natural gas pipeline specifications are reported in Tab. 7.4. The business opportunity for membranes lies in processing the gas to meet these specifications. The total worldwide market for new natural gas separation equipment is probably $5 billion per year. Currently, membrane processes have less than 1% of this market, almost all for the removal of carbon dioxide.

The massive scale of this opportunity is illustrated by the Qadirpur carbon dioxide plant shown in Fig. 7.11 [25]. This is probably the world's largest membrane gas-separation unit after the U^{235}/U^{237} gas diffusion plants built as part of the Manhattan Project.

Carbon dioxide. The design of a typical membrane system to remove carbon dioxide from natural gas is shown in Fig. 7.12. Two-stage designs are often used to reduce methane loss. The first carbon dioxide plants installed treated only a few million scfd of gas; more recently, a number of large systems have been installed. Membrane-based units are particularly favored over amine absorption plants (the usual competitive technology) in offshore installations because of their smaller footprint and reduced weight.

Numerous membrane materials with selectivities and permeabilities far in excess of those used by the industrial membrane suppliers have been reported in the academic and patent literature. Unfortunately, when evaluated under real-world conditions (high-pressure gas containing heavy hydrocarbons, aromatics and water), these membranes are seldom able to match their laboratory performance. Having said this, cellulose acetate, the current industry workhorse

Table 7.4 Composition specifications for natural gas delivery to the US national pipeline grid.

Component	Specification
Carbon dioxide	<2%
Water	<120 ppm
Hydrogen sulfide	<4 ppm
C_{3+} content	950–1050 Btu/scf; dew point: –20 °C
Total inert gases (N_2, CO_2, He, etc.)	<4%

Fig. 7.11 A section of the Qadirpur, Pakistan, carbon dioxide removal plant built by UOP (Separex). The plant reduces the carbon dioxide content of natural gas from 6.5% to less than 2%. The original plant bulit in 1995 had a capacity of 265 MMscfd but was expanded in 2004 to a capacity of 500 MMscfd of gas. This is the largest natural gas membrane plant built to date [25].

CO_2 permeance: 50 gpu
CO_2/CH_4 selectivity: 15

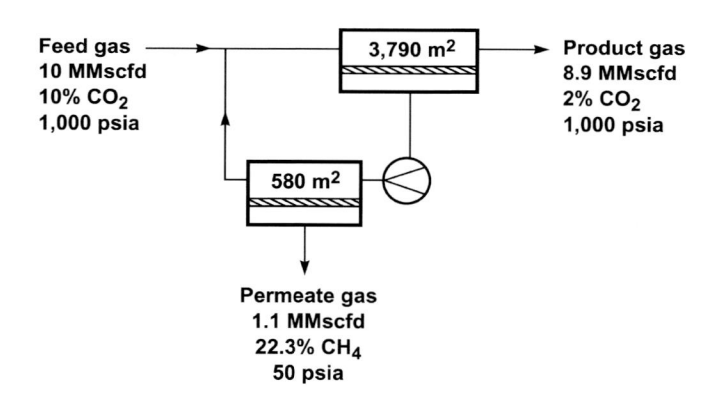

Feed gas
10 MMscfd
10% CO_2
1,000 psia

3,790 m^2

Product gas
8.9 MMscfd
2% CO_2
1,000 psia

580 m^2

Permeate gas
1.1 MMscfd
22.3% CH_4
50 psia

Membrane area: 4,290 m^2
Compressor load: 547 hp (theoretical)
Methane loss: 2.6% of feed methane

Fig. 7.12 Two-stage membrane system to remove carbon dioxide from natural gas. Membrane properties shown are similar to those of current good-quality cellulose acetate membranes.

membrane, is gradually being displaced by better polyaramide and perfluoro polymer membrane materials, making prospects for substantial growth in this area good.

Other natural gas separations. Although carbon dioxide represents the bulk of the current natural gas processing market, there are a number of opportunities for membrane processes to remove other impurities. Natural gas dehydration represents a significant opportunity for membrane technology. Water is a small, condensable compound, so many membranes have high water permeabilities, combined with water/methane selectivities of several hundred. However, it has proved difficult thus far for membranes to compete with glycol absorption plants – the current competitive technology – and only a handful of plants have been installed [26, 27]. Another potential application is the separation of nitrogen from natural gas. The challenge is to develop membranes with the necessary nitrogen/methane separation characteristics. Either glassy polymers, usually nitrogen-selective, or rubbery polymers, usually methane-selective, could be used. Figure 7.13 compares trial calculations performed for methane-selective and nitrogen-selective membranes. The figure shows the calculated membrane selectivity required to separate a gas stream containing 10% nitrogen and 90% methane into two streams: one, containing 4% nitrogen, to be delivered to the pipeline and the other, containing 50% nitrogen, to be used as on-site fuel. The separation corresponds to 93% recovery of methane in the product gas stream –

(a) Methane-selective membrane

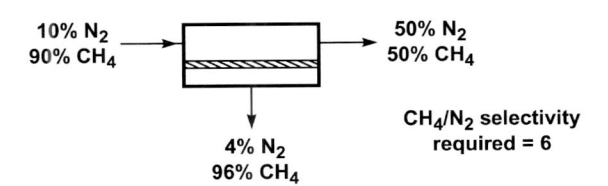

10% N_2 / 90% CH_4 → [membrane] → 50% N_2 / 50% CH_4

↓ 4% N_2 / 96% CH_4

CH_4/N_2 selectivity required = 6

(b) Nitrogen-selective membrane

10% N_2 / 90% CH_4 → [membrane] → 4% N_2 / 96% CH_4

↓ 50% N_2 / 50% CH_4

N_2/CH_4 selectivity required = 17

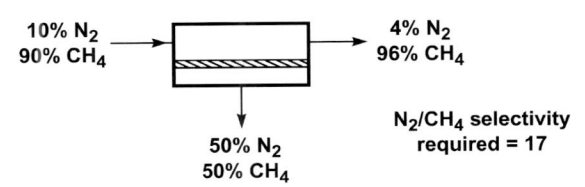

Fig. 7.13 One-stage membrane separations of 10% nitrogen/90% methane feed gas to produce a 50% nitrogen/50% methane reject stream and a 4% nitrogen/96% methane product gas. This target can be achieved by (a) a methane-selective membrane with a selectivity of 6, close to today's membranes, or (b) a nitrogen-selective membrane with a selectivity of 17, far higher than that of the best current membranes.

a very acceptable target for a nitrogen-removal process. A methane-permeable membrane with a methane/nitrogen selectivity of 6 can achieve the target separation. In contrast, a nitrogen-permeable membrane must have a nitrogen/methane selectivity of 17 to achieve the target separation. The best polymeric nitrogen-selective membrane currently known has a selectivity of 2 to 3 – far below the value required. This is why the most commercially developed membrane process uses methane-selective membranes. This approach requires recompression of the permeate gas for delivery to the pipeline, but the recompression cost is not high enough to significantly impact the process economics. The first membrane nitrogen removal plant was installed in 2003; three more plants were installed in 2004–early 2005. This application could grow, because the competitive technology is cryogenic distillation, which is expensive and only suited to large gas fields.

7.2.4.4 Vapor/Gas, Vapor/Vapor Separations

Separation of light hydrocarbon vapors from each other (for example, propylene from propane) or separation of vapors from gases (for example, propane and butane from hydrogen or propylene from nitrogen) are major commercial opportunities for membranes. A number of membrane plants have already been installed for such separations. For example, hydrogen-permeable membranes are being used in a number of refineries to recover hydrogen. Reliability problems caused by hydrocarbon condensation on the membranes have occurred, but many carefully designed and operated plants have worked well for years. More recently, hydrocarbon-selective membranes have been used to separate valuable hydrocarbons from vent streams [28, 29]. More than 20 units to separate and recover vinyl chloride monomer from nitrogen in polyvinyl chloride plants and 50 units to separate and recover ethylene and propylene from nitrogen in polyolefin plants are operating. The units installed thus far have only tapped the potential membrane opportunity.

An example of the very large potential opportunity for vapor separation processes is in the production of ethylene. Ethylene is produced in larger volume than any other petrochemical: about 200 large crackers are in operation worldwide, each with an average production of about 1 billion pounds per year of ethylene. The separation and purification of ethylene and other products from the cracked gas represents the majority of the capital and energy used in those plants. Currently, distillation is used, but the volatility of the gas mixtures to be separated means that high-pressure/low-temperature distillation columns are required.

Olefin plants do not use distillation as their product separation method because this is a good application of distillation – quite the contrary. The initial demethanizer tower, for example, operates at 450 psig and uses –100 °C ethylene as a coolant. These conditions require special metallurgies, and essentially complete removal of water, carbon dioxide and hydrogen sulfide from the feed gas. Similarly, the C_2 splitter and C_3 splitter are massive towers containing 150

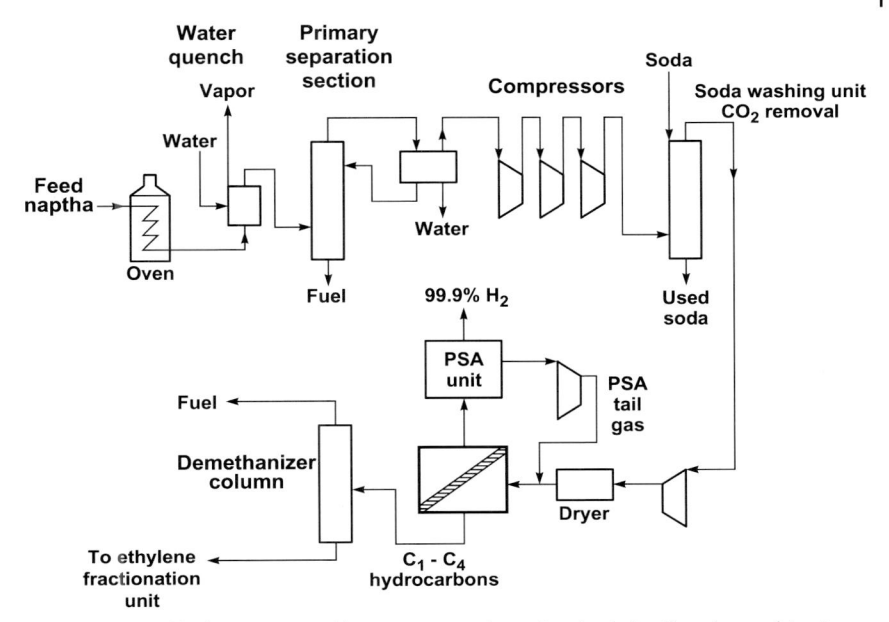

Fig. 7.14 Use of hydrogen-permeable membranes to separate hydrogen from light hydrocarbons in a cracker before the gas is sent to the demethanizer column, the first step in the cracker cold train. The operating temperature of the first demethanizer column is raised significantly, resulting in operating and capital cost savings. A PSA unit is used to upgrade the hydrogen-rich gas from the membrane unit to 99.9% hydrogen and to recycle any ethylene that permeates the membrane [30, 31].

to 200 trays, because of the very close boiling points of the ethylene/ethane and propylene/propane mixtures to be separated. Distillation is used for these separations because it is currently the only separation technique that works.

A number of opportunities exist for membrane separation units in olefin plants. The first large opportunities are likely to be in debottlenecking processes by using a membrane unit to perform an initial separation. This will reduce the load on the refrigeration plant or on subsequent distillation operations. A number of authors have proposed schemes to use membranes to remove hydrogen and methane from the feed gas to the cracker cold train. This reduces the refrigeration load of the plant and, because the demethanizer can then operate at a higher temperature, use of special construction materials can be avoided. One possible design is shown in Fig. 7.14.

In the design shown in Fig. 7.14, a polysulfone or polyaramide hydrogen-permeable membrane is used to remove hydrogen from the gas sent to the low-temperature demethanizer [30, 31]. Hydrogen represents only a small weight fraction of the feed gas, but because its molecular weight is low it may represent 20–30% or more of the volume of the cracked gas. Removing the hydrogen prior to the demethanizer makes the gas much more condensable, reducing the refrigeration load as well as producing a valuable byproduct stream.

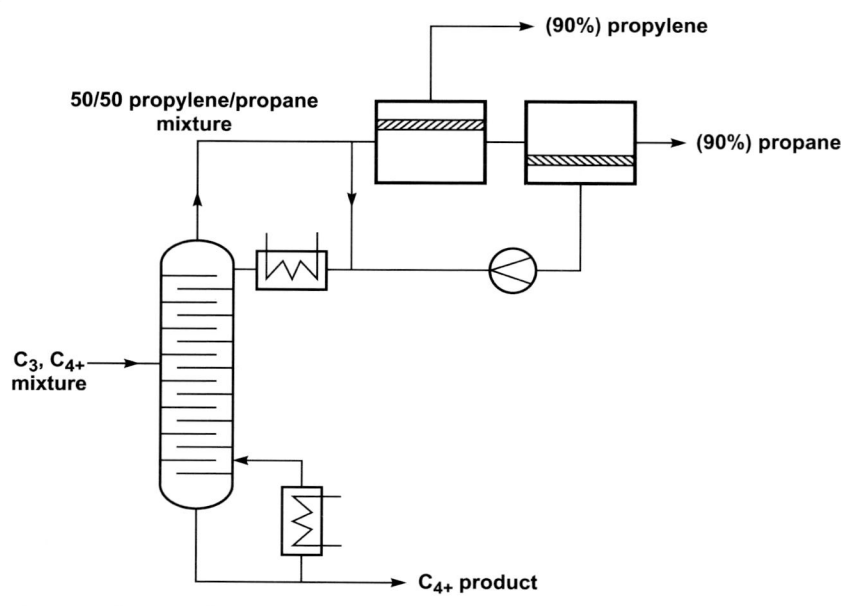

Fig. 7.15 Schematic of a propylene/propane membrane separation system installed to treat the overhead vapor from a refinery depropanizer column.

Another very large potential application of membranes in ethylene plants is replacing the C_2 and C_3 splitters. An example of a possible process design is shown in Fig. 7.15. In this example, a two-step membrane system equipped with propylene-permeable membranes is used to split a 50/50 propylene/propane overhead stream from a depropanizer column into a 90% propylene stream and a 90% propane stream. Both streams could then be sent to distillation units for polishing, but the size of columns required would be much reduced. For this design to be feasible, membranes with an olefin/paraffin selectivity of 5 to 10 are required. Many other designs that combine membranes and distillation columns to achieve good separation are possible [23].

Olefin/paraffin selectivities of 5 to 10 appear quite modest compared to the claims of some reports [32, 33]. However, much of the literature selectivity data has been calculated from the ratio of the permeabilities of pure olefin to pure paraffin. Olefin/paraffin selectivities measured with gas mixtures under conditions likely in a real process show that using pure gas permeabilities overestimates the membrane selectivity by a factor of 2 to 10. Therefore, it will be some time before olefin/paraffin-selective membranes are used in ethylene plants, although some nearer-term applications exist in petrochemical and refinery operations.

7.2.5
Pervaporation

The commercial success of pervaporation has been a disappointment to many process developers. Current pervaporation sales worldwide are probably less than US $10 million; almost all are for dehydration of ethanol or isopropanol solutions using water-permeable poly(vinyl alcohol) or equivalent membranes. A smaller market also exists for the separation of volatile organics from water using silicone-rubber membranes.

The historical development of pervaporation technology can be tracked by the number of US patents issued each year that relate to pervaporation, as illustrated by the plot in Fig. 7.16. Prior to 1960, only a handful of patents had been issued on pervaporation, but beginning in 1960 a series of patents were issued to Binning, Lee and others at American Oil covering the use of pervaporation membranes to separate organic mixtures, dissolved organics from water, and so on. Membrane technology was not sufficiently advanced to make these applications practical at that time, and American Oil abandoned the program after a few years. Through the rest of the 1960s and into the 1970s, a low level of interest was maintained in the process at Monsanto, Exxon and Standard Oil (Indiana), principally with the hope of separating organic mixtures such as styrene/ethylbenzene. Interest in the process surged in the 1980s following the installation of the first solvent dehydration plants by GFT. At that time, many researchers thought it would only be a matter of time before pervaporation would begin to replace distillation in large

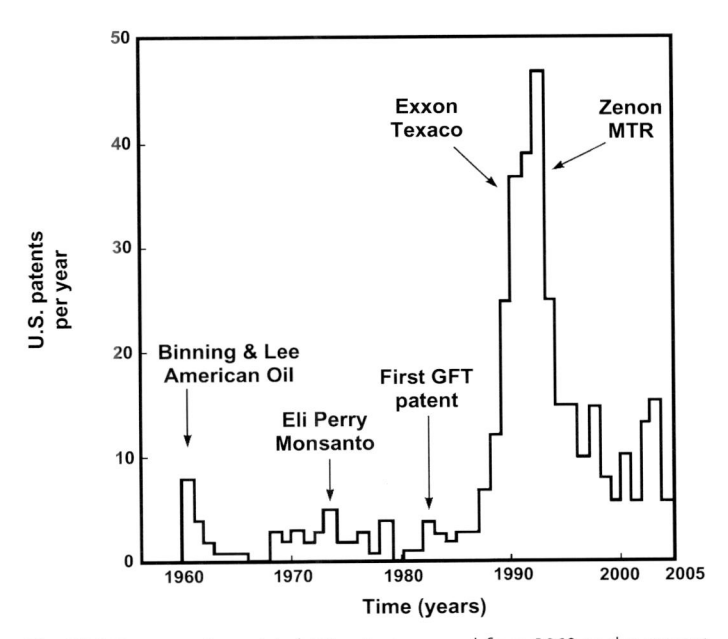

Fig. 7.16 Pervaporation-related US patents ussued from 1960 to the present.

refinery and petrochemical applications. As a result research groups were established at Exxon, Texaco, Standard Oil (Indiana), and elsewhere to develop the technology. The Exxon group in particular devoted considerable resources to developing membranes able to separate close-boiling aromatic/aliphatic refinery mixtures. Membranes with toluene/n-octane separation factors of up to 10 were obtained, and the process was taken to the pilot scale. In practice, however, the technology was still not competitive with distillation, and by the mid-1990s most of the oil companies were closing down their pervaporation research groups.

Currently, only a handful of companies – Sulzer Chemtech, Ube, Mitsui and MTR – remain active in the pervaporation area. The current market is likely to expand over the next years, particularly in the food industry, where interesting opportunities to recover high-value flavor and aroma compounds from process and waste streams exist, and in bioethanol plants, where improved dehydration membranes might compete with molecular sieve dehydration units. Another promising large-scale application is the separation of organic azeotropes and close boiling mixtures, but significant improvements in membrane selectivity will be required to make the process economically viable.

7.2.6
Ion-conducting Membranes

In a discussion of future developments of membranes in the chemical industry, a mention of the considerable effort now being made to develop ion-conducting membranes is in order. The overall concept is to use ceramic membranes that conduct oxygen ions at high temperatures. Materials that can conduct both oxygen ions and electrons are called mixed-conducting matrices. Transport through such materials is shown schematically in Fig. 7.17. Various complex metal oxide compositions, including some better known for their properties as superconductors, have mixed-conducting properties; recent efforts in this field appear to be focused on these materials [34, 35]. An example of this type of material is perovskites having the structure $La_xA_{1-x}Co_yFe_{1-y}O_{3-z}$ where A is barium, strontium or calcium, x and y are 0 to 1, and the value of z makes the material charge neutral overall. Passage of oxygen ions and electrons is related to the defect structure of these materials; at temperatures of 800 to 1000 °C, discs of these materials have been shown to be extraordinarily permeable.

Two large consortia, one headed by Air Products and the other by Praxair/BP-Amoco, are developing the membranes. To date, the work has used small ceramic tubes coated with 5–10-μm thick films of perovskites or other ion-conducting materials. Typical sample membrane areas are 20–30 cm^2. At the appropriate operating temperatures of 800–1000 °C, the membranes are perfectly selective for oxygen over nitrogen, and oxygen permeabilities of 10 000 Barrer can be obtained. This means that if the membrane thickness can be reduced to 1 μm, permeances of 10 000 gpu are possible. On this basis, a plant to produce 1 million scfh of oxygen will require about 4000 m^2 of membrane tube area – a large, but not inconceivably large, membrane area.

Mixed ion-conducting membrane

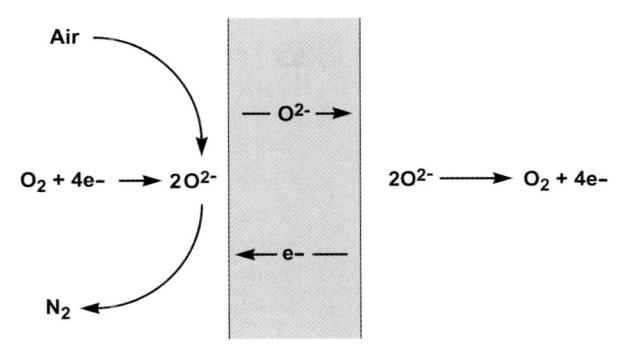

Fig. 7.17 Transport of oxygen ions and electrons through mixed ion-conducting membranes.

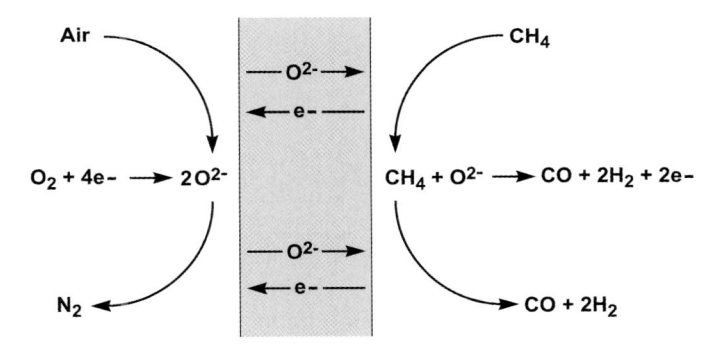

Fig. 7.18 Partial oxidation of methane using a mixed ion-conducting membrane.

Although various schemes to produce oxygen from air using heat integration with power plants or steel mills have been proposed, the most practical application and the principal driving force behind the development of these membranes is the production of synthesis gas by the partial oxidation of methane, as shown in Fig. 7.18 [35]. Oxygen ions diffusing through the membrane react with methane to form carbon monoxide and hydrogen. This gas can then be used without further separation to form methanol or other petrochemicals. Will it work? It is difficult to know: certainly enough money is dedicated to the problem. The membrane areas involved are reasonable but, on the other hand, the membranes must be operated at 800–1000 °C; must be resistant to poisoning by other gases such as carbon dioxide, water, or sulfur compounds; must be defect free; and must be able to withstand repeated thermal cycles from ambient to 1000 °C. These are all serious technical challenges, but a major effort is being made to solve them. A 70-scfm oxygen-from-air pilot plant is supposed to be turned on by the end of 2005. We will see.

7.3
The Future: Predictions for 2020

In 1983, I was hired as a consultant to predict development in the membrane separation area for the next 20 years. Looking at these predictions today, I find I was generally far too optimistic – sometimes ludicrously optimistic. I predicted that many refineries would be using pervaporation to separate organic mixtures

Ultrafiltration/Reverse Osmosis/Microfiltration

- Truly chlorine-resistant interfacial composite reverse osmosis membrane finally developed
- Reverse osmosis/nanofiltration of organic liquid mixtures is accepted in refinery and food oil processing operations. Total annual market U.S. $5 million/year in 2010, $20 million/year in 2020.
- Installed reverse osmosis capacity > 1 billion gal/day (2000)
- Installed reverse osmosis capacity > 3 billion gal/day (2010)
- Installed reverse osmosis capacity > 5 billion gal/day; half municipal/half ultrapure water (2020)
- Ceramic module and vibrating/rotating module costs reduced to U.S. $400/m². Begin to be widely used to treat chemical plant process and waste streams
- Microfiltration tertiary treatment units and bioreactors capture more than 50% of the new water treatment plant market.

Pervaporation

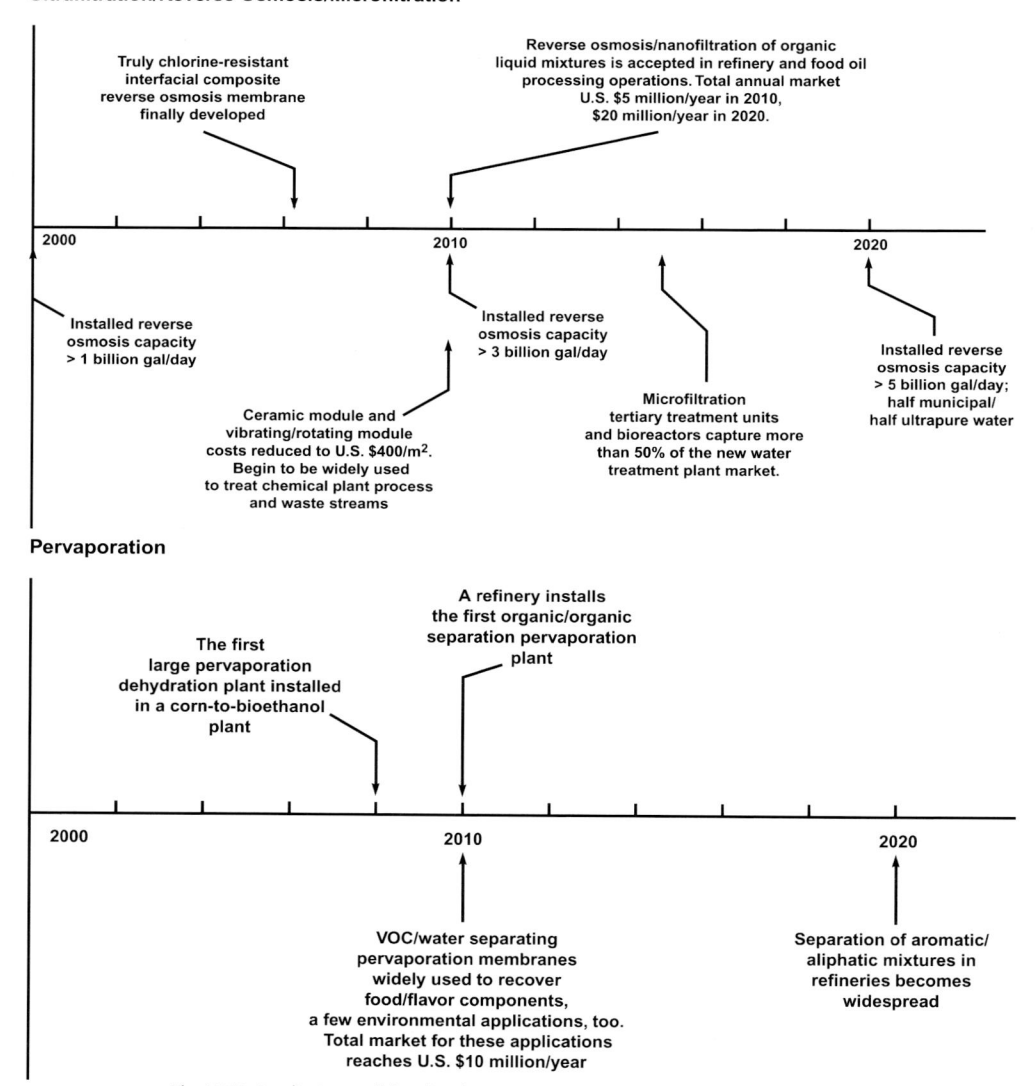

- The first large pervaporation dehydration plant installed in a corn-to-bioethanol plant
- A refinery installs the first organic/organic separation pervaporation plant
- VOC/water separating pervaporation membranes widely used to recover food/flavor components, a few environmental applications, too. Total market for these applications reaches U.S. $10 million/year (2010)
- Separation of aromatic/aliphatic mixtures in refineries becomes widespread (2020)

Fig. 7.19 Predictions of the development of membrane technology by 2020.

Gas Separation

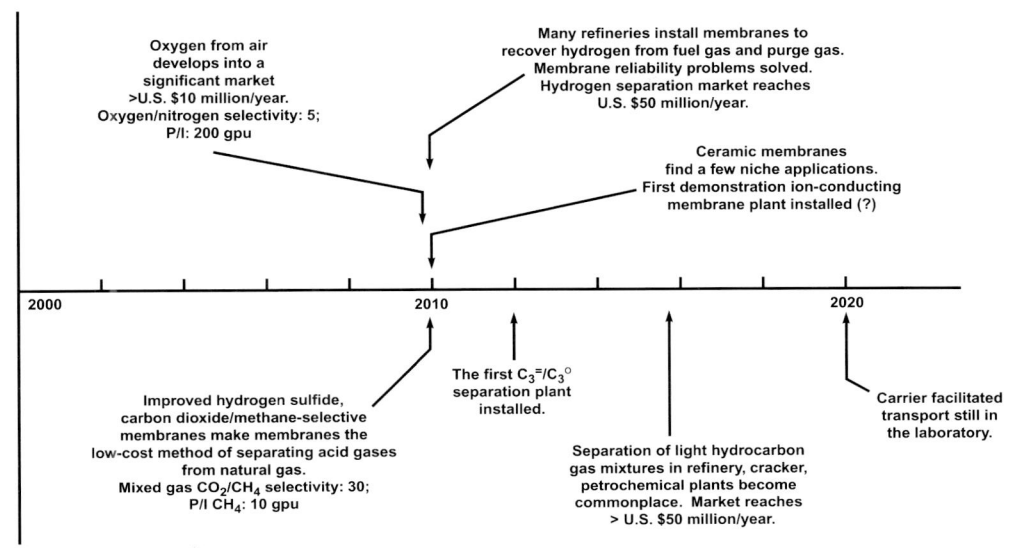

Fig. 7.19 (continued)

by the year 2000 and that carrier-facilitated transport would be in use to separate oxygen from air. Neither of these has happened or is likely to happen in the near future. On the other hand, I grossly underestimated the success of membranes for the production of nitrogen from air. Despite this track record, I venture to show my predictions for 2020 in Fig. 7.19, confident that opportunities for new applications are indicators of a bright future for membranes.

References

1 S. Loeb, S. Sourirajan, "Seawater Demineralization by Means of an Osmotic Membrane," in *Water Conversion II Advances in Chemistry* Series Number **28**, American Chemical Soc., Washington, DC (1963).

2 L. M. Robeson, "Correlation of Separation Factors Versus Permeability for Polymeric Membranes," *J. Memb. Sci.* **62**, 165 (1991).

3 L. T. Rozelle, J. E. Cadotte, K. E. Cobian, C. V. Kopp, Jr., "Nonpolysaccharide Membranes for Reverse Osmosis: NS-100 Membranes," in *Reverse Osmosis and Synthetic Membranes*, S. Sourirajan, (Ed.), National Research Council Canada, Ottawa (1977).

4 J. E. Cadotte, "Interfacially Synthesized Reverse Osmosis Membranes," US Patent 4277344 (July 1981).

5 R. J. Petersen, "Composite Reverse Osmosis and Nano Filtration Membranes", *J. Memb. Sci.* **83**, 81 (1993).

6 B. Culkin, A. Plotkin, M. Monroe, "Solve Membrane Fouling Problems With High-Shear Filtration," *Chem. Eng. Prog.*, (January 1998).

7 L. S. White, "Transport Properties of a Polyimide Solvent-resistant Nanofiltration Membrane", *J. Memb. Sci.* **205**, 191 (2002).

8 X. L. Yang, A. G. Livingston, L. M. Freitas dos Santos, "Experimental Observa-

tions of Nanofiltration With Organic Solvents," *J. Memb. Sci.* **190**, 45 (2001).
9   R. M. Gould, A. R. Nitsch, "Lubricating Oil Dewaxing With Membrane Separation of Cold Solvent", US Patent 5,494,566 (February 1996).
10 N. Bhore, R. M. Gould, S. M. Jacob, P. O. Staffeld, D. McNally, P. H. Smiley, C. R. Wildemuth, "New Membrane Process Debottlenecks Solvent Dewaxing Unit", *Oil and Gas Journal* **97**, 67 (1999).
11 J. T. Scarpello, D. Nair, L. M. Freitas dos Santos, L. G. White, L. G. Livingston, "The Separation of Homogenous Organometallic Catalysts Using Solvent-resistant Nanofiltration", *J. Memb. Sci.* **203**, 71 (2002).
12 M. Kolega, G. S. Grohmann, R. F. Chiew, A. W. Day, "Disinfection and Clarification of Treated Sewage by Advanced Microfiltration", *Water Sci. Technol.* **23**, 1609 (1991).
13 O. Futamura, M. Katoh, K. Takeuchi, "Organic Waste Treatment by Activated Sludge Process Using Integrated Type Membrane Separation", *Desalination* **98**, 17 (1994).
14 H. Shino, K. Nishimori, S. Kawakami, T. Uesaka, K. Izumi, "Wastewater: Future Applications for Submerged Flat-sheet membranes", *Ultrapure Water*, **37** (November 2004).
15 W. J. Henshaw, M. Mahendran, B. Bechmann, "Vertical Cylindrical Skien of Hollow Fiber Membranes and Method of Maintaining Clean Fiber Surfaces", US Patent 5783083 (July 1998).
16 P. L. Cote, B. M. Smith, A. A. Deutchmann, C. F. F. Rodrigues, S. K. Pedersen, "Frameless Array of Hollow Fiber Membranes and Method of Cleaning Fiber Surfaces While Filtering a Substrate to Withdraw a Permeate", US Patent 5248424 (September 1993).
17 R. W. Okey and P. L. Stavenger, "Reverse Osmosis Applications in Industrial Wastewater Treatment", in Membrane Processes for Industry Proceedings of the Symposium, Southern Research Institute, (May 19–20 1966).
18 K. Yamamoto, M. Hiasa, T. Mahmood, T. Matsuo, "Direct Solid-Liquid Separation Using Hollow Fiber Membrane in an Activated Sludge Aeration Tank", *Water Sci. Technol.* **21**, 43 (1989).
19 C. Chiemchaisri, K. Yamamoto, S. Vigreswaran, "Household Membrane Bioreactor in Domestic Wastewater Treatment", *Water Sci. & Tech.* **27**, 171 (1993).
20 B. D. Cho, A. G. Fane, "Fouling Transients in Nominally Sub-critical Flux Operation of a Bioreactor", *J. Memb. Sci.* **209**, 391 (2002).
21 Z. F. Cui, S. Chang, A. G. Fane, "The Use of Bubbling to Enhance Membrane Processes", *J. Memb. Sci.* **221**, 1 (2003).
22 R. Barchas, T. Hickley, "Membrane-separation process for Cracked Gas", US Patent 5082481 (January 1992).
23 D. E. Gottschlich, D. L. Roberts, "Energy Minimization of Separation Processes Using a Conventional/Membrane Hybrid System", Report to the Dept. of Energy/DE910047-10 (September 1990).
24 B. O. Bhide, S. A. Stern, "A New Evaluation of Membrane Processes for Oxygen-enrichment of Air", *J. Memb. Sci.* **62**, 87 (1991).
25 D. Dortmundt, "Qadirpur Water or CO_2 Removal Membrane Plant Expansion", Proceedings of the Gas Processors Association Annual Meeting (March 2004).
26 R. J. Arrowsmith, K. Jones, "Process for the Dehydration of a Gas," US Patent 5641337 (June 1997).
27 D. J. Stookey, K. Jones, D. G. Kalthod, T. Johannessen, B. C. C. Membrane Technology Planning Conference, Newton, MA (October 1996).
28 R. W. Baker, M. Jacobs, "Improved Monomer Recovery from Polyolefin Resin Degassing", *Hydrocarbon Processing* **75** (March 1996).
29 R. W. Baker, J. G. Wijmans, "Membrane Separation of Organic Vapors from Gas Streams", in *Polymeric Gas-separation membranes*, D. R. Paul, Y. P. Yampol'skii (Eds.), CRC Press, Boca Raton, FL (1994).
30 L. J. Howard, H. C. Rowles, "Olefin Recovery from Olefin Hydrogen Mixtures", US Patent 5634354 (June 1997).
31 E. Schmidt, I. Sinclair, "Maximize Hydrogen Production and Increase Ethyl-

ene Capacity With the Opal System", Presented at the Second European Petrochemical Technology Conference, Prague, Czech Republic (June 2000). See also: "Process for Recovering Olefins", US Patent 6141988 (November 2000).

32 O. M. ILINITCH, G. L. SEMIN, M. V. CHERTOVA, K. I. ZAMARAEV, "Novel Polymeric Membranes for Separation of Hydrocarbons", *J. Memb. Sci.* **66**, 1 (1992).

33 K. TANAKA, A. TAGUCHI, J. HAO, H. KITA, K. OKAMOTO, "Permeation and Separation Properties of Polyimide Membranes to Olefins and Paraffins", *J. Memb. Sci.* **121**, 197 (1996).

34 T. MAZANEC, T. L. CABLE, "Coated Membranes", US Patent 5273035 (March 1998).

35 T. MAZANEC, T. L. CABLE, "Process for Partial Oxidation of Hydrocarbons", US Patent 5714091 (February 1998).

Subject Index

a
acid-gas separations 143
air separation 64

c
carbon membranes 62, 63
– carbon molecular sieve membranes 62
– gas-separation membranes 62
cellulose 32
– cellophane 32
– cuprophane 32
concentration polarization 208

d
dehydrogenation
– Catofin 248
– Oleflex 248

e
electrochemical processing
– chlor-alkali industry 291
– electroanalysis 287
– electrochemistry 286
– electrohydrodimerization (EHD) 299
electromembrane processes 259, 274
– bipolar membrane 283
– desalination 285
– electrochemical oxidation 300
– electrochemical regeneration 282
– electrochemistry 286
– electrodeionization (EDI) 280, 281
– electrodialysis 274, 275, 285
– electrolysis 288
– electroosmotic phenomenon 275
– fuel cells 297
– HYDRINA® process 289, 290
– ion-exchange membranes 259
– zero-gap electrode 294

f
free volume polymers 58
– butane/methane selectivity 61
– free volume (FFV) 59
– PDMS 61
– PMP 59, 60, 61
– poly(1-trimethylsilyl-1-propyne) 58
– polymers of intrinsic microporosity (PIMs) 61
– PTMSP 59, 60, 61
– Teflon AF2400 59, 60, 61
fuel cells 45 ff, 297
– direct methanol (DMFC) 46
– electrolytes 45
– ion exchange membranes 45
– membranes
– – for high temperatures 51
– – – phosphonated polymers 51
– – – polybenzimidazole (PBI) 51
– – Gore-Select 299
– – Nafion 299
– – perfluorinated membranes 291, 292, 298
– proton conductivity 45
future directions 305 ff

g
gasoline vapor recovery 103
– PVC production 104
gas-separation 53 ff, 64, 119, 315
– air dryers 145
– ammonia synthesis 137
– carbon dioxide 323, 324
– carbon membranes 62, 63
– critical oxygen concentration 141
– helium separation 140
– hydrogen recovery 138, 318
– hydrogen separation 137
– ion-conduction membranes 330
– membranes 54, 59

Membrane Technology in the Chemical Industry.
Edited by Suzana Pereira Nunes and Klaus-Viktor Peinemann
Copyright © 2006 WILEY-VCH Verlag GmbH & Co. KGaA, Weinheim
ISBN: 3-527-31316-8

– – carbon molecular sieve membranes   62
– – cellulose acetate   57
– – polydimethylsiloxane   54, 59
– – polyimide   54, 57, 59
– – polysulfone   54, 57, 59
– – polyvinyltrimethylsilane   54
– methane-selective   325
– methanol synthesis   138
– mixed ion-conducting   331
– natural gas   147, 323, 324, 325
– natural gas dehydration   325
– nitrogen (and oxygen) separation from air   319
– nitrogen-enriched air (NEA)   147
– nitrogen rejection   147
– nitrogen-separation   140
– olefin/paraffin separation   328
– oxygen-enriched air   146
– ozone generator   146
– permeability   121
– perovskite-type   64
– process design   69
– propylene/propane separation   328
– refinery hydrogen applications   317
– Robertson's plot   309

h

hollow fibers   126
hydrogen separation   137
hydrogen/hydrocarbon separation
– off-gas streams   114
– petrochemical industry   114
– process   114
hydrophilic membranes   189

i

ion exchange membranes   45, 259, 260, 273
– anion-exchange   261, 263
– bipolar membranes   260, 283, 290
– cation-exchange   261, 263
– crosslinked   260
– diffusion permeability   268
– electrical conductivity   263
– electrochemical performance   267
– electrodialysis   261
– electromembrane process   274
– electroosmotic permeability   270
– hydraulic permeability   269
– ion-exchange   261, 265
– limiting current   272
– osmotic permeability   269
– polarization   271, 273

– properties   262
– styrene crosslinked   261
– swelling   262
– three-conductor model   265

m

market   5 ff
– hemodialysis   5
– microfiltration   6
– reverse osmosis (RO)   5
– ultrafiltration   6
Matrimid   57
membrane modules   98, 306
– anisotropic membranes   306
– capillary fibers   306, 307
– envelope-type GS module   98
– hollow fine fiber   306, 307, 308
– module configuration   306
– module geometry   125
– module manufacture   130
– organic/organic separation   311
– plate and frame   307, 311
– plate and frame design   125
– process design   132
– sealing materials   129
– spiral wound   99, 306, 307
– tubular   307
membrane preparation   9
membrane reactors   229, 232, 241
– bioprocess engineering   230
– bioreactor   254, 238, 316
– catalyst retention   241
– catalytic esterification   246
– catalytic membrane reactors   231
– catalytically active membrane   241
– Catofin   248, 249
– ceramic membrane   248
– dehydrogenation   234, 248, 249
– extraction   253
– gas-liquid reactions   237
– gas-phase reaction   233, 239
– heterogeneous catalysis   231
– homogeneous catalysis   231
– liquid-phase reactions   235, 240
– nanofiltration (NF)   240
– Oleflex   248, 249
– OTM Syngas   250, 251
– perovskites   233
– pervaporation   242, 243, 244, 248
– reverse osmosis (RO)   240
– ultrafiltration (UF)   240
membrane transport, models   207
microfiltration   34, 332

– Celgard® 34, 35
– Gore-Tex® 34, 37
– membrane bioreactors 314
– poly(ethylene terephthalate 37, 38
– polycarbonate 37
– polyethylene 34
– Polypore 35
– polypropylene 34
– polytetrafluoroethylene 36
– PTFE 37
mixed-matrix membranes 67
– Maxwell 67
– silicalite 67
– zeolites 67
models 207
module configuration 306
monomer recovery 110

n
nanofiltration 18, 203
– Desal-5 25
– Desal 19
– Docosane-Toluene 216
– FILMTEC 19
– MAX DEWAX 204, 205
– pH Stable SelRO® 22
– polysulfone 23, 25
– SelRO® 20, 21
– solvent dewaxing 204
– solvent-resistant membranes 20
– STARMEN122 210, 212, 213
– TOABr-Toluene 219
natural gas treatment 113
nitrogen generator 133
nitrogen separation 140
nonfluorinated membranes 48 ff
– poly(ether ketone ketone 49, 50
– polyetherketone 49
– polyimide 49
– polyphosphazene 50
– polysulfone 48
– sPEEK 49, 50
– sulfonated polyetherketones 49
– sulfonated polyimide 49
– sulfonated polysulfone 48

o
organic solvent nanofiltration 203 ff
organic vapors 93, 96
– gasoline vapor recovery 103
– hydrocarbon dewpointing 94
– hydrogen 94
– hydrogen/hydrocarbon separation 114

– natural gas treatment 113
– off-gas 102
– petrochemical industry 114
– petrol station 111
– polydimethylsiloxane (PDMS) 96, 106
– polyoctylmethylsiloxane (POMS) 96, 106
– polyolefin production 109
– pressure swing adsorption 106
– process gas 102
organic-inorganic membranes for fuel cells 52
– DMFC 52
– heteropolyacids 52
– montmorillonite 52
– SiO_2 52
– ZrO_2 52
organic-organic separation 197
organophilic membranes 188
oxygen-enriched air 146
– ozone generator 146

p
perfluorinated membranes 46 ff, 291
– Gore-select® 47
– Nafion® 46
– radiation grafting 47, 48
– $α,ββ$-Trifluorostyrene 47
pervaporation 151, 329, 332
– "cushion" modules 185
– desorption 155
– diffusion 155
– hydrophilic membranes 189
– modules 182
– organophilic membranes 188
– plates modules 183
– polydimethylsiloxane (PDMS) 177
– polymethyloctylsiloxane (POMS) 177
– principles 168
– sorption 155
– spiral-wound modules 185
– tubular modules 186
– vapor permeation 154, 171
– zeolites 178, 180
phase inversion 10, 11
– demixing 11
– finger-like cavities 12
– gelation 14
– macrovoid formation 13
– nucleation and growth mechanism (NG) 11
– pore structure 10
– spinodal decomposition (SD) 11
– sponge-like 12

poly(1-trimethylsilyl-1-propyne) 58
poly(tetrafluoroethylene) 36
polyacrylonitrile (PAN) 22, 30
polybenzimidazole (PBI) 51
polyetherimide (PEI) 14, 28
polyetherketones 33
polyethersulfone 23
polyimide 49
polysulfone 23
process design 132

r

reverse osmosis 15, 332
– FILMTEC® FT-30 16, 17
– hollow-fine-fiber 309
– interfacial polymerization 15
– Loeb-Sourirajan 309
– municipal drinking 310
– spiral-wound 312
– thin-film composites 15

s

spiral-wound modules 185
surface modification 39 ff, 42
– chemical oxidation 39
– corona discharge 39
– crosslinking 40
– grafting 40, 41, 42
– hydrophilic 41
– hydrophobic 42
– plasma treatment 40
– polysulfone 40, 41
– reverse osmosis 40
– UV irradiation 43

t

Teflon AF2400 61
transport mechanisms 53, 55
– diffusion 56, 57
– permeability 58
– selectivity 58
– silicone rubber 56
– solubility 56, 57
transport process 208
– concentration polarization 208, 213, 214, 221, 222, 223
– Hagen-Poiseuille equation 205, 212, 213
– mass-transfer 215
– – coeffcient 221
– MAX-DEWAX 205
– OSN 205, 207, 209
– solution-diffusion model 206, 211, 213, 215
– Stefan-Maxwell equation 205, 206

u

ultrafiltration 23, 24, 332
– cellulose 32
– ceramic membranes 313
– GE Osmonics 30
– poly(acrylo nitrile) 31
– poly(phenylene sulfide) (PPS) 33
– poly(vinylidene fluoride) 26
– polyacrylonitrile 30
– polyetherimide 28
– polyetherketones 33
– solvent-resistant membranes 32
– UltraFilic 30
ultrathin membranes 305
– anisotropic membrane 305, 306
– cellulose acetate 305
– Loeb-Sourirajan 305, 306

v

vapor permeation 154, 171, 191
vapor separation
– vapor/gas 326
– vapor/vapor 326